Advances in *Cordyceps* Research

Prospects and Avenues

Series: Progress in Mycological Research

Advances in *Cordyceps* Research

Prospects and Avenues

Editors

Kandikere R. Sridhar
Department of Biosciences
Mangalore University, Mangalore, India

Sunil K. Deshmukh
Scientific Advisor
Greenvention Biotech Pvt. Ltd., Pune, India

Shin-Yee Fung
Professor, Department of Molecular Medicine
Faculty of Medicine, Universiti Malaya, Malaysia

Shivannegowda Mahadevakumar
Scientist C, Botanical Survey of India
Andaman and Nicobar Regional Center,
Port Blair, India

CRC Press
Taylor & Francis Group
Boca Raton London New York

CRC Press is an imprint of the
Taylor & Francis Group, an **informa** business

A SCIENCE PUBLISHERS BOOK

First edition published 2025
by CRC Press
2385 NW Executive Center Drive, Suite 320, Boca Raton FL 33431

and by CRC Press
4 Park Square, Milton Park, Abingdon, Oxon, OX14 4RN

Library of Congress Cataloging-in-Publication Data (applied for)

ISBN: 978-1-032-71698-5 (hbk)
ISBN: 978-1-032-73870-3 (pbk)
ISBN: 978-1-003-46642-0 (ebk)

DOI: 10.1201/9781003466420

Typeset in Palatino Linotype
by Prime Publishing Services

Preface

Mycology is one of the most fascinating branches of biology and has influenced almost all facets of human life owing to its significance or threat to agriculture, animals, the environment, industry and medicine. Fungi have the power to contribute to the global economy in all spheres through ecological and environmental services to benefit humans in nutrition, health, biomaterials and economically valued products (e.g., antibiotics, beverages, biopesticides, cosmetics, drugs, enzymes, fermented foods, metabolites, nutraceuticals, organic acids and pigments) (Niego et al., 2023).

Among fungi, entomopathogens are valuable sources of biological control agents as they serve as natural broad-spectrum insecticides. *Cordyceps* and allied species are well known to parasitize a wide range of herbivorous insects and other arthropods in varied ecological habitats, which qualifies them as biopesticides in the agricultural sector (Bamisile et al., 2021). Among more than 600 species of *Cordyceps*, *Cordyceps sinensis* is the most studied fungus. *Cordyceps* are known to be abundant in humid climatic habitats and tropical forests in North America, Europe, East Asia and Southeast Asia (Qu et al., 2022). The hotspot of the Himalayan ecoregions is well known for the occurrence of *Cordyceps* (popularly referred to as 'Himalayan Gold' or 'Himalayan Viagra') (Lin and Li, 2011), while another hotspot in the Western Ghats is also showing promising avenues for tropically adapted *Cordyceps* and allies (Karun and Sridhar, 2017; Dattaraj et al., 2018). *Cordyceps* species are the most employed species in Traditional Chinese Medicine which has attracted global attention for their utilization in drug discovery and drug development (Li et al., 2021). Currently, up to 35 *Cordyceps* species are known to produce metabolites of therapeutic value against many lifestyle diseases (Zhang et al., 2023). Owing to human interference and climate change, the ecological niches of *Cordyceps* are dwindling, which has resulted in artificial cultivation to utilize the mycelia as well as fruit bodies in therapeutics (Dong et al., 2016).

The book *Advances in Cordyceps Research: Prospects and Avenues* offers different facets of investigation on *Cordyceps* and allied fungi. Chapters written by 41 researchers from eight countries delve into different aspects of *Cordyceps* research. This volume highlights seven subdivisions in basic and applied areas: 1) Overview; 2) Taxonomy, Phylogeny and Genetics;

3) Diversity and Distribution; 4) Bioactive Potential; 5) Nutraceutical Prospects; 6) Plant Protection; 7) Safety and Products. The first subsection, in one chapter, presents a world scenario of studies on *Cordyceps* along with distribution, ecology, impact of the environment, pathogenicity, cultivation, methods of study and world market. The second and third subsections deal with taxonomy, phylogeny, diversity and distribution in two chapters each. In the fourth subsection, biotechnological potential, along with traditional knowledge and pharmaceutical and cosmeceutical applications, were dealt with in four chapters. The fifth subsection, in two chapters, presents nutraceutical prospects from diet to medicine. The sixth subsection delves into the significance of plant protection in one chapter. The last subsection in two chapters records the applications, safety and marketed products.

We are hopeful that the intentions of the authors of chapters and editors in following various aspects of *Cordyceps* research and developments will be useful for readers with a wide interest in mycology and allied fields (applied biology, biochemistry, biotechnology, botany, ecology, entomology, environmental biology, field biology, forestry and zoology).

The authors appreciate the kind gestures of the contributors as well as the reviewers for the submission of chapters and meticulous evaluation on time, respectively. We are grateful to the CRC Press, Boca Raton, USA, for its fruitful collaboration to fulfil many official formalities to be observed to publish this book on time.

Manglore, India   **Kandikere R. Sridhar**
Pune, India   **Sunil K. Deshmukh**
Kaula Lumpur, Malaysia   **Shin-Yee Fung**
Port Blair, India   **Shivannegowda Mahadevakumar**
February 16, 2024

References

Bamisile, B.S., Akutse, K.S., Siddiqui, J.A. and Xu, Y. (2021). Model application of entomopathogenic fungi as alternatives to chemical pesticides: Prospects, challenges, and insights for next-generation sustainable agriculture. Front. Plant Sci., 12: 741804. https://doi.org/10.3389/fpls.2021.741804.

Dattaraj, H.R., Jagadish, B.R., Sridhar, K.R. and Ghate, S.D. (2018). Are the scrub jungles of southwest India potential habitats of *Cordyceps*? KAVAKA—Trans. Mycol. Soc. India, 51: 20–22.

Dong, C.H., Li, W.J., Li, Z.Z., Yan, W.J., Li, T.H. et al. (2016). *Cordyceps* industry in China: Current status, challenges and perspectives-Jinhu declaration for *Cordyceps* industry development. Mycosystema, 35(1): 1–15. https://doi.org/10.13346/j.mycosystema.150207.

Karun, N.C. and Sridhar, K.R. (2013). The stink bug fungus *Ophiocordyceps nutans*—A proposal for conservation and flagship status in the Western Ghats of India. Fungal Conservation, 3: 43–49.

Li, Z.Z., Luan, F.G., Hywel-jones, N.L., Zhang, S.L., Chen, M.J. et al. (2021). Biodiversity of cordycipitoid fungi associated with *Isaria cicadae* Miquel II: Teleomorph discovery and nomenclature of chanhua, an important medicinal fungus in China. Mycosystema, 40(1): 95–107. https://doi.org/10.13346/j.mycosystema.200119.

Lin, B.-Q. and Li, S.-P. (2011). *Cordyceps* as an herbal drug. pp. 73–105. *In*: Benzie, F.F. and Wachtel-Galor, S. (eds.). Herbal Medicine: Biomolecular and Clinical Aspects. 2nd Edition, CRC Press, Boca Raton, USA. http://dx.doi.org/10.1201/b10787-6.

Neigo, A.G.T., Lambert, C., Mortimer, P., Thongklang, N., Rapior, S. et al. (2023). The contribution of fungi to the global economy. Fungal Divers., 121: 95–137. https://doi.org/10.1007/s13225-023-00520-9.

Qu, S.-L., Li, S.-S., Li, D. and Zhao, P.-J. (2022). Metabolites and their bioactivities from the genus *Cordyceps*. Microorganisms, 10(8): 1489. https://doi.org/10.3390/microorganisms10081489.

Zhang, Y., Li, K., Zhang, C., Liao, H. and Li, R. (2023). Research progress of *Cordyceps sinensis* and its fermented mycelium products on ameliorating renal fibrosis by reducing epithelial-to-mesenchymal transition. J. Inflamm. Res., 16: 2817–2830. https://doi.org/10.2147/JIR.S413374.

Contents

List of Contributors

Anju, MS,
Forest Health Division, KSCSTE-Kerala Forest Research Institute, Peechi 680653, Kerala, India.

Chaiwut, Phanuphong
School of Cosmetic Science, Mae Fah Luang University, 333 M.1 Thasud, Muang, Chiang Rai, Thailand.
Green Cosmetic Technology Research Group, Mae Fah Luang University, 333 M.1 Thasud, Muang, Chiang Rai, Thailand.

Chhipa, Hemraj
College of Horticulture and Forestry, Agriculture University, Kota, Jhalawar (Agriculture University, Kota) Rajasthan-326023, India.

Datta, Sonal
1 Department of Bio-Sciences and Technology, Maharishi Markandeshwar (Deemed to be University), Mullana (Ambala) Haryana-133207, India.

Devika Lal, MR
Forest Health Division, KSCSTE-Kerala Forest Research Institute, Peechi 680653, Kerala, India.

Jędrejko, Karol
Department of Pharmaceutical Botany, Faculty of Pharmacy, Jagiellonian University Medical College, 9 Medyczna Street, 30-688 Kraków, Poland.

Jen, Koh Gui
School of Biosciences, Faculty of Health and Medical Sciences Taylor's University, Subang Jaya 47500, Malaysia.

Kahraman, Çiğdem
Hacettepe University, Faculty of Pharmacy, Department of Pharmacognosy, 06100, Ankara, Türkiye.

Kała, Katarzyna
Department of Pharmaceutical Botany, Faculty of Pharmacy, Jagiellonian University Medical, College, 9 Medyczna Street, 30-688 Kraków, Poland.

Karunarathna, Samantha C.
Center for Yunnan Plateau Biological Resources Protection and Utilization, College of Biological Resource and Food Engineering, Qujing Normal University, Qujing 655011, China.

Krishnan, Jithu U
Forest Health Division, KSCSTE-Kerala Forest Research Institute, Peechi 680653, Kerala, India.

Lim, Lek Teng
BioFact Life Sdn Bhd, Parit Jawa, Muar, Johor, Malaysia.

Lumyong, Saisamorn
Department of Biology, Faculty of Science, Chiang Mai University, Chiang Mai 50200, Thailand.

Mahadevakumar, Shivannegowda
Botanical Survey of India, Andaman and Nicobar Regional Center, Haddo 744 102, Port Blair, Andaman, India.

Murali, R,
Forest Health Division, KSCSTE-Kerala Forest Research Institute, Peechi 680653, Kerala, India.

Muszyńska, Bożena
Department of Pharmaceutical Botany, Faculty of Pharmacy, Jagiellonian University Medical, College, 9 Medyczna Street, 30-688 Kraków, Poland.

Nahar, Lutfun
Laboratory of Growth Regulators, Palacký University and Institute of Experimental Botany, The Czech Academy of Sciences, Šlechtitelů 27, 78371 Olomouc, Czech Republic.

Özenver, Nadire
Hacettepe University, Faculty of Pharmacy, Department of Pharmacognosy, 06100, Ankara, Türkiye.

Paul, Purnendu
West Bengal Biodiversity Board, Prani Sampad Bhawan, 5th Floor, LB-2, Sector-III, Salt Lake, Kolkata 700 106, West Bengal, India.

Pintathong, Punyawatt
School of Cosmetic Science, Mae Fah Luang University, 333 M.1 Thasud, Muang, Chiang Rai, Thailand.
Green Cosmetic Technology Research Group, Mae Fah Luang University, 333 M.1 Thasud, Muang, Chiang Rai, Thailand.

Pradhan, Prakash
West Bengal Biodiversity Board, Prani Sampad Bhawan, 5th Floor, LB-2, Sector-III, Salt Lake, Kolkata 700 106, West Bengal, India.

Priyashnatha, A.K Hasith
Department of Biology, Faculty of Science, Chiang Mai University, Chiang Mai 50200, Thailand.
Center for Yunnan Plateau Biological Resources Protection and Utilization, College of Biological Resource and Food Engineering, Qujing Normal University, Qujing 655011, China.

Quan, Tang Yin
School of Biosciences, Faculty of Health and Medical Sciences Taylor's University, Subang Jaya 47500, Malaysia.
Medical Advancement for Better Quality of Life Impact Lab, Taylor's University, 47500 Subang Jaya, Selangor Darul Ehsan, Malaysia.

Rakhi, KR
Forest Health Division, KSCSTE-Kerala Forest Research Institute, Peechi 680653, Kerala, India.

Saini, Shallu
Department of Bio-Sciences and Technology, Maharishi Markandeshwar (Deemed to be University), Mullana (Ambala) Haryana-133207, India.

Sangthong, Sarita
School of Cosmetic Science, Mae Fah Luang University, 333 M.1 Thasud, Muang, Chiang Rai, Thailand.
Green Cosmetic Technology Research Group, Mae Fah Luang University, 333 M.1 Thasud, Muang, Chiang Rai, Thailand.

Sarker, Satyajit D.
Centre for Natural Products Discovery, School of Pharmacy and Biomolecular Sciences, Liverpool John Moores University, James Parsons Building, Byrom Street, Liverpool L3 3AF, United Kingdom.

Shambhu Kumar,
Forest Health Division, KSCSTE-Kerala Forest Research Institute, Peechi 680653, Kerala, India.

Sharma, Anil Kumar
Department of Biotechnology, Amity University, Sector 82 A, IT City Rd, Block D, Sahibzada Ajit Singh Nagar, Mohali, Punjab 140306, India.

Shukla, Kamlesh Kumar
School of Studies in Biotechnology, Pt. Ravi Shankar Shukla University, Raipur (C.G.) India.

Sridhar, Kandikere R.
Department of Biosciences, Mangalore University, Mangalagangotri, Mangalore 574 199, Karnataka, India.

Sułkowska-Ziaja, Katarzyna
Department of Pharmaceutical Botany, Faculty of Pharmacy, Jagiellonian University Medical, College, 9 Medyczna Street, 30-688 Kraków, Poland.

Tandon, Chanderdeep
Department of Biotechnology, Amity University, Sector 82 A, IT City Rd, Block D, Sahibzada Ajit Singh Nagar, Mohali, Punjab 140306, India.

Tandon, Simran
Department of Health Sciences, Amity University, Sector 82 A, IT City Rd, Block D, Mohali, Punjab 140306, India.

Tirkey, Samay
School of Studies in Biotechnology, Pt. Ravi Shankar Shukla University, Raipur (C.G.) India.

Tuli, Hardeep Singh
Department of Bio-Sciences and Technology, Maharishi Markandeshwar (Deemed to be University), Mullana (Ambala) Haryana-133207, India.

Verma, Srishti
School of Studies in Biotechnology, Pt. Ravi Shankar Shukla University, Raipur (C.G.) India.

Wanhao, Chen
Center for Mycomedicine Research, Basic Medical School, Guizhou University of Traditional Chinese Medicine, Guiyang 550025, Guizhou, PR China.

Yanfeng, Han
Institute of Fungus Resources, Department of Ecology, College of Life Sciences, Guizhou University, Guiyang 550025, Guizhou, PR China.

Yi, Emilyn Yapp Wye
School of Biosciences, Faculty of Health and Medical Sciences Taylor's University, Subang Jaya 47500, Malaysia.

Yi, Sabrina Khor Xin
School of Biosciences, Faculty of Health and Medical Sciences Taylor's University, Subang Jaya 47500, Malaysia.

Overview

1

Research on *Cordyceps*—An Appraisal

A.K. Hasith Priyashanatha,[1,2,*]
Samantha C. Karunarathna[2,3,*] and *Saisamorn Lumyong*[1,*]

1. Introduction

Entomopathogenic fungi constitute a group of fungi known for infecting a diverse range of insects. The infections caused by these fungi are often lethal, leading to the death of the host organism within a relatively short period, typically ranging from several days to a few weeks. The group encompasses over 1,200 identified species distributed across approximately 100 genera (Jaber and Ownley, 2018; Ashraf et al., 2020). Due to their role as natural antagonists of herbivorous insects, entomopathogenic fungi are increasingly employed as biocontrol agents in the agriculture sector (Bamisile et al., 2021; Reason et al., 2022; Irsad et al., 2023). In recent decades, entomopathogenic fungi, particularly *Cordyceps*, have garnered attention, for their potential in the modern healthcare system.

The term *Cordyceps* is derived from the Latin words, "cord" meaning club and "ceps" meaning head (Tuli et al., 2014). Popularly known as winter-worm-summer grass, or Dong Chong Xia Cao in Chinese (Guo et al., 2017a), and Yarsagumba (fungus cum larvae) in Nepali

[1] Department of Biology, Faculty of Science, Chiang Mai University, Chiang Mai 50200, Thailand.

[2] Center for Yunnan Plateau Biological Resources Protection and Utilization, College of Biological Resource and Food Engineering, Qujing Normal University, Qujing 655011, China

[3] National Institute of Fundamental Studies (NIFS), Kandy 20000, Sri Lanka.

* Corresponding authors: priyashanthahasith@gmail.com; samanthakarunarathna@gmail.com; scboi009@gmail.com

(Baral, 2017), *Cordyceps sinensis*, a species within the genus, has been the focus of numerous pharmaceutical developments. *Cordyceps sinensis* is renowned for its rich composition of bioactive components including nucleosides (cordycepin, adenosine guanosine, cordysinin A-E), polysaccharides (exopolysaccharide fraction, acid polysaccharide, CPS-1, CPS-2, mannoglucan, CME-1, PS-A, cordyglucan, D-mannitol), sterols (ergosterol, H1-A, β-sitosterol, cerevisterol), amino acids and polypeptides (cordymin, eyclodipeptides A, cordycemides A and B). These components contribute to the various therapeutic effects associated with *Cordyceps* (Li et al., 2021; Liu et al., 2022; Zhang et al., 2022a,b). Due to its remarkable medicinal properties, *Cordyceps* is colloquially referred to as "soft gold" in China (Liu et al., 2020).

In this chapter, our focus is to provide a brief introduction to *C. sinensis*, highlighting its medicinal value, and other relevant background information. The medicinal aspects we will cover include its potential in enhancing respiratory health, anti-aging effects, lifespan extension, anti-tumor properties, attenuation of diabetes, improvement of heart health, and its role in anti-obesity. We aim to emphasize these selected medicinal values of *C. sinensis* based on recent studies. Before delving into the medicinal aspects, we will provide an overview of the fungus from various perspectives, including current research trends. Additionally, we will discuss taxonomic placement, habitat, distribution, ethnopharmacological uses, the current status of the world market for the fungus, morphological features, isolation, cultivation, and infection mechanisms.

2. World Attention

Species of *Cordyceps* have been known for a long time in various countries, with a few exceptions where they remain relatively unknown. However, the global awareness of these fungi and their benefits significantly increased in 1993. This breakthrough occurred when Chinese women athletes, during the National Games, achieved remarkable results in running events. The success was attributed, at least in part, to the consumption of *C. sinensis* (Guo et al., 2018). Following this initial report public interest in *C. sinensis* has grown significantly over the past decade (Buenz et al., 2005).

The commercial value of *C. sinensis* is noteworthy, with wild-collected fungi fetching high prices. In western Nepal, they are sold for over 100,000 Indian Rupees (approx. 1,200 US$) per kilogram (Tuli et al., 2014). The demand for *C. sinensis* surged significantly in 2003 during the severe acute respiratory syndrome (SARS) outbreak in China (Yan et al., 2014). While about 35 species of *Cordyceps* are recognized for their therapeutic properties (Zhang et al., 2023b), only a few, including *C. cicadicola*,

C. liangshanesis, C. militaris, C. ophioglossoides, and *C. sobolifera* are being cultivated alongside the majorly cultured *C. sinensis* (Ashraf et al., 2020).

3. Ethnopharmacological Uses of *Cordyceps*

The historical usage of *Cordyceps* is rooted in China where Chinese folk medicine recognized its medicinal value over three centuries ago. It was first reported by Wang Ang in 1694 in the compendium of Materia Medica-Ben Cao Bei Yao (Li et al., 2006). In the Western region, it gained attention in the the 17th century when the British mycologist Berkely first described the fungus in 1843 as *Sphaeria sinensis* Berk (Devkota, 2006). However, it was officially named *Cordyceps sinensis* in 1878 by Italian scholar Saccardo. (Li et al., 2006). *Cordyceps* has been utilized not only in China but also in India, Nepal, and Tibet for over a thousand years (Sen et al., 2023). In Nepal, where a significant portion of the population relies on traditional medicine (Giri et al., 2023), Western Nepalis refer to *C. sinensis* as "Himalayan Viagra" or "Himalayan Gold" (Lin and Li, 2011). The locals use this fungus medicine to treat various ailments, including liver disease, rheumatism, coughing, diarrhea, and headaches. In Sikkim, India, particularly in the Lachung and Lachen areas of North Sikkim, *C. sinensis* is popular in traditional practices, considered a remedy for various health issues. Traditional healers use it as a tonic to improve appetite, endurance, energy, libido, sleeping patterns, and stamina. Chinese traditional medicine also commonly employs *C. sinensis* to treat conditions such as cough, asthenia after severe illness, renal dysfunction, and renal failure (Lin and Li, 2011).

The local knowledge and usage of *Cordyceps* in Sikkim reveal an interesting cultural and traditional understanding of the fungus. Observing the robustness of animals like yaks, goats, and sheep that consumed *Cordyceps* while grazing in the forest led early inhabitants to experiment with its medicinal properties. They tested the fungus by feeding it to their cattle, powdered it with jaggery and observed improvements in their reproduction and vitality. Recognizing the significance of *Cordyceps* , locals started collecting it in the wild, separating the stroma, drying them under the sun, and consuming them. Early dwellers discovered that *Cordyceps* enhanced vigor and vitality. People used to drink a cup of milk with a piece of *Cordyceps* and it became a part of local traditions. Additionally, locals incorporated *Cordyceps* into homemade alcohol by placing a piece of fungus into an alcohol bottle and used it to take a sip. Drinking this concoction in the morning and evening became a traditional practice serving as a tonic for the community though some people got addicted to it (Panda and Swain, 2011).

The utilization of *Cordyceps* in Tibet dates back approximately 1,500 years, reminiscent of the practices in Sikkim. Similar to the locals in Sikkim

Tibetans observed enhanced energy in their livestock after consuming *Codryceps*. About 1,000 years ago, during the Ming Dynasty, the emperor's physicians learned about the Tibetan knowledge use of *Cordyceps*. They integrated the knowledge with their traditional practices creating potent medicines (Sharma, 2004). The traditional knowledge of Tibetans in finding wild *Cordyceps* is noteworthy. They recognized that this fungus could primarily be found in areas where yaks graze. Additionally, Tibetans fed their pack animals which carried loads in the mountainous regions of the Himalayas with *Cordyceps*. This practice aimed to enhance the animals' ability to withstand the low oxygen pressure at high latitudes (Sharma, 2004).

4. Current Scenario of *C. sinensis Studies*

The progress of studies on *C. sinensis* has been analyzed based on the Scopus database (https://www.scopus.com). Over the last 30 years (since 1994), a total of 371 articles have been retrieved, with a notable increase in the number of studies since 2000. The trend shows overall growth, although there are fluctuations in the number of studies over time (Fig. 1). The majority of these studies are concentrated in the fields of pharmacology, toxicology, and pharmaceutics (70.9%), followed by medicine (24.9%); and biochemistry, genetics, and molecular biology (21.0%) (Fig. 2). Geographically, more than half of the studies originate from mainland China (56.2%), with Taiwan

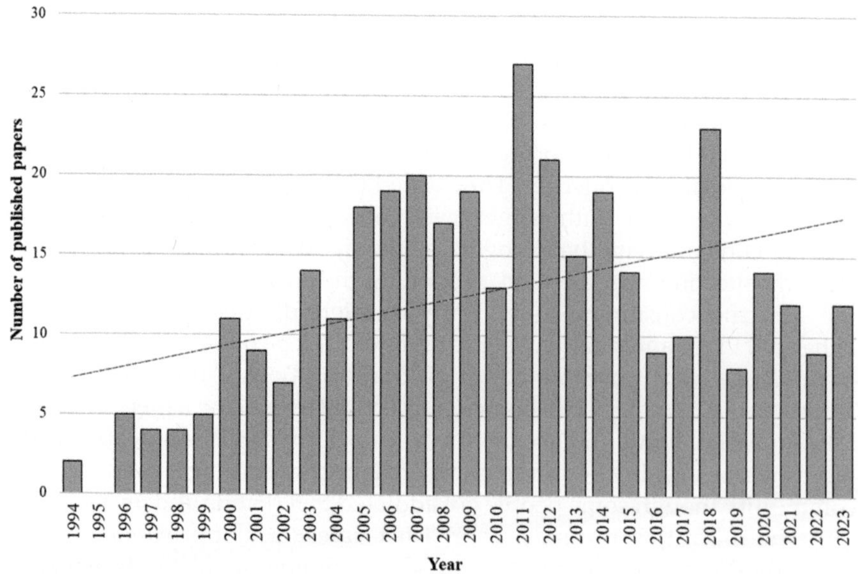

Fig. 1. The number of studies or article publications during 1994–2023 on *C. sinensis*.

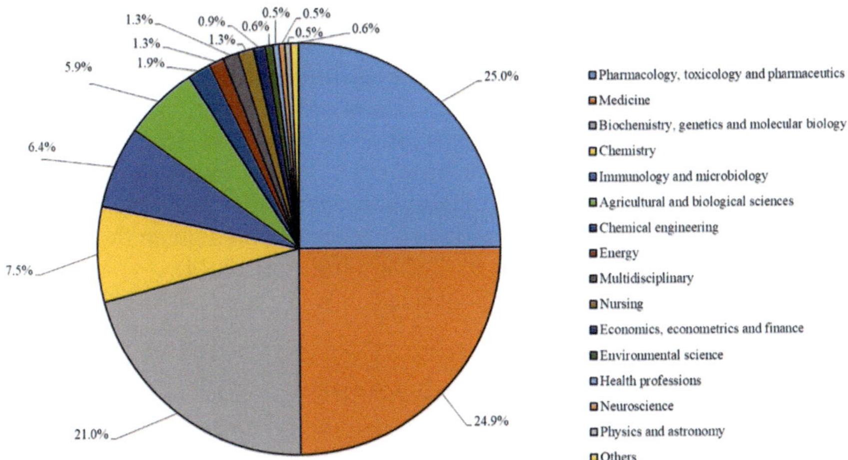

Fig. 2. Percentage of studies with the related areas: The study areas of *C. sinensis* are wider, and a plethora of studies have fallen into pharmacology, toxicology, and pharmaceutics. The least number of studies were conducted on environmental science (0.6%), health professions (0.5%), neuroscience (0.5%), physics (0.5%) and others (0.6%) (articles published in computer science, engineering, materials science, and veterinary) (the search was conducted using the Scopus database accessed on December 20, 2023).

(8.9%), Hong Kong (6.2%), the United States (4.2%), Japan (4%), India (3.3%), Canada (2.8%), Macao (2.5%), the United Kingdom (2.5%), and South Korea (2.3%) also contributing significantly. Several other countries have published three or fewer articles each constituting equal to or less than 0.5% of the data (data not presented).

5. Morphology, Taxonomy, Habitat and Distribution

5.1 *Morphology*

The fruiting bodies of *C. sinensis* are stipitate, cylindrical to slightly clavate exhibit a color range from yellowish orange to orange to reddish-orange. These structures typically feature stripes with varying thickness (1.5 to 3 mm) and a fertile clava terminal that is 2.0 to 6.0 mm wide. The overall height of the stroma is within the range of 1.5 and 7.0 cm, with variations based on the size of the host (Tuli et al., 2014). The ascocarp is of the perithecioid type, and the ascospores are filamentous and multi-septate (Shrestha et al., 2017).

5.2 *Taxonomy*

Cordyceps belongs to the Ascomycota division, specifically falling under the class Sordariomycetes, order Hypocreales, and Family Clavicipitaceae

(Dworecka-Kaszak, 2014). Within the Clavicipitaceae family, *Cordyceps* is the most diverse genus, with numerous species exhibiting a higher degree of host specificity. The genus is widely distributed globally, with a notable concentration of species in subtropical and tropical regions, particularly East and Southeast Asia. *Cordyceps* is found in various regions worldwide, excluding the harsh conditions of Antarctica. As of current records, Index Fungorum lists 598 *Cordyceps* species, while Mycobank provides details on 641 species, though this number may be underestimated given the ongoing discovery of new species (Sung et al., 2007; Tuli et al., 2014).

5.3 *Habitat and Distribution*

The ancestral origins of *C. sinensis* are believed to trace back to north-western Yunnan, China as indicated by an evolutionary biogeographical study (Dai et al., 2020). The fungus is commonly found in habitats characterized by higher latitudes, featuring cold and arid environments. The native occurrence of fungus is observed in regions such as the Himalayan Mountains in (3,000–4,000 meters above sea level) and China (3,500–4,500 meters above sea level, most commonly seen) (Sharma, 2004). While the current distribution of the fungus is not fully understood studies have reported its presence in various regions. Apart from Yunan, *C. sinensis* has been identified in the Tibetan Plateau and its surrounding areas including Gansu, Qinghai, Sichuan, and Tibet. Additionally, the southern flank of the Himalayas, encompassing Bhutan, India, and Nepal,is known to host the fungus. In India, Himachal Pradesh, Sikkim (located in the northeastern part of the country, bordered by Bhutan, Tibet, and Nepal) and Gori Valley, in the Pithoragarh district of Uttarakhand, are notable locations. In Nepal, the fungus can be found in 27 districts with Darchula and Dolpa being among the top-headed districts (Lin and Li, 2011; Laha et al., 2018; Baral, 2017).

6. Wild Collection of *C. sinensis*: Dead or Alive!

For medicinal purposes, *C. sinensis* is collected as a whole, including the stroma and the sclerotia (Wei et al., 2021). In Nepal, the wild collection of *C. sinensis.* for commercial purposes began approximately 20 years ago. About 10 years after the world realized the significance of this fungus. Legal regulations for collecting the fungus in the country were established about two years later. Barun and other inner valleys of the Makalu-Barun National Park in Nepal have become renowned sites for *Cordyceps* hunters, since the government granted permission for collection. Annually, over 3,000 local and non-local collectors embark on a mission to gather *Cordyceps* between May 15 and July 15 (in general) for one mission-collecting *Cordyceps* as much as they can. In 2014, reports, indicated that 45–50 kg of fungi were collected from Barun Valley. Notably, it requires the collection of 8,000–10,000 individuals to obtain one kilogram of fungi, illustrating the extensive effort involved in harvesting (Byers et al., 2020). The harvesting of *Cordyceps* raises concerns about sustainability, but the significant income generated from this activity often outweighs such considerations (Paterson, 2008). In Gori Valle, India, 60–78% of the annual household income is derived from collecting the fungus, highlighting the impact of *C. sinensis* on livelihood security (Laha et al., 2018).

The wild collection of *C. sinensis* is a challenging and labor-intensive task. In Nepal, non-local collectors travel to the Barun Valley and stay there until the season concludes. Collectors walk hard, climbing and digging the cliffs from dawn to dusk in pursuit of these valuable fungi (see Fig. 3). The task is perilous, involving the risk of climbing cliffs, and unfortumately, there have been instances of collectors dying by falling. Additionally, social conflicts between different groups can arise, ranging from verbal arguments to occasional incidents of violence including murders. The degradation of traditional values in the local culture is also documented, which is expected when there is an uncontrolled flux of people from diverse regions and backgrounds who stay continuously for an extended period. Moreover, the environmental impact is significant as the collecting activities contribute to alpine shrub destruction, wildlife poaching, and improper garbage disposal (Byers et al., 2020).

Fig. 3. Collecting *C. sinensis* in the wild: A, A high level of concentration and patience is required, keeping the head closer to the ground, crawling, and searching for the small hidden fungus (shown in the arrow); B and C, Careful digging and collecting of the fungal stroma with the whole sclerotia are required for quality harvesting; D, Freshly collected fungus before cleaning (photo credits: www.inaturalist.org).

7. World Market for *C. sinensis*: Current Scenario

7.1 Growth of the Demand

Cordyceps sinensis is commercially significant, with well-known companies located in various countries such as China, Bhutan, India, Singapore, Malaysia, Nepal, and the United States. China stands out as the leading country in the production of *Cordyceps*-based bioactive nutrients and other products (Bibi et al., 2020). The price of *Cordyceps* has shown a gradual increase, with significant growth observed from 1998 to 2008, where the price of *C. sinensis* rose by 900% (Chen et al., 2013). According to the analysis by Data Bridge Market Research, (https://www.databridgemarketresearch. com/), the global market for *C. sinensis*, valued at USD 1.07 billion in 2022, is expected to reach USD 2.44 billion by 2030. The compound annual growth rate is expected to be 5.01% from 2023 to 2030 (Li et al., 2019).

Intriguingly, the use of *C. sinensis* has led to an accelerated demand during the recent global pandemic of coronavirus disease 2019 (COVID-19) (Liu et al., 2023). This surge in demand reflects a growing public trust in the therapeutic value of the fungus. The high prices associated with *Cordyceps* are not only attributed to its therapeutic properties but also to its unique growth environment (Li et al., 2019). China stands out as one of the leading countries in the use of *Cordyceps* as a herbal drug. The most popular species is *C. sinensis*, officially recognized by the Chinese government as a herbal remedy in 1964. It took another 30 years for the USA to approve the utilization of the species. In 1994, after the USA passed the Dietary Supplement Health and Education Act, a large market of websites selling capsules and tonics made from independepently grown *Cordyceps* emerged. Despite this, there is no supplement developed by the world's leading pharmaceutical companies that surpasses *Cordyceps*.

7.2 What Matters?

The failure to meet the global demand for *C. sinensis* can be attributed to various factors. One significant factor is the geographically restricted distribution of the fungus. *C. sinensis* needs specific environmental or climatic conditions leading to a limited geographic range, which in turn, restricts wild collection (Tong et al., 2020; Wei et al., 2021). Additionally, artificial cultivation of the fungus poses challenges and may not accelerate production rates as expected. Cultivating the fruiting bodies, for example, is difficult, as they take one to two years to mature under optimal artificial conditions, limiting their widespread use (Tong et al., 2020). Moreover, the seasonal emergence of the fungus, following a specific life-cycle strategy, makes year-round collection impossible. This seasonality restricts the continuous collection of the fungus to several months in the year (Laha et al., 2018).

Ensuring sustainable production in the *C. sinensis* industry requires addressing several crucial aspects, particularly focusing on policy regulations and quality control measures in both production and marketing. Gaps in policies and testing have resulted in the circulation of products claiming to be *C. sinensis* but actually based on different strains or species. For instance, the CS-4 strain sold in China is reported to contain mycelia of *Paecilomyces hepialid*, a species distinct from *C. sinensis*. Some products produced by Wang Fong Pharmaceutical factory (Zheijang, China) claimed to be based on *C. sinensis* have been found to use *Cephalosporium sinensis* Chen, another species entirely (Chen et al., 2013). Additionally, *C. militaris* fruiting bodies are sometimes used as an alternative source for *C. sinensis* due to their easier production (Ko et al., 2017). Traditionally, the authentication of wild *C. sinensis* has relied on the morphology, color, and odor of the entire fungus-larva complex, especially when consumed directly in medicinal soups or liquors. However, the modern pharmaceutical industries often process raw fungi into various forms such as tablets, capsules, and purified oral liquids, making traditional authentication methods less applicable. (Guo et al., 2018).

On the other distinguishing between wild and cultivated *C. sinensis* or differentiating between species can be challenging due to the lack of morphological differences (Zhou et al., 2019). However, various analytical techniques have been propsed to address the issue and support the authentication process. Qian and Li (2017) highlighted the significance of the multi-column liquid chromatography (MC-LC) system, particularly in the qualitative analysis of macromolecules and micromolecules in *Cordyceps*. This system could potentially aid in the characterization of *Cordyceps* and differentiate between wild and cultivated varieties. Zhang et al. (2018) explored the use of ultra-performance liquid chromatography combined with Orbitrap Fusion mass spectrometry and chemometrics. This analytical approach demonstrated promising results in distinguishing between wild and cultivated fungus mycelia, providing a potential method for authentication. Guo et al. (2017b) suggested the use of stable carbon isotope ratios for verifying the authenticity of wild *Ophicordyceps sinensis* products.

7.3 *Hazards*

The contamination of *Cordyceps*-based products with hazardous trace metals is indeed a significant concern. The presence of elevated levels of arsenic (As), cadmium (Cd), lead (Pb), and mercury (Hg) in traditional Asian medicines, including herbal-based prducts, has been reported in various studies (Jairoun et al., 2020). Long-term consumption of products contaminated with these heavy metals can have detrimental effects on various organs (e.g., brain, kidney, heart) and systems (e.g., central nervous system, reproductive system, digestive system) leading to serious health issues. Studies have highlighted the potential toxicity associated with the consumption of contaminated dietary supplements, and the presence of heavy metals in these products can pose health risks. The carcinogenic properties of certain heavy metals further emphasize the importance of addressing and monitoring contamination in herbal and dietary supplements (Sarker et al., 2022). The study by Ćwieląg-Drabek et al. (2020) which tested 41 dietary supplements produced from terrestrial plants or microalgae, revealed a concerning level of contamination. Accordingly, 79.2% of supplement samples were made from terrestrial plants, and 88.2% were made from microalgae contaminated with at least one of the three heavy metals. Further, of all the supplements, 68.3% of the samples had Cd and Pb contamination (not necessarily from the same samples), and 29.3% had Hg contamination. This underscores the need for strict quality control measures and regular testing to ensure the safety of herbal and dietary supplements, including those containing *Cordyceps*.

The issue of heavy metal contamination is not unique to *Cordyceps*-based products; it has been observed in various mushroom-based products as well. Studies by Melgar et al. (2014) and Llorente-Mirandes et al. (2014) have reported As contamination in both edible mushrooms and mushroom supplements. In the case of *Cordyceps*, the problem of heavy metal contamination, particularly Pb has been documented. Wu et al. (1996) reported two cases of Pb poisoning after the consumption of *Cordyceps* in China, marking one of the early instances of such contamination being associated with *Cordyceps* consumption. Subsequently, the Chinese Food and Drug Administration issued a warning about lead contamination in *C. sinensis* on February 15, 2016 (Wei et al., 2017) In a limited number of studies examining heavy metal contamination in *C. sinensis* Wei et al. (2017) identified elevated levels of As, Cd, and Pb contamination with *C. sinensis*. In a subsequent study, Li et al. (2019), an accumulation of As with *C. sinensis*. In another comprehensive study, Zhou et al. (2017) reported the accumulation of five heavy metals, namely, copper (Cu), Pb, As, Cd, and Hg. Their findings demonstrated that all tested batches (17) are at a safe level of accumulation of Cu, Pb, Cd, and Hg, which meets international standards of the Chinese Medicine-Chinese Herbal Medicine Heavy Metal Limit. Nevertheless, the contamination is higher in the stroma than in the body of caterpillars. In contrast, the troublesome effect is with As, where excess accumulation of As was reported mainly in the larvae body part of the fungus (88.24%), which is also seven to twelve times higher than that of the stroma. Li et al. (2019) emphasized the heightened toxicity of the unknown As) in *C. sinensis*, perhaps converted into free, highly toxic inorganic As. Xiao et al. (2021) highlighted a significant accumulation of As in *Cordyceps*, specifically within the digestive tract of the host larva. The findings, underscore the necessity for quality assessment of *Cordyceps*-based products, a domain that is still in its early stages of development (Guo et al., 2018).

8. Isolation and Cultivation of *C. sinensis*

8.1 *Isolation*

Indeed, artificial cultivation of *C. sinensis* has become necessary to meet the growing demand for the fungus. This method allows for controlled and efficient production, ensuring a more sustainable supply to meet market needs. Despite the generally lower cost of products derived from cultivated *Cordyceps* compared to natural ones, consumer acceptance has been limited. This reluctance stems from the belief that effective components may vary between the two types, with artificially grown fungi potentially containing lower levels of these components (Zhou et al., 2019). Due to the lack of comprehensive studies, it is challenging to alter public perception. However, it has been recognized that certain fungal strains isolated from *C. sinensis* demonstrate pharmacological efficacy similar to that of wild fungal material, even when cultivated in a laboratory setting (Yan et al., 2014).

8.2 *Cultivation*

The fungi can be isolated using the fruiting body in the laboratory. To do this, the fruiting body needs to be cut into small pieces after washing with 1% sodium hypochlorite (NaOCl) for 1 min, followed by a series of washings with distilled water, placed on a PDA medium at a low temperature (e.g., 10–20°C). Then the hyphal tip method can be used to get a pure culture. The fungus mycelium is hyaline, branched, and smooth-walled and produces diffuse, white colonies with dense aerial mycelium and regular margins. This is the asexual stage (anamorph) of the fungus, also called *Hirsutella sinensis* (Ko et al., 2017).

Due to the scarcity of natural *Cordyceps* and its inability to meet the growing demand, mass-scale production of the species has emerged. The fungal mycelia can also be grown on media other than the PDA; therefore, in commercial cultivation, it uses different nutritional media such as insect larvae residuals (silkworms) and various cereal grains. However, the least fungus growth success has been reported when it was attempted to culture it using insect larvae (Baral., 2017). The infection rate of the fungus under artificial conditions is considerably lower, maybe due to the absence of companion microflora (like the natural environment has) to stimulate their infection (Sun et al., 2023). There is also a concern that the medicinal properties of the fungus could vary based on the cultured medium; nonetheless, there are no studies that prove such a variation. Therefore, it has been suggested that the feasibility of the usage of any of the availble media is favored (Tuli et al., 2014). However, after generating the spawn in the media, it is incubated on another disinfected substrate and permitted to colonize the substrate. The mycelium grows throughout the substrate utilizing nutrients during the brooding stage. This process is commonly known as the "spawn run" or "production run".

Typically, the temperature is maintained at 25°C during this spawn run. The mycelia then enter the regenerative stage and are prepared for mushrooming. This process is triggered by a sudden drop in temperature (5–10°C), and exposure to carbon dioxide (CO_2, below 500 lux and 2,000 lux for 12 hours). Relative humidity is also kept at 85%–92% (Memoria et al., 2023). There might be slight variations in the specified conditions when cultivating the fungus in different laboratories. Notably, Li et al. (2023) endeavoured to replicate conditions as closely as possible to the natural habitat of the fungus. They kept the laboratory windows open to allow for adequate sunshine and breeze, and the cultivated samples were placed in paper cups. The room temperature naturally fluctuates; during the night and the morning, it ranges from 18–19°C, increasing to between 22–23°C in the early afternoon. Water spraying (using an atomizer, twice a day) is applied to maintain humidity, with an absolute humidity of about 90%–100% on the surface of stromata and a relative humidity of about 30%–40%.

8.3 Mass Production

The mass production of *C. sinensis* poses significant challenges, with low nutrient infection rates and unclear infection mechanisms being key issues that need to be addressed. The inoculation method for the fungus entails substantial labor costs, time, and material resources. Moreover, there is an increased mortality rate of the host larvae attributed to mechanical injuries (Wu et al., 2022b). In the process of fungal biomass production, various fermentation techniques have been employed, with solid and submerged techniques being the most widely used globally (Yan et al., 2014). In the USA and Japan, a significant number of manufacturers utilize solid-medium cultivation, primarily because of its lower production costs. However, a drawback is that fungal mycelia contain a higher content of grain matter compared to their active components or substances (Ghatnur et al., 2015). Compared with the solid fermentation method, submerged fermentation is considered the most reliable as culture conditions and other relevant parameters such as growth supplements and additives can be easily controlled (Kaushik et al., 2020). In their latest paper, Chellapandi and Saranya (2023) provided additional insights into these two methods; readers can find more details by referring to their paper.

9. Mechanism of Infection

Like many other entomopathogenic fungi, *Cordyceps* shares the same infection mechanism. The initial symptoms may appear after a year, as they exhibit 5–12 months of symptomless infection (Meng et al., 2021). *Cordyceps* go through the teleomorph and anamorph phases to complete their life cycle. The fungi can persist in the soil for an extended period until they encounter a suitable host. In this case, the conidia (or mycelia) deposited in deeper soils can infect the soil-dwelling insects in autumn (Li et al., 2011). The fungus can infect approximately 30 species of *Thitarodes* caterpillars (Lepidoptera: Hepialidae) (Baral, 2017). The female adult insect *Thitarodes* lays the eggs in grassland and the larvae emerge after the incubation. Then they find a suitable plant substrate and habitat in the root system (topsoil). The entire larval stage lives under ground, which may take three to four years or more, and feed on the caudexes and roots of the host, for example, the alpine plant. The larvae eventually get infected by the *Cordyceps* (Liu et al., 2021). The fungal infection may induce a larva to sluggishly crawl upwards at a position near the ground to facilitate their respiration (Guo et al., 2015). The fungus generally infects the newly molted larva, and some scholars suggest that the infection could prominently happen in the neck region of the host insects (Guo et al., 2017a). However, in the infection process, initially, the fungus needs to go through the larval cuticle, which is a protective barrier for the larvae against desiccation by sunlight (UV) and protects their floppy bodies from physical damage (Baral, 2017). Immediately after the attachment of the spores to the hard cuticle, the spores start germinating. First, they produce the germ tube, appressoria, and penetration peg. The mechanical pressure and the extracellular enzymes produced by appressoria facilitate penetration (Zhou et al., 2021).

On the other hand, penetration may occur possible through the opening of the larva-spiracle, providing an accessible pathway into the host's body. Spores can also enter the larvae through the digestive tract. If the fungus spores adhere to the tender plant roots, larvae may un-intentionally ingest them while consuming plant parts. The surviving spores in the digestive system can then germinate and initiate the infection (Guo et al., 2017a). Additionally, the exact pathway of fungal entry into the host larvae lacks definitive evidence, and various possibilities, including those mentioned earlier, are considered (Wu et al., 2022b). Its crucial to emphasize that factors such as humidity, temperature,and nutritional and chemical cues, play a significant role in successful infection (Skinner et al., 2014). After penetration, fungi subsequently grow as discrete yeast-like cells (hyphal bodies) inside the insect body and hemocoel. Those cell types start multiplying and invade many parts of the host, including the alimentary canal and midgut. Its worth noting that *C. sinensis* deviates from other entomopathogenic fungi by spending an extended period of more than five months in the larval hemocoel during its vegetative growth (Li et al., 2023).

Increased nutrient acquisition by the fungus weakens the host's body inducing paralysis (Zhou et al., 2021). Subsequently, the larva undergoes a gradual transformation, becoming stiff and ultimately mummified by the fungus's mycelium. This process results in the formation of a caterpillar-shaped sclerotium known as *Chongcao* in Chinese, a state reached during winter dormancy (Wei et al., 2021). Later, the fungal fruiting body erupts from the dead host larvae in the spring or summer of the following year, generally from the head region and its stroma breaks through the topmost soil layer, continuously growing and appearing above ground. Eventually, it forms a stalked fruiting body-ascostroma, baring millions of sexual spores (Guo et al., 2015; Liu et al., 2021). The mechanism behind fruiting body formation is still unclear, nonetheless, generally, this may happen owing to some severe stressors such as sudden temperature fluctuations, fire, floods or submerged conditions, or nutrient deficiency (Feng et al., 2017). Eventually these spores will infect the new host and continue the life cycle.

10 Environmental Factors in the Growth and Emergence of *C. sinensis*

10.1 Temperature

In order to germinate a stroma from the muscardine, a cadaver generally needs temperature changes. The optimum temperature ranges for the growth of the hyphae are between 15 and 18°C. However, the hyphal growth can still be seen even at 25°C. They can also grow even at 2°C but cannot retain it below 0°C (Li et al., 2011). It was noticed that under the range of −4 to 4°C, the fungus produces multiple stromata (Guo et al., 2015).

10.2 Light Intensity

Reports indicate that the stromata of *C. sinensis* exhibit rapid upward growth during the spring season reaching a length of 3 mm per day below the soil surface. However,this growth rate may decrease by half (1.5 mm/day) upon exposure to sunlight as the fungus emerges from the soil (Guo et al., 2015).

10.3 Soil Pressure

Typically, the wild-collected *Cordyceps* exhibits the emergence of a single stroma. However, under laboratory cultivation conditions, where the soil factor is eliminated, multiple stromata have been observed to germinate at different positions in the larvae's dead bodies. This observation highlights the influence of the soil factor on the sporulating position of the fungus. Therefore, there is a distinct difference between the appearance of wild-harvested and indoor-cultivated fungi (Guo et al., 2015).

11. Nutritional and Medicinal Value of *C. sinensis*

Nutrition is is a crucial factor when selecting food and *Cordyceps* is no exception. It has been reported to contain various nutrients, including a range of essential amino acids and vitamins such as B1, B2, B12, and K. Additionally, *Cordyceps* comprises carbohydrates such as monosaccharides, oligosaccharides, polysaccharides alongwith proteins, sterols, and nucleosides like N6-4-methylbutyrat-adenosine, 3'-deoxy-6-*O*-methylinosine) (Tuli et al., 2014; Qu et al., 2022). However, public interest in the fungus does not stem from its nutritional significance but rather from its medicinal value. As mentioned earlier, the medicinal properties of *C. sinensis* are well known, and ongoing research by scientists continues to unveil additional applications for this fungus. Moreover, modern pharmaceutical experiments indicate heightened effectiveness of the *Cordyceps*-based treatments, particularly in a dose-dependent manner (Li et al., 2023). In this section, we aim to discuss the medicinal values of the fungus incorporating novel findings alongside background knowledge from previous understandings.

11.1 Improvement of the Respiratory System

The impact of *C. sinensis* on lung injury, particularly in reducing the pulmonary infiltrate, hypoxemia, and pulmonary capillary wedge pressure deficiency, is crucial, attributed to its anti-inflammatory and antioxidant properties (Fu et al., 2019). However, in their study, Wang et al. (2016) analyzed the effect of *C. sinensis* Corbrin capsule (1.2 g) on asthma. The researchers administered the capsule to a selected group of individuals (60) in various groups (but over 18 years old), three times per day for 3 months. Despite a few incidents (five patients), Wang et al. (2016) discovered that the *Cordyceps* treatment significantly improves the quality of life for the majority of patients, reducing asthma symptoms and inflammation and while improving lung function. In a susequent study, Chen et al. (2020a) tested the effect of fungal extraction on ovalbumin-induced allergic rhinitis and asthma using a rat model experiment. It was found that there was a reduction of immunoglobulin E, in the plasma and bronchoalveolar fluid. Furthermore, eosinophil peroxidase levels in the bronchoalveolar fluid were also reduced. The treatment also suppressed the increase in the level of interleukin (IL-4, IL-5, and IL-13) in rat lung tissue. It also minimized the level of immunoglobulin Eovalbumin- immunoglobulin E and interleukin (IL-4, IL-13) in the nasal fluid.

Additionally, the treatment also reduced airway responsiveness in rats. Yang et al. (2018) demonstrated the potential of *C. sinensis* to alleviate chronic inflammatory airway diseases using a rat model of chronic obstructive pulmonary disease, induced by cigarette smoking and an intratracheal injection of lipopolysaccharide.

Accordingly, the researchers observed that the fungal treatment significantly reduced the thickening of the airway wall, including collagen deposition, fibrosis of the airway wall, hypertrophy of smooth muscles, and hyperplasia of the epithelium in diseased rats. Moreover, Yang et al. (2018) reported a reduction in the accumulation of inflammatory cells and the synthesis of inflammatory cytokines in bronchoalveolar lavage fluid. Other than that, in a supportive study, Chen et al. (2012) showed that fungal treatment can minimize the formation of reactive oxygen species (ROS) and also minimize lung fibrosis.

In a separate study, Liu et al. (2016) demonstrated the inhibition of smoke extract-induced senescence in human bronchial epithelial cells using *C. sinensis*. They have found a decrease in cellular senescence upon fungal treatment. They further observed a decrease in ROS and the phosphatidylinositol 3-kinase (PI3K)/AKT-rapamycin complex (mTOR) signaling pathways after the treatment. Note that the activation of ROS and PI3K-AKT-mTOR signaling pathways is involved with cell senescence. Furthermore, the authors highlighted the importance of their study directed towards the slowing of the human aging process as well as aging-related diseases.

11.2 *Antiaging Effect and Lifespan Increase*

Aging is an inevitable phenomenon in human life, leading to a gradual reduction in the functioning of organs, tissues, and cells over time. Memory dysfunction, decline in sexual function, and skin shrinking are among theeffects of aging (Ganceviciene et al., 2012). In a study, Ji et al. (2009) conducyed experiments on mice by feedingthem *C. sinensis* extract and observed morphological and biochemical changes. They observed improved learning and memory in the tested animals, along with the increased activity of antioxidative enzymes in aged mice. The researchers also noted that the level of superoxide dismutase activity in the liver, brain, and erythrocytes in treated mice was compared to the control group. Additionally, there was a significant decrease in catalase activity in the blood. Moreover, a noteworthy reduction in the activity of GSH-Px, which is a more efficient metabolizer of hydrogen peroxide than catalase in glutathione depletion, was reported in the blood.

Further, Ji et al. (2009) observed a significant increase in spontaneous lipid peroxidation in the brains and livers coupled with a decline in monoamine oxidase activity as a result of the treatments. Nevertheless, the overall results of their study evidence the delaying of the aging of the treated mice. In another study, Zou et al. (2015) observed a prolonged lifespan of the fruit fly, *Drosophila melanogaster*, after being fed with the *C. sinensis* oral liquid. They were fed with this liquid at concentrations of 0.02, 0.06, and 0.20 mg/ml, and the results showed significant increase in the lifespan of the fruit flies compared to the control by 25, 31, and 32%, respectively. They further recognized that this occurs through an anti-oxidative stress pathway, which involves the upregulation of copper-zinc-containing superoxide dismutase 1 and catalase activity and the inhibition of lipofuscin accumulation. Tan et al. (2011) examined the gene expression profiles of neocortex and gastrocnemius from young (five months of age), old (25 months), and old fungi mycelia fermentation products of *C. sinensis*-treated (0.3 g/kg) C57Bl/6 mice. The lifespan was extended to 10–66 days at 50% survival with treatment and 45–153 days at 10% survival. The age of the oldest surviving mice was extended to 152 days (1.5 g/kg) and > 210 days (both 0.5 and 1.0 g/kg), showing anti-aging properties related to humans.

11.3 Potential Anti-Tumor Effects

It has been documented that *Cordyceps* has anticancer properties in different tumor conditions (Luo et al., 2019). In a study, Nakamura et al. (2015) investigated the hot water extraction (70°C for 5 min) of *C. sinensis* for anticancer and antimetastatic effects against B16 mouse melanoma (B16) and Lewis lung carcinoma cells. They have obtained positive results, as the tested extraction has a considerable ability to inhibit the cancer-causing and spreading. This action is found to be antagonized by stimulating the adenosine A3 receptor, and further promoted by an adenosine deaminase inhibitor. Moreover, glycogen synthase kinase (GSK)-3β activation and cyclin D1 suppression help to mitigate the carcinogenic effect. The platelet aggregation (induced by cancer cells) and the invasiveness of cancer cells are also inhibited by blocking the activity of matrix metalloproteinase (MMP)-2 and MMP-9 and hastening the release of tissue inhibitors of metalloproteinase (TIMP)-1 and TIMP-2 from cancer cells.

In a similar study, Huo et al. (2017) also worked with Lewis lung cancer cells, where the effect of extraction from *C. sinensis* gave positive results in mouse testing. They noted the mitigation of tumor growth, toxicity, and a prolongation of the survival time. In their most recent bioinformatics analysis and in vivo validation study, Zhang et al. (2023a) also came up with the analogous funding for effective inhibition of Lewis lung cancer cells. In contrast, through the orthotopic xenograft mouse model, Li et al. (2020) investigated the effect of *C. sinensis* on breast tumors and the change of the macrophage phenotype in the tumor. They also obtained positive results, as the fungal extracts inhibited the 4T1 tumor growth depending on the dosage. Li et al. (2020) further recognized the positive correlation between tumor growth inhibition and macrophage M1-like polarized phenotype, as the fungal extract boosted macrophage differentiation towards the M1 phenotype. Tan et al. (2023) recognize that fungal treatment can also be used to minimize liver cancer by inhibiting the proliferation of H22 tumors and improving immune functions. In line with these studies, several other studies have also shown the tremendous ability of fungal extraction in different cancer reduction studies (e.g., Fong et al., 2019; Guan et al., 2020; Arul et al., 2023).

11.4 Diabetes Attenuation

Lo et al. (2006) tested the effect of *C. sinensis* on nicotinamide and streptozotocin-induced diabetes in rats. An interesting aspect of their study is that the researchers have compared the anti-hyperglycemic activity of the fermented fungal mycelia and broth (normally discarded as waste products of industrial fermentation processes with that of the natural products from the fruiting bodies). According to their findings, the diabetic rats fed with *C. sinensis* fruiting bodies, fermented mycelia, and/or broth showed a significantly improved glucose response in the oral glucose tolerance test and the 2-hour postprandial period, along with a significantly decreased serum fructosamine concentration. El Ashry et al. (2012) also studied streptozotocin-induced diabetes in rats; oral administration of *Cordyceps* decreased serum glucose, fructosamine, total cholesterol, triglycerides levels, insulin resistance index, and pancreatic malondialdehyde content.

Furthermore, the treatment improved the levels of serum insulin, HDL-cholesterol, total antioxidant capacity levels, β cell function percent, and pancreatic-reduced glutathione. Despite these, Choi (2011) demonstrated *C. sinensis* against type II diabetes. The mice were treated with *Cordyceps* doses of 200 mg/kg body weight, which resulted in a 47% drop in fasting blood glucose after seven weeks compared with the control. The study further proves that the *Cordyceps* treatment works effectively by maintaining normal blood glucose levels through the increase in plasma insulin and leptin expression. Dong et al. (2019) showed that the fungus treatment works against diabetic nephropathy by inhibiting the epithelial–mesenchymal transition and safeguarding against renal fibrosis. A similar effect may also be present even in the pharmaceutically available fermented *C. sinensis* powder (Bailing capsule), as proved by Zhang et al. (2022a).

On the other hand, clinical testing of *C. sinensis* against kidney disease caused by diabetics has shown over 20 years of history, yet studies in this field are continuously growing. The efficacy of this fungus in this regard is still not fully understood, however, available findings show its significant ability to control diabetic kidney disease (Luo et al., 2015). One of the most recent studies carried out by Zhang et al. (2023c) indicated that *C. sinensis* minimizes diabetic kidney disease, possibly through the nephroprotective effects achieved by promoting proliferation and inhibiting apoptosis of renal proximal tubular cells.

11.5 Minimize Heart Diseases

Yan et al. (2013) found that *C. sinensis* shows cardio-protection via enhanced adenosine receptor activation, followed by a multiple-step pathway. A later study conducted by Liu et al. (2014) used a rat model system to recognize the effect of this fungus on minimizing heart disease and the results were positive. Nevertheless, one of the most recent studies conducted by Fang et al. (2021) has conducted a comprehensive analysis of the importance of *C. sinensis* (*H. sinensis*) in enhancing cardiac function in a heart failure mouse model. The authors have further found that the fungus treatment works well by protecting myocardial tissue to prevent ventricular remodeling and collagen hyperplasia after transverse aortic constriction. The results further showed that fungal treatment is significant even when the tested individuals are at a high risk of heart failure. Wu et al. (2018) evaluated the protective effect of fermented *C. sinensis* over Doxorubicin-induced cardiotoxicity in a rat-based model. Note that doxorubicin is recognized as an antitumor drug applied in chemotherapy, however, it also has side effects such as cardiotoxicity. The results of Wu et al. (2018) indicated that fermented *C. sinensis* reduced the left ventricular weight index, heart weight index, and mortality rates.

On the other hand, the treatment enhanced the diastolic blood pressure and arterial pressure. Nevertheless, results further showed that *C. sinensis* reduces cardiotoxicity by inhibiting myocardial hypertrophy and myocardial damage, ameliorating systolic function, and the antioxidant enzyme system. Upon further treatment, it showed an increase in cardiac energy metabolism. Another aspect of heart-related issues is irregular heartbeat arrhythmia, which can also be overcome with the *Cordyceps* treatment. Wang et al. (2022) found that *C. sinensis* can be used to treat such conditions based on a meta-analysis study. In contrast to the above positive results, Liao et al. (2019) found that combined treatment of the plants, *Rhodiola crenulata* and *C. sinensis* did not change the lipid profile of the blood in young sedentary participants. It is also noteworthy that the lipid content of the blood is also an important factor in heart disease, as increased lipids can narrow and block the arteries.

11.6 *Anti-Obesity*

The effects of *Cordyceps* in preventing excess weight, fat accumulation, and lowering triglyceride levels and related issues have been extensively studied by several researchers and their outcome in dealing with the obesity is intriguing (Jang et al., 2022; Wu et al., 2022a). The action of *C. sinensis*, in mitigating the lipidemic effect and developing and accumulating as adipose is an essential factor in minimizing obesity. Jang et al. (2022) observed that *Cordyceps* extracts have a considerable ability to prevent excess weight gain, fat accumulation, liver hypertrophy, and lowered triglyceride levels, thus overcoming obesity. Tiamyom et al. (2019) tested the *C. sinensis* extracts on adipogenesis and lipase activity. It was recorded that the treatment inhibited intracellular lipid accumulation in 3T3-L1 adipocytes and pancreatic lipase activity. Wu et al. (2019) isolated the polysaccharides from *C. sinensis* and further tested their ability to mitigate diet-induced obesity. They also used the mouse model and found anti-obesogenic effects.

Most interestingly, they have found that the treatment promoted the growth of *Parabacteroides goldsteinii*, a commensal gut bacterium that plays a role in reducing the impact of a high-fat diet. On the other hand, Chen et al. (2020b) also came up with the positive and negative results of feeding *C. sinensis* polysaccharides to mice. Their study also highlighted the fungus's ability to prevent body weight gain; in contrast, it may accelerate liver fat accumulation, fibrosis, hepatic steatosis, and steatohepatitis in treated animals. Furthermore, Chen et al. (2020b) also recorded the changes in the natural gut microflora population upon treatment in mice, which also had negative effects (progression from obesity to nonalcoholic steatohepatitis and related diseases). However, findings suggest that positive changes in gut microflora (drop-down abundance of the *Clostriduim* genus) facilitate the reduction of obesity. The researchers also observed the negative changes in the abundance of various microbes (e.g., *Anaerostipes*, *Bactereroides*, *Dehalobacterium*, *Dorea*, and *Odoribacter*), which are associated with escalating metabolic syndrome (Li et al., 2022).

12. Medicare for Animal Husbandry

Widespread factory farming is one of the current trends in the world due to its inexpensive production costs compared with traditional farming. Nonetheless, many critics have been raised owing to their poor animal welfare, with intensive farming undoubtedly causing animals to suffer from stress and various diseases throughout their lives. There is a strong link between the food that they get and the food that we consume. More precisely, we can explain this, as the animals may receive various foods, treated drinks, antibiotics, and many other chemicals from birth to death. Upon consumption of those animal products (e.g., milk, eggs, and meat), consumers may introduce those compounds into their bodies as there is a possibility of bioaccumulation of heavy metals and other chemicals, posing the health risks. Thus, it is clear that animal farming should be done with care (Anomaly, 2015; Hayek, 2022).

However, according to the literature survey, there are only a few studies conducted on the importance of *Cordyceps* for the perception of animals, and they are mainly focused on poultry farming. The study conducted by Shihab and Nafea (2023) was quite impressive as it sought to determine the impact of adding *C. sinensis* extract and a probiotic to broiler diets on assessing their productivity (body weight). They fed the one-day-old, unsexed chicks with *C. sinensis* extract at a level of 300 and 600 mg/kg feed as well as probiotics (3 and 6 g/kg feed). They also tested the combination of both fungal and probiotic feeds. Shihab and Nafea (2023) obtained superior results with the combined application of the ratio (*Cordyceps:* Probiotics) 300 mg/kg feed: 3 g/kg feed, and 600 mg/kg feed: 3 g/kg fodder and 6 g/kg feed, except for fungal treatment at 600 mg/kg feed. Their study further shows the effectiveness of *C. sinensis* extract utilization without causing any side effects to the animals and can also minimize the utilization of a probiotic application. In a separate study, Han et al. (2015) proved that *Cordyceps* (*C. militaris*) treatment can increase the body weight of the broiler chicks, by minimizing unhealthier feeds and treatments. In addition to chicken weight gain, Park's (2011) findings revealed that *Cordyceps* treatment can lower the pathogenic *E. coli* and *Salmonella* and increase the beneficial *Bifidobacterium*.

Conclusion

Owing to the immense popularity of the medical value of *C. sinensis*, it has begun to to be cultivated artificially. It is also worthwhile, as the current wild collection is not enough to satisfy the demand, and it also minimizes the ruthless wild collection. However, the culturing of the fungus is still not at the expected level of success, mainly due to the requirement of specific conditions such as the long lifecycle of the fungi, and unknown factors behind the infection mechanism. Therefore, researchers need to make more attempts on this aspect as well. Under these circumstances, many other successful studies have recently been conducted to reveal many more therapeutic properties, including enhancement of respiratory health, anti-aging effects, lifespan increase, anti-tumor effects, diabetes attenuation, heart disease, and anti-obesity. However, unfortunately, most of those experiments have been conducted based on animal trials; thus, human-based studies need to be done in the future. On the other hand, the importance of fungal-based treatments in animal husbandry has also been raised, mostly along with poultry farming. Moreover, this also could be the new direction of studies, aiming at safer foods and a healthy life.

References

Anomaly, J. (2015). What's wrong with factory farming? Public Health Ethics, 8(3): 246–254. https://doi.org/10.1093/phe/phu001

Arul, V., Kandasamy, R. and Adharshna, S.P.M. (2023). Antioxidant and cytotoxic potential of potentized preparation of *Cordyceps sinensis in vitro* in cancer cell lines. Ind. J. Res. Homoeopathy, 17(1): 3–10.

Ashraf, S.A., Elkhalifa, A.E.O., Siddiqui, A.J., Patel, M., Awadelkareem, A.M. et al. (2020). Cordycepin for health and wellbeing: A potent bioactive metabolite of an entomopathogenic *Cordyceps* medicinal fungus and its nutraceutical and therapeutic potential. Molecules, 25(12): 2735. https://doi.org/10.3390/molecules25122735

Bamisile, B.S., Akutse, K.S., Siddiqui, J.A. and Xu, Y. (2021). Model application of entomopathogenic fungi as alternatives to chemical pesticides: Prospects, challenges, and insights for next-generation sustainable agriculture. Front. Plant Sci., 12: 741804. https://doi.org/10.3389/fpls.2021.741804

Baral, B. (2017). Entomopathogenicity and biological attributes of Himalayan treasured fungus *Ophiocordyceps sinensis* (Yarsagumba). J. Fungi, 3(1): 4. https://doi.org/10.3390/jof3010004

Bibi, S., Wang, Y.-B., Tang, D.-X., Kamal, M.A. and Yu, H. (2020). Prospects for discovering the secondary metabolites of *Cordyceps senensis* Lato by the integrated strategy. Med. Chem., 17(2): 97–120. https://doi.org/10.2174/1573406416666191227120425

Buenz, E.J., Bauer, B.A., Osmundson, T.W. and Motley, T.J. (2005). The traditional Chinese medicine *Cordyceps sinensis* and its effects on apoptotic homeostasis. J. Ethnopharmacol., 96(1–2): 19–29. https://doi.org/10.1016/j.jep.2004.09.029

Byers, A.C., Byers, E., Shrestha, M., Thapa, D. and Sharma, B. (2020). Impacts of *Yartsa gunbu* harvesting on alpine ecosystems in the Barun valley, Makalu-Barun National Park, Nepal. Himalaya, 39(2): 44–59.

Chellapandi, P. and Saranya, S. (2023). *Ophiocordyceps sinensis*: A potential caterpillar fungus for the production of bioactive compounds. Explor. Res. Hypothesis Med., 1–14. https://doi.org/10.14218/erhm.2023.00040

Chen, J., Chan, W.M., Leung, H.Y., Leong, P.K., Yan, C.T.M. and Ko, K.M. (2020a). Anti-inflammatory effects of a *Cordyceps sinensis* mycelium culture extract (Cs-4) on rodent models of allergic rhinitis and asthma. Molecules, 25(18): 4051. https://doi.org/10.3390/molecules25184051

Chen, L., Zhang, L., Wang, W., Qiu, W., Liu, L. et al. (2020b). Polysaccharides isolated from *Cordyceps sinensis* contribute to the progression of NASH by modifying the gut microbiota in mice fed a high-fat diet. PloS One, 15(6): e0232972. https://doi.org/10.1371/journal.pone.0232972

Chen, M., Cheung, F.W., Chan, M.H., Hui, P.K., Ip, S.P. et al. (2012). Protective roles of *Cordyceps* on lung fibrosis in cellular and rat models. J. Ethnopharmacol., 143(2): 448–454. https://doi.org/10.1016/j.jep.2012.06.033

Chen, P.X., Wang, S., Nie, S. and Marcone, M. (2013). Properties of *Cordyceps sinensis*: A review. J. Funct. Foods, 5(2): 550–569. https://doi.org/10.1016/j.jff.2013.01.034

Choi, J.-W. (2011). Anti-diabetic effect of the exopolysaccharides (EPS) produced from *Cordyceps sinensis* on ob/ob Mice. Kor. Soc. Biotech. Bioeng., 26(1), 33–40. https://doi.org/10.7841/ksbbj.2011.26.1.033

Cordyceps sinensis market size, demand, scope and revenue analysis by 2030 (2023). Databridgemarketresearch.com (https://www.databridgemarketresearch.com/reports/global-cordyceps-sinensis-market) (Accessed: December 20, 2023).

Ćwieląg-Drabek, M., Piekut, A., Szymala, I., Oleksiuk, K., Razzaghi, M. et al. (2020). Health risks from consumption of medicinal plant dietary supplements. Food Sci. Nutr., 8(7): 3535–3544. https://doi.org/10.1002/fsn3.1636

Dai, Y., Wu, C., Yuan, F., Wang, Y., Huang, L. et al. (2020). Evolutionary biogeography on *Ophiocordyceps sinensis*: An indicator of molecular phylogeny to geochronological and ecological exchanges. Geosci. Front., 11(3): 807–820. https://doi.org/10.1016/j.gsf.2019.09.001

Devkota, S. (2006). Yarsagumba [*Cordyceps sinensis* (Berk.) Sacc.]; Traditional utilization in Dolpa District, Western Nepal. Our Nature, 4(1), 48–52. https://doi.org/10.3126/on.v4i1.502

Dong, Z., Sun, Y., Wei, G., Li, S. and Zhao, Z. (2019). A Nucleoside/nucleobase-rich extract from *Cordyceps sinensis* inhibits the epithelial-mesenchymal transition and protects against renal fibrosis in diabetic nephropathy. Molecules, 24(22): 4119. https://doi.org/10.3390/molecules24224119

Dworecka-Kaszak, B. (2014). *Cordyceps* fungi as natural killers, new hopes for medicine and biological control factors. Ann. Parasitol., 60(3): 151–158.

El Ashry, F.E.Z.Z., Mahmoud, M.F., El Maraghy, N.N. and Ahmed, A.F. (2012). Effect of *Cordyceps sinensis* and taurine either alone or in combination on streptozotocin induced diabetes. Food Chem. Toxicol., 50(3–4): 1159–1165. https://doi.org/10.1016/j.fct.2011.12.020

Fang, M., Jin, L., Mao, W., Jin, L., Cai, Y. et al. (2021). *Hirsutella sinensis* fungus improves cardiac function in mouse model of heart failure. Biomed. Pharmacother., 142: 111885. https://doi.org/10.1016/j.biopha.2021.111885

Feng, K., Wang, L.Y., Liao, D.J., Lu, X.P., Hu, D.J. et al. (2017). Potential molecular mechanisms for fruiting body formation of *Cordyceps* illustrated in the case of *Cordyceps sinensis*. Mycology, 8(4): 231–258. https://doi.org/10.1080/21501203.2017.1365314

Fong, P., Loi, C.-I., Wan-Fong, U., Choi, C.-I., Yi, T. and Meng, L.-R. (2019). Antitumor effects of MRS5698, a cordycepin derivative, on endometrial cancer cells. Nat. Prod. Comm. 14(10): 1934578X1988156. https://doi.org/10.1177/1934578x19881564

Fu, S., Lu, W., Yu, W. and Hu, J. (2019). Protective effect of *Cordyceps sinensis* extract on lipopolysaccharide-induced acute lung injury in mice. Biosci. Rep., 39(6): BSR20190789. https://doi.org/10.1042/BSR20190789

Ganceviciene, R., Liakou, A.I., Theodoridis, A., Makrantonaki, E. and Zouboulis, C.C. (2012). Skin anti-aging strategies. Dermatoendocrinol., 4(3): 308–319. https://doi.org/10.4161/derm.22804

Ghatnur, S.M., Parvatam, G. and Balaraman, M. (2015). Culture conditions for production of biomass, adenosine, and Cordycepin from *Cordyceps sinensis* CS1197: optimization by desirability function method. Pharmacogn. Mag., 11(Suppl 3): S448–S456. https://doi.org/10.4103/0973-1296.168946

Giri, S., Ojha, N., Subedi, S., Rana, S., Bhandari, Y. and Khanal, A. (2023). Ethnobotany of the medicinal plants: Case of *Ophiocordyceps sinensis* (Yarsagumba) and its benefits for Nepal, India, and Bhutan. J. Environ. Info. Lett., 9(2): 78–86. https://doi.org/10.3808/jeil.202300103

Guan, J., Han, L., Shi, N., Zhu, H. and Wang, J. (2020). Development, in vitro biocompatibility, and antitumor efficacy of acetic acid-modified *Cordyceps sinensis* polysaccharide nanoparticle drug delivery system. Braz. J. Pharm. Sci., 56. https://doi.org/10.1590/s2175-97902019000418470

Guo, L.X., Hong, Y.H., Zhou, Q.Z., Zhu, Q., Xu, X.M. and Wang, J.H. (2017a). Fungus-larva relation in the formation of *Cordyceps sinensis* as revealed by stable carbon isotope analysis. Sci. Rep., 7(1): 7789. https://doi.org/10.1038/s41598-017-08198-1

Guo, L.-X., Xu, X.-M., Hong, Y.-H., Li, Y. and Wang, J.-H. (2017b). Stable carbon isotope composition of the lipids in natural *Ophiocordyceps sinensis* from major habitats in China and its substitutes. Molecules, 22(9): 1567. https://doi.org/10.3390/molecules22091567

Guo, L.-X., Xu, X.-M., Liang, F.-R., Yuan, J.-P., Peng, J. et al. (2015). Morphological observations and fatty acid composition of indoor-cultivated *Cordyceps sinensis* at a high-altitude laboratory on Sejila mountain, Tibet. PloS One, 10(5): e0126095. https://doi.org/10.1371/journal.pone.0126095

Guo, L.X., Zhang, G.W., Li, Q.Q., Xu, X.M. and Wang, J.H. (2018). Novel arsenic markers for discriminating wild and cultivated *Cordyceps*. Molecules. 23(11): 2804. https://doi.org/10.3390/molecules23112804.

Han, J.C., Qu, H.X., Wang, J.G., Yan, Y.F., Zhang, J.L. et al. (2015). Effects of fermentation products of *Cordyceps militaris* on growth performance and bone mineralization of broiler chicks. J. Appl. Anim., 43(2): 236–241. https://doi.org/10.1080/09712119.2014.928630

Hayek, M.N. (2022). The infectious disease trap of animal agriculture. Sci. Adv., 8(44): eadd6681. https://doi.org/10.1126/sciadv.add6681

Huo, X., Liu, C., Bai, X., Li, W., Li, J. et al. (2017). Aqueous extract of *Cordyceps sinensis* potentiates the antitumor effect of DDP and attenuates therapy-associated toxicity in non-small cell lung cancer via IκBα/NFκB and AKT/MMP2/MMP9 pathways. RSC Adv., 7(66): 41796–41798. https://doi.org/10.1039/c7ra90091k

Irsad, Shahid, M., Haq, E., Mohamed, A., Rizvi, P.Q. and Kolanthasamy, E. (2023). Entomopathogen-based biopesticides: Insights into unraveling their potential in insect pest management. Front. Microbiol., 26(14): 1208237. https://doi.org/10.3389/fmicb.2023.1208237

Jaber, L.R. and Ownley, B.H. (2018). Can we use entomopathogenic fungi as endophytes for dual biological control of insect pests and plant pathogens? Biol. Control., 116: 36–45. https://doi.org/10.1016/j.biocontrol.2017.01.018

Jairoun, A.A., Shahwan, M. and Zyoud, S.H. (2020). Heavy metal contamination of dietary supplements products available in the UAE markets and the associated risk. Sci. Rep., 10(1): 18824. https://doi.org/10.1038/s41598-020-76000-w

Jang, D., Lee, E., Lee, S., Kwon, Y., Kang, K.S. et al. (2022). System-level investigation of anti-obesity effects and the potential pathways of *Cordyceps militaris* in ovariectomized rats. Med. Ther., 22(1): 132. https://doi.org/10.1186/s12906-022-03608-y

Ji, D.-B., Ye, J., Li, C.-L., Wang, Y.-H., Zhao, J. and Cai, S.-Q. (2009). Antiaging effect of *Cordyceps sinensis* extract. Phytother. Res., 23(1): 116–122. https://doi.org/10.1002/ptr.2576

Kaushik, V., Singh, A., Arya, A., Sindhu, S.C., Sindhu, A. and Singh, A. (2020). Enhanced production of cordycepin in *Ophiocordyceps sinensis* using growth supplements under submerged conditions. Biotechnol. Rep., 28: e00557. https://doi.org/10.1016/j.btre.2020.e00557

Ko, Y.F., Liau, J.C., Lee, C.S., Chiu, C.Y., Martel, J. et al. (2017). Isolation, culture and characterization of *Hirsutella sinensis* mycelium from caterpillar fungus fruiting body. PloS One, 12(1): e0168734. https://doi.org/10.1371/journal.pone.0168734

Laha, A., Badola, R. and Hussain, S.A. (2018). Earning a livelihood from Himalayan caterpillar fungus in kumaon Himalaya: Opportunities, uncertainties, and implications. Mount. Res. Dev., 38(4): 323–331. https://doi.org/10.1659/mrd-journal-d-17-00063.1

Li, J., Cai, H., Sun, H., Qu, J., Zhao, B. et al. (2020). Extracts of *Cordyceps sinensis* inhibit breast cancer growth through promoting M1 macrophage polarization via NF-κB pathway activation. J. Ethnopharmacol., 260(112969): 112969. https://doi.org/10.1016/j.jep.2020.112969

Li, M., Zhang, J., Qin, Q., Zhang, H., Li, X. et al. (2023). Transcriptome and metabolome analyses of *Thitarodes xiaojinensis* in response to *Ophiocordyceps sinensis* infection. Microorganisms, 11(9): 2361. https://doi.org/10.3390/microorganisms11092361

Li, S.P., Yang, F.Q. and Tsim, K.W.K. (2006). Quality control of *Cordyceps sinensis*, a valued traditional Chinese medicine. J. Pharm. Biomed. Anal., 41(5): 1571–1584. https://doi.org/10.1016/j.jpba.2006.01.046

Li, Y., Liu, Y., Han, X., Jin, H. and Ma, S. (2019). Arsenic species in *Cordyceps sinensis* and its potential health risks. Front. Pharmacol., 10: 1471. https://doi.org/10.3389/fphar.2019.01471

Li, Y., Talbot, C.L., Chandravanshi, B., Ksiazek, A., Sood, A. et al. (2022). *Cordyceps* inhibits ceramide biosynthesis and improves insulin resistance and hepatic steatosis. Sci. Rep., 12(1): 7273. https://doi.org/10.1038/s41598-022-11219-3

Li, Y., Wang, L., Xu, B., Zhao, L., Li, L. et al. (2021). Based on network pharmacology tools to investigate the molecular mechanism of *Cordyceps sinensis* on the treatment of diabetic nephropathy. J. Diabetes Res., 2021: 1–12 8891093. https://doi.org/10.1155/2021/8891093

Li, Y., Wang, X.-L., Jiao, L., Jiang, Y., Li, H. et al. (2011). A survey of the geographic distribution of *Ophiocordyceps sinensis*. J. Microbiol., 49(6): 913–919. https://doi.org/10.1007/s12275-011-1193-z

Liao, Y.H., Chao, Y.C., Sim, B.Y., Lin, H.M., Chen, M.T. and Chen, C.Y. (2019). *Rhodiola/Cordyceps*-based herbal supplement promotes endurance training-improved body composition but not oxidative stress and metabolic biomarkers: A preliminary randomized controlled study. Nutrients, 11(10): 2357. https://doi.org/10.3390/nu11102357

Lin, B.-Q. and Li, S.-P. (2011). *Cordyceps* as an herbal drug. pp. 73–105. *In*: Benzie, F.F. and Wachtel-Galor, S. (eds.). Herbal Medicine: Biomolecular and Clinical Aspects. 2nd Edition, CRC Press, Boca Raton, USA. http://dx.doi.org/10.1201/b10787-6

Liu, A., Wu, J., Li, A., Bi, W., Liu, T. et al. (2016). The inhibitory mechanism of *Cordyceps sinensis* on cigarette smoke extract-induced senescence in human bronchial epithelial cells. Int. J. Chron. Obstruct. Pulmon. Dis., 11: 1721–1731. https://doi.org/10.2147/COPD.S107396

Liu, J., Guo, L., Li, Z., Zhou, Z., Li, Z. et al. (2020). Genomic analyses reveal evolutionary and geologic context for the plateau fungus *Ophiocordyceps sinensis*. Chin. Med., 15: 107. https://doi.org/10.1186/s13020-020-00365-3

Liu, Q.-B., Liu, J., Lu, J.-G., Yang, M.-R., Zhang, W. et al. (2023). Quantitative 1H NMR with global spectral deconvolution approach for quality assessment of natural and cultured *Cordyceps sinensis*. J. Pharm. Biomed. Anal., 235(115603): 115603. https://doi.org/10.1016/j.jpba.2023.115603

Liu, W., Gao, Y., Zhou, Y., Yu, F., Li, X. and Zhang, N. (2022). Mechanism of *Cordyceps sinensis* and its extracts in the treatment of diabetic kidney disease: A review. Front. Pharmacol., 13: 881835. https://doi.org/10.3389/fphar.2022.881835

Liu, X., Zhong, F., Tang, X.L., Lian, F.L., Zhou, Q. et al. (2014). *Cordyceps sinensis* protects against liver and heart injuries in a rat model of chronic kidney disease: A metabolomic analysis. Acta Pharmacol. Sin., 35(5): 697–706. https://doi.org/10.1038/aps.2013.186

Liu, Y., Shi, M., Liu, X., Xie, J., Yang, R. et al. (2021). Arsenic transfer along the soil-sclerotium-stroma chain in Chinese *Cordyceps* and the related health risk assessment. Peer J., 9(e11023): e11023. https://doi.org/10.7717/peerj.11023

Llorente-Mirandes, T., Barbero, M., Rubio, R. and López-Sánchez, J.F. (2014). Occurrence of inorganic arsenic in edible Shiitake (*Lentinula edodes*) products. Food Chem., 158: 207–215. https://doi.org/10.1016/j.foodchem.2014.02.081

Lo, H.-C., Hsu, T.-H., Tu, S.-T. and Lin, K.-C. (2006). Anti-hyperglycemic activity of natural and fermented *Cordyceps sinensis* in rats with diabetes induced by nicotinamide and streptozotocin. Am. J. Chin. Med., 34(05): 819–832. https://doi.org/10.1142/s0192415x06004314

Luo, L., Ran, R., Yao, J., Zhang, F., Xing, M. et al. (2019). Se-enriched *Cordyceps militaris* inhibits cell proliferation, induces cell apoptosis, and causes G2/M phase arrest in human non-small cell lung cancer cells. OncoTargets Ther., 12: 8751–8763. https://doi.org/10.2147/OTT.S217017

Luo, Y., Yang, S.-K., Zhou, X., Wang, M., Tang, D. et al. (2015). Use of *Ophiocordyceps sinensis* (syn. *Cordyceps sinensis*) combined with angiotensin-converting enzyme inhibitors (ACEI)/angiotensin receptor blockers (ARB) versus ACEI/ARB alone in the treatment of diabetic kidney disease: A meta-analysis. Renal Failure, 37(4): 614–634. https://doi.org/10.3109/0886022x.2015.1009820

Melgar, M.J., Alonso, J. and García, M.A. (2014). Total contents of arsenic and associated health risks in edible mushrooms, mushroom supplements and growth substrates from Galicia (NW Spain). Food Chem. Toxicol., 73: 44–50. https://doi.org/10.1016/j.fct.2014.08.003

Memoria, M., Shah, S.K., Anandaram, H., Ali, A., Joshi, K. et al. (2023). An internet of things enabled framework to monitor the lifecycle of *Cordyceps sinensis* mushrooms. Int. J. Electr. Comput. Eng., 13(1): 1142. https://doi.org/10.11591/ijece.v13i1.pp1142-1151

Meng, Q., Wu, P.-P., Li, M.-M., Shu, R.-H., Zhou, G.-L. et al. (2021). Distinct responses of *Thitarodes xiaojinensis* β-1,3-Glucan recognition protein-1 and immulectin-8 to *Ophiocordyceps sinensis* and *Cordyceps militaris* infection. J. Immunol., 207(1): 200–209. https://doi.org/10.4049/jimmunol.2000447

Nakamura, K., Shinozuka, K. and Yoshikawa, N. (2015). Anticancer and antimetastatic effects of cordycepin, an active component of *Cordyceps sinensis*. J. Pharmacol. Sci., 127(1): 53–56. https://doi.org/10.1016/j.jphs.2014.09.001

Panda, A.K. and Swain, K.C. (2011). Traditional uses and medicinal potential of *Cordyceps sinensis* of Sikkim. J. Ayurveda Integr. Med., 2(1): 9–13.

Park, B.-S. (2011). Effect of feeding *Cordyceps* with fly pupa on growth performance in broiler chickens. J. Life Sci., 21(11), 1541–1548. https://doi.org/10.5352/jls.2011.21.11.1541

Paterson R.R. (2008). *Cordyceps*: A traditional Chinese medicine and another fungal therapeutic biofactory? Phytochemistry, 69(7): 1469–1495. https://doi.org/10.1016/j.phytochem.2008.01.027

Qian, Z. and Li, S. (2017). Analysis of *Cordyceps* by multi-column liquid chromatography. Acta Pharm. Sin. B., 7(2): 202–207. https://doi.org/10.1016/j.apsb.2016.10.002

Qu, S.L., Li, S.S., Li, D. and Zhao, P.J. (2022). Metabolites and their bioactivities from the genus *Cordyceps*. Microorganisms, 10(8), 1489. https://doi.org/10.3390/microorganisms10081489

Reason, A., Bulgarella, M. and Lester, P.J. (2022). Identity, prevalence, and pathogenicity of entomopathogenic fungi infecting invasive Polistes (Vespidae: Polistinae) paper wasps in New Zealand. Insects, 13(10): 922. https://doi.org/10.3390/insects13100922

Sarker, A., Kim, J.-E., Islam, A.R.M.T., Bilal, M., Rakib, M.R.J. et al. (2022). Heavy metals contamination and associated health risks in food webs-a review focuses on food safety and environmental sustainability in Bangladesh. Environ. Sci. Pollut. Res. Int., 29(3): 3230–3245. https://doi.org/10.1007/s11356-021-17153-7

Sen, S., Karati, D., Priyadarshini, R., Dua, T.K., Paul, P. et al. (2023). *Cordyceps sinensis* (yarsagumba): pharmacological properties of a mushroom. Pharmacol. Res. Mod. Chin. Med., 8(100294): 100294. https://doi.org/10.1016/j.prmcm.2023.100294

Sharma, S. (2004). Trade of *Cordyceps sinensis* from high altitudes of the Indian Himalaya: Conservation and biotechnological priorities. Curr. Sci., 86(12): 1614–1619.

Shihab, S.K. and Nafea, H.H. (2023). Effect of Adding *Cordyceps sinensis* extract and probiotic to the diet on productive performance of broiler. Arch. Razi Inst., 78(2): 659–666. https://doi.org/10.22092/ARI.2022.359478.2430

Shrestha, B., Sung, G.H. and Sung, J.-M. (2017). Current nomenclatural changes in *Cordyceps sensu lato* and its multidisciplinary impacts. Mycology, 8(4): 293–302. https://doi.org/10.1080/21501203.2017.1386242

Skinner, M., Parker, B.L. and Kim, J.S. (2014). Role of entomopathogenic fungi in integrated pest management. pp. 169–191. *In*: Abrol, D.P. (ed.). Integrated Pest Management, Academic Press.

Sun, T., Zou, W., Luo, R., Li, C., Zhang, C. and Yu, H. (2023). Compositional and functional diversities of core microbial communities in wild and artificial *Ophiocordyceps sinensis*. Int. Microbiol., 26(4): 791–806. https://doi.org/10.1007/s10123-023-00333-5

Sung, G.-H., Hywel-Jones, N.L., Sung, J.-M., Luangsa-Ard, J.J., Shrestha, B. and Spatafora, J.W. (2007). Phylogenetic classification of *Cordyceps* and the clavicipitaceous fungi. Stud. Mycol., 57: 5–59. https://doi.org/10.3114/sim.2007.57.01

Tan, L., Liu, S., Li, X., He, J., He, L. et al. (2023). The large molecular weight polysaccharide from wild *Cordyceps* and its antitumor activity on H22 Tumor-bearing mice. Molecules, 28(8): 3351. https://doi.org/10.3390/molecules28083351

Tan, N.-Z., Barger, J.L., Zhang, Y., Ferguson, S.B., Wu, Z.-M. et al. (2011). *Cordyceps sinensis* Cs-4 restores aging-associated changes in gene expression and extends lifespan in normal aged mice. Fed. Am. Soc. Exp. Biol., 25(S1): https://doi.org/10.1096/fasebj.25.1_supplement.1090.13

Tiamyom, K., Sirichaiwetchakoon, K., Hengpratom, T., Kupittayanant, S., Srisawat, R. et al. (2019). The Effects of *Cordyceps sinensis* (Berk.) Sacc. and *Gymnema inodorum* (Lour.) decne. extracts on adipogenesis and lipase activity in vitro. Evid. Based. Complement. Alternat. Med., 2019: 5370473. https://doi.org/10.1155/2019/5370473

Tong, X., Zhang, H., Wang, F., Xue, Z., Cao, J. et al. (2020). Comparative transcriptome analysis revealed genes involved in the fruiting body development of *Ophiocordyceps sinensis*. Peer J, 8: e8379. https://doi.org/10.7717/peerj.8379

Tuli, H.S., Sandhu, S.S. and Sharma, A.K. (2014). Pharmacological and therapeutic potential of *Cordyceps* with special reference to Cordycepin. 3 Biotech, 4(1): 1–12. https://doi.org/10.1007/s13205-013-0121-9

Wang, L., Sun, H., Yang, M., Xu, Y., Hou, L. et al. (2022). Bidirectional regulatory effects of *Cordyceps* on arrhythmia: Clinical evaluations and network pharmacology. Front. Pharmacol., 13: 948173. https://doi.org/10.3389/fphar.2022.948173

Wang, N., Li, J., Huang, X., Chen, W. and Chen, Y. (2016). Herbal medicine *Cordyceps sinensis* improves health-related quality of life in moderate-to-severe asthma. Alternat. Med. 2016: 6134593. https://doi.org/10.1155/2016/6134593

Wei, X., Hu, H., Zheng, B., Arslan, Z., Huang, H.C. et al. (2017). Profiling metals in *Cordyceps sinensis* by using inductively coupled plasma mass spectrometry. Anal. Methods, 9(4): 724–728. https://doi.org/10.1039/C6AY02524B

Wei, Y., Zhang, L., Wang, J., Wang, W., Niyati, N. et al. (2021). Chinese caterpillar fungus (*Ophiocordyceps sinensis*) in China: Current distribution, trading, and futures under climate change and overexploitation. Sci. Total Environ., 755(Pt 1): 142548. https://doi.org/10.1016/j.scitotenv.2020.142548

Wu, G.D., Pan, A., Zhang, X., Cai, Y.-Y., Wang, Q. et al. (2022a). *Cordyceps* improves obesity and its related inflammation via modulation of *Enterococcus cecorum* abundance and bile acid metabolism. Am. J. Chin. Med., 50(3): 817–838. https://doi.org/10.1142/S0192415X22500343

Wu, P., Qin, Q., Zhang, J., Zhang, H., Li, X. et al. (2022b). The invasion process of the entomopathogenic fungus *Ophiocordyceps sinensis* into the larvae of ghost moths (*Thitarodes xiaojinensis*) using a GFP-labeled strain. Front. Microbiol., 13: 974323. https://doi.org/10.3389/fmicb.2022.974323

Wu, R., Yao, P.-A., Wang, H.-L., Gao, Y., Yu, H.-L. et al. (2018). Effect of fermented *Cordyceps sinensis* on doxorubicin-induced cardiotoxicity in rats. Mol. Med. Rep., 18(3): 3229–3241. https://doi.org/10.3892/mmr.2018.9310

Wu, T.-N., Yang, K.-C., Wang, C.-M., Lai, J.-S., Ko, K.-N. et al. (1996). Lead poisoning caused by contaminated *Cordyceps*, a Chinese herbal medicine: Two case reports. Sci. Total Environ., 182(1–3): 193–195. https://doi.org/10.1016/0048-9697(96)05054-1

Wu, T.-R., Lin, C.-S., Chang, C.-J., Lin, T.-L., Martel, J. et al. (2019). Gut commensal *Parabacteroides goldsteinii* plays a predominant role in the anti-obesity effects of polysaccharides isolated from *Hirsutella sinensis*. Gut, 68(2): 248–262. https://doi.org/10.1136/gutjnl-2017-315458

Xiao, Y., Li, C., Xu, W., Du, Y., Zhang, M. et al. (2021). Arsenic content, speciation, and distribution in wild *Cordyceps sinensis*. Evid. Based. Complement. Alternat. Med., 2021. 6651498. https://doi.org/10.1155/2021/6651498

Yan, J.K., Wang, W.Q. and Wu, J.Y. (2014). Recent advances in *Cordyceps sinensis* polysaccharides: Mycelial fermentation, isolation, structure, and bioactivities: A review. J. Func. Foods, 6: 33–47. https://doi.org/10.1016/j.jff.2013.11.024

Yan, X.-F., Zhang, Z.-M., Yao, H.-Y., Guan, Y., Zhu, J.-P. et al. (2013). Cardiovascular protection and antioxidant activity of the extracts from the mycelia of *Cordyceps sinensis* act partially via adenosine receptors. Phytother. Res., 27(11): 1597–1604. https://doi.org/10.1002/ptr.4899

Yang, L., Jiao, X., Wu, J., Zhao, J., Liu, T. et al. (2018). *Cordyceps sinensis* inhibits airway remodeling in rats with chronic obstructive pulmonary disease. Exp. Ther. Med., 15(3): 2731–2738. https://doi.org/10.3892/etm.2018.5777

Zhang, P., Li, S., Li, J., Wei, F., Cheng, X. et al. (2018). Identification of *Ophiocordyceps sinensis* and its artificially cultured *Ophiocordyceps* mycelia by ultra-performance liquid chromatography/orbitrap fusion mass spectrometry and chemometrics. Molecules, 23(5): 1013. https://doi.org/10.3390/molecules23051013

Zhang, Q., Xiao, X., Li, M., Yu, M. and Ping, F. (2022a). Bailing capsule (*Cordyceps sinensis*) ameliorates renal triglyceride accumulation through the PPARα pathway in diabetic rats. Front. Pharmacol., 13: 915592. https://doi.org/10.3389/fphar.2022.915592

Zhang, X., Wang, M., Liang, X., Wang, J., Zhang, M. et al. (2023a). Bioinformatics analysis and in vivo validation study of *Ophiocordyceps sinensis* (Berk.) G.H. Sungetal against lung adenocarcinoma. J. Ethnopharmacol., 317(116739): 116739. https://doi.org/10.1016/j.jep.2023.116739

Zhang, X., Wang, M., Qiao, Y., Shan, Z., Yang, M. et al. (2022b). Exploring the mechanisms of action of *Cordyceps sinensis* for the treatment of depression using network pharmacology and molecular docking. Ann. Transl. Med., 10(6): 282. https://doi.org/10.21037/atm-22-762

Zhang, Y., Li, K., Zhang, C., Liao, H. and Li, R. (2023b). Research progress of *Cordyceps sinensis* and its fermented mycelium products on ameliorating renal fibrosis by reducing epithelial-to-mesenchymal transition. J. Inflamm. Res., 16: 2817–2830. https://doi.org/10.2147/JIR.S413374

Zhang, Y., Xu, L., Lu, Y., Zhang, J., Yang, M. et al. (2023c). Protective effect of *Cordyceps sinensis* against diabetic kidney disease through promoting proliferation and inhibiting apoptosis of renal proximal tubular cells. BMC Complement. Med. Ther., 23(1): doi:10.1186/s12906-023-03901-4

Zhou, L., Hao, Q.X., Wang, S., Yang, Q., Kang, C.Z. et al. (2017). Study on distribution of five heavy metal elements in different parts of *Cordyceps sinensis* by microwave digestion ICP-MS. China J. Chinese Mat. Med., 42(15): 2934–2938. https://doi.org/10.19540/j.cnki.cjcmm.20170714.016

Zhou, Q., Yu, L., Ying, S.-H. and Feng, M.-G. (2021). Comparative roles of three adhesin genes (*adh1-3*) in insect-pathogenic lifecycle of *Beauveria bassiana*. Appl. Microbiol. Biotechnol., 105(13): 5491–5502. https://doi.org/10.1007/s00253-021-11420-w

Zhou, Y., Wang, M., Zhang, H., Huang, Z. and Ma, J. (2019). Comparative study of the composition of cultivated, naturally grown *Cordyceps sinensis*, and stiff worms across different sampling years. PloS One, 14(12): e0225750. https://doi.org/10.1371/journal.pone.0225750

Zou, Y., Liu, Y., Ruan, M., Feng, X., Wang, J. et al. (2015). *Cordyceps sinensis* oral liquid prolongs the lifespan of the fruit fly, *Drosophila melanogaster*, by inhibiting oxidative stress. Int. J. Mol. Med., 36(4): 939–946. https://doi.org/10.3892/ijmm.2015.2296

Taxonomy, Phylogeny and Genetics

2

Taxonomy, Phylogeny and Genetics of *Cordyceps*

Chen Wanhao[1] and *Han Yanfeng*[2,]*

1. Introduction

In the traditional classification system of epigenetic features (macro and micro), "*Cordyceps*" refers to the genus *Cordyceps* (Fr.) Link and belongs to the fungi kingdom, Ascomycota, Ascomycetes, Sortariomycetidae, Hyporales, and Clavicepitaceae (Kirk et al., 2001). In this classification system, *Cordyceps* includes three subgenera, *Ophiocordyceps* (Petch) Y. Kobayasi, *Eucordyceps* Y. Kobayasi and *Neocordyceps* Y. Kobayasi (Kobayasi, 1983). Phylogenetic analysis has been widely applied in taxonomic research with the development of molecular biology and sequencing technology. Four genera have been recognized in the genus *Cordyceps*, including, *Cordyceps* Fr., *Metacordyceps* G.H. Sung, J.M. Sung, Hywel-Jones & Spatafora, *Ophiocordyceps* Petch and *Elaphocordyceps* G.H. Sung & Spatafora, belonging to the families Cordycipitaceae Kreisel ex G.H. Sung, J.M. Sung, Hywel-Jones & Spatafora, Clavicipitaceae Rogerson and Ophiocordycipitaceae G.H. Sung, J.M. Sung, Hywel-Jones & Spatafora, respectively. Meanwhile, among the taxonomic units related to the names of *Cordyceps* in the three families mentioned above, the morphology of *Cordyceps* is similar to that of many other fungi. Currently, *Cordyceps* sensu lato contains all the species in the

[1] Center for Mycomedicine Research, Basic Medical School, Guizhou University of Traditional Chinese Medicine, Guiyang 550025, Guizhou, PR China.

[2] Institute of Fungus Resources, Department of Ecology, College of Life Sciences, Guizhou University, Guiyang 550025, Guizhou, PR China.

*Corresponding author: swallow1128@126.com

families Cordycipitaceae, Ophiocordycipitaceae, Polycephalomycetaceae and some species in the family Clavicipitaceae. In this chapter, a review of the taxonomy, phylogeny, and genetics of *Cordyceps* is presented, along with a prediction for the future of the species.

2. Taxonomy of *Cordyceps*

2.1 *Morphological Traits*

For taxonomic purposes, *Cordyceps* is primarily classified based on morphological characteristics. The first taxonomic definition of *Cordyceps* was based on the concept of phenotypic species (Mayr, 1942), which is a morphological feature. In the early 19th century, the description of *Cordyceps* was mainly based on macroscopic morphological features, such as the type of host insect or fungus, the shape of the perithecium, and the position of the perithecium in the fertile part of the fruiting body. Schroeter (1894) classified the species of *Cordyceps* that were partially buried in the perithecium as the subgenus *Racemella* Ces., while the species that were completely buried in the perithecium were classified as *Cordylia* Tul. Subsequently, the subgenus *Cordylia* was classified into the subgenus *Eucordyceps* Sacc. based on the host species and insect species.

In the mid-19th century, a microscopic morphological classification of *Cordyceps* was also performed, such as the number, arrangement, shape, color, length, and the state after rupture of ascospores in the ascus. Kobayasi (1941) divided the genus *Cordyceps* into three subgenera *Ophiocordyceps*, *Eucordyceps* and *Neocordyceps*. The classification system of *Cordyceps* was established by Kobayasi (1983). It divided the genus into three subgenera and seven groups.

With the increasing number of *Cordyceps* and their asexual forms, other methods were added to the taxonomy of *Cordyceps* such as the DELTA (description language for taxonomy) system, a flexible and powerful information processing tool for taxonomic description using computers. The main function of DELTA is to standardize the morphological description of taxonomy, perform automatic descriptions of classification groups, facilitate interactive identification, manage classification information, and process this information into a data structure suitable for analyzing representative classification systems and branch classification systems (Zhang, 2009). Han et al. (2007, 2009) studied *Paecilomyces* sensu lato in China using the DELTA system for the first time and reported two new species *Taifanglania berberidis* Y.E Han & Z.Q. Liang and *Taifanglania jiangsuensis* Y.F Han & Z.Q. Liang (Han et al. 2010). Chen et al. (2014) applied the DELTA system to the taxonomy of the genus *Gibellula* Cavara and successfully expanded its functions, such as search tables clustering trees and interactive identification, based on the database of phenotypic features. Three specimens were quickly identified by the DELTA system.

Currently, three cordyceps families can be distinguished: Cordycipitaceae, Ophiocordycipitaceae, and Clavicipitaceae. Subsequently, Polycephalomycetaceae was added by Xiao et al. (2023). Nevertheless, applying "One Fungus = One Name" (Taylor, 2011) and relating teleomorph and anamorph species to the taxonomy of *Cordyceps* has resulted in difficulties in identifying *Cordyceps* species based on morphological characteristics.

2.2 *Biochemical and Physiological Traits*

Cordyceps has also been classified by biochemical and physiological characteristics. There have been several attempts to discover chemotaxonomic markers that could be used to isolate and identify *Cordyceps* species. According to Stadler et al. (2003), *Pochonia* Bat. & O.M. Fonseca and other conidial fungi with anamorphs similar to *Verticillium* can be classified according to chemotaxonomy. *Pochonia* has a different profile of secondary metabolites compared to *Lecanicillium* W. Gams & Zare, *Haptocillium* W. Gams & Zare, and *Rotiferophthora* G.L. Barron. The resorcylic acid lactones could be used as a marker for the taxonomy of *Pochonia*. Watanabe et al. (2006) showed that dipicolinic acid could be used to distinguish *Cordyceps*, Isaria Pers., *Beauveria* Vuill. and *Lecanicillium*. Luangsa-ard et al. (2009) showed that it is possible to use beauvericin as a chemotaxonomic marker for these *Isaria tenuipes* complex species. Isaka et al. (2011) conducted the chemical taxonomy of *Torrubiella sensu* lato and suggested that the zeorin could be used as a marker for *Conoideocrella* D. Johnson, G.H. Sung, Hywel-Jones & Spatafora.

Metabolite profiles are used to taxonomize fungal species, which include chromatographic (mainly HPLC) data on many exometabolites. Although many secondary metabolites have species-specific features, the synthesis of these compounds is unstable. The use of MALDI-TOF mass spectrometry (MS) for comparing microorganisms based on their protein fingerprints appears to be a more effective method for proteomic profiling. In this way, strain-specific mass spectra can be obtained, which can then be used to identify and type microorganisms intraspecifically (Aebersold, 2003; Domon and Aebersold, 2006; Pinel et al., 2011). *Metarhizium anisopliae* complex has been identified using MALDI-TOF mass spectrometry by Lopes et al. (2014). As a result, this approach was found to be extremely useful for identifying the *M. anisopliae* complex and identifying new cultures. A MALDI-TOF mass spectrometry method was used by Hricakova et al. (2018) to identify two *Beauveria* species, and the result proved that the MALDI-TOF technique was reliable for entomopathogenic fungi identification, particularly for those that belong to the genus *Beauveria*.

3. Taxonomy of the Major Genera of *Cordyceps* Sensu Lato

3.1 *Cordyceps Sensu Stricto*

In the family Cordycipitaceae, the genus *Cordyceps* was the genus type established by Vaillant in 1727. *Cordyceps* sensu lato was classified into four genera, i.e., *Cordyceps* sensu stricto, *Ophiocordyceps*, *Metacordyceps* and *Elaphocordyceps* (Sung et al., 2007a). *Cordyceps* consisted of around 50 species of fungi that parasitized insects at that time (Sung et al., 2007a). The rejections of *Isaria*, *Microhilum* H.Y. Yip & A.C. Rath, *Phytocordyceps* C.H. Su & H.H. Wang and *Evlachovaea* B.A. Borisov & Tarasov were proposed by Kepler et al. (2017). Some new *Cordyceps* species were reported later, and *Cordyceps* now consists of over 100 species. Among those species, both sexual and asexual species were present. The taxonomy of *Cordyceps* needs a combined analysis of multi-gene phylogeny and morphological characteristics.

3.2 *Akanthomyces*

Akanthomyces Lebert was established by Lebert for the species, *A. aculeatus* Lebert in 1858. Mains (1950) reported three additional species: *A. ampullifer* (Petch) Mains, *A. angustisporus* Mains and *A. aranearum* (Petch) Mains. Subsequently, several other species have been reported. Chen et al. (2023b) summarized that there were 34 species in the genus *Akanthomyces*. Several *Akanthomyces* species were transferred to *Hevansia* Luangsa-ard et al. based mostly on phylogenetic analyses (Kepler et al., 2017). In addition, several *Lecancillium* species were transferred to *Akanthomyces*. Currently, there are 36 species in the genus *Akanthomyces*.

Akanthomyces species were originally divided according to their morphology. The rejection of the genera *Torrubiella* Boud. and *Lecanicillium* was proposed, and the species were recombined into *Akanthomyces* by Kepler et al. (2017). As a result, the morphological characteristics and phylogenetic analyses were combined to determine the taxonomy of *Akanthomyces*. The PHI tests and base difference rates were introduced by Chen et al. (2023b), which might help delimit cryptic species taxonomically in *Akanthomyces*. For the cryptic *Akanthomyces* species, at least two loci were recommended, as well as multiple methods for comparing cryptic species with their relatives.

3.3 *Beauveria*

Among the abundant entomopathogenic fungi, *Beauveria* includes species with ecological and economic importance (Ownley et al., 2008). Nevertheless, *Beauveria* can be endophytes or saprobes (Moonjely et al.,

2016). *Beauveria* species were characterized by branched, penicillate, or trichodermoid conidiophores; the hyaline conidia originate from clusters of small sympodial and globose conidiogenous cells with apical denticulate rachis on conidiophores. (Rehner et al., 2011).

As a result of an analysis on four protein code genes (Bloc, TEF, RPB1, and RPB2) by Rehner et al. (2011), *Beauveria* was divided into 12 species. Subsequently, seven new species were reported later by Zhang et al. (2013), Chen et al. (2013), Agrawal et al. (2014), Ariyawansa et al. (2015), Robène-Soustrade et al. (2015) and Imoulan et al. (2016).

In a phylogenetic analysis conducted by Kepler et al. (2017), the Cordycipitaceae were reevaluated and transferred to *Cordyceps acridophila* T. Sanjuan & Franco-Mol., *C. diapheromeriphila* T. Sanjuan & S. Restrepo, *C. locustiphila* Henn., *C. scarabaeidicola* Kobayasi and *C. staphylinidicola* Kobayasi & Shimizu into the genus *Beauveria*. Moreover, *B. blattidicola* M. Chen, Aime, T.W. Henkel & Spatafora was reported. *B. araneola* Wan H. Chen, Y.F. Han, Z.Q. Liang & D.C. Jin, *B. majiangensis* W.H. Chen, M. Liu, Z.X. Huang, G.M. Yang, Y.F. Han, J.D. Liang & Z.Q. Liang, *B. baoshanensis* Z.H. Chen & L. Xu, *B. yunnanensis* Z.H. Chen & L. Xu, *B. peruviensis* D.E. Bustamante, M.S. Calderon, M. Oliva & S. Leiva and *B. mimosiformis* Khonsanit, Thanakitppattana, Kobmoo & Luangsa-ard were reported later by Chen et al. (2017a, 2018a, 2019a), Bustamante et al. (2019) and Khonsanit et al. (2020), respectively. Eight new *Beauveria* species were reported later (Kobmoo et al., 2021; Wang et al., 2022a; Sato et al., 2022). Currently, *Beauveria* consists of 36 species.

In the genus *Beauveria*, species delineation has mainly relied on morphological characteristics in combination with multi-locus phylogenetic analyses. Bustamante et al. (2019) demonstrated that polyphasic approaches, where different methods align, provide a reliable indication of *Beauveria* species boundaries. Establishing these boundaries in *Beauveria* involves considering genetic distance, coalescence approaches, and multi-locus phylogenies. Kobmoo et al. (2021) showed that whole-genome sequence data could provide strong support for intra-species genetic groups and prove powerful for delimiting species of *Beauveria*. Besides, the phylogenetic network was applied to the phylogenetic relationship among *Beauveria* species by Khonsanit et al. (2020).

3.4 *Metarhizium*

The genus *Metarhizium* was proposed with the type species *M. anisopliae* (Metschn.) Sorokīn by Sorokin (1883). Subsequently, several species and varieties of this genus have been described in recent years due to seemingly minor morphological differences from *M. anisopliae* (Latch, 1965; Petch, 1931, 1935). Tulloch (1976) established only two species, *M. anisopliae* and *M. flavoviride* W. Gams & Rozsypal, as well as a variety, *M. anisopliae* var.

majus (J.R. Johnst.) M. C. Tulloch (as var. *major*). *M. flavoviride* var. *flavoviride*, as well as *M. flavoviride* var. *minus* Rombach, Humber & D.W. Roberts, which were classified as two varieties by Rombach et al. (1986). Two new species, *M. taii* Z.Q. Liang & A.Y. Liu and *Cordyceps taii* Z.Q. Liang & A.Y. Liu, were proposed by Liang et al. (1991) and connected by microcycle conidiation.

DNA sequencing and other molecular techniques have been developed to study phylogenetic relationships and clarify the taxonomic positions of confusing species. *Metarhizium* was phylogenetically classified using single and multi-locus techniques (Bischoff et al., 2006, 2009; Curran et al., 1994; Driver et al., 2000; Rakotonirainy et al., 1994). Phylogenetic approaches based on multiple genes are superior for identifying *Metarhizium* species and their relationships. According to Kepler et al. (2014), *Metarhizium* is a synonym for *Metacordyceps*; the genus *Pochonia* was disassembled, and only 24 have been recognized. A total of 22 new species or recombined species have been reported since then (Montalva et al., 2016; Chu et al., 2016; Nishi et al., 2017; Luangsa-ard et al., 2017; Chen et al., 2017b, 2018b, 2019b; Luz et al., 2019; Gutierrez et al., 2019; Yamamoto et al., 2020). Mongkolsamrit et al. (2020) re-evaluated the genus *Metarhizium*, described some new species from Thailand, and recognized 65 *Metarhizium* species. *M. lymantriidarum* Z.H. Chen & L. Xu and *Metarhizium macrosemiae* M.J. Chen & B. Huang were reported by Chen et al. (2022a, 2023c). Currently, *Metarhizium* consists of 67 species.

3.5 *Ophiocordyceps*

As the major genus in the family Ophiocordycipitaceae, *Ophiocordyceps* Petch includes invertebrates' pathogenic fungi (Luangsa-ard et al., 2018). It was originally established by Petch (1931) to accommodate a type of cordyceps species with a clear top cover in the ascospores, a clear septum in the ascospores, and no detachment into secondary spores at maturity. In the genus *Ophiocordyceps*, *O. blattae* (Petch) Petch was known as the type species, which is a rarely collected cockroach with no culture or molecular data available (Petch, 1931). As a result of the morphological data that Kobayasi (1941) examined, the genus *Ophiocordyceps* was considered a subdivision of *Cordyceps* sensu lato. However, according to Sung et al. (2007b), *Ophiocordyceps* belongs to the family Ophiocordycipitaceae, and it forms sister branches with *Elaphocordyceps*. The main characteristics were as follows: species with the *Hirsutella* asexual forms, fibrous, pliable, and sometimes wiry, with aperithecial apices and side pads. The perithecia vary in arrangement from superficial to completely immersed, ordinal or oblique. Species with *Hymenostilbe* asexual forms produce brightly colored and fleshy stroma. The perithecium is fully buried and inclined. The ascospores are cylindrical, multi-septate, and mature into 64 secondary ascospores (Sung et al., 2007b).

There are many species of asexual forms related to the genus *Ophiocordyceps*, including *Hirsutella, Hymenostelbe, Paraisaria,* and *Syngliocladium* Petch. However, according to the latest requirement of "ICN", the dual naming system for fungi should be abolished and a naming convention of one fungal name should be adopted. According to Quandt et al. (2014), the genera *Hymenostilbe, Paraisaria,* and *Syngliocladium* should be discouraged, as they were proposed later and contain fewer taxa than *Hirsutella* or *Ophiocordyceps*. The genus *Ophiocordyceps* is remarkably diverse with 235 accepted species were recorded until 2018 (Luangsa-ard et al., 2018; Spatafora et al., 2015; Khonsanit et al., 2019), and 56 new species were reported later (Tang et al., 2023; Fan et al., 2021; Yang et al., 2021; Khao-ngam et al., 2021; Tasanathai et al., 2019; Zou et al., 2022; Araújo et al., 2020; Tasanathai et al., 2020; Clifton et al., 2021; Mongkolsamrit et al., 2021; Wang et al., 2021a; Xiao et al., 2019; Khonsanit et al., 2019; Sharma et al., 2023; Crous et al., 2023; Lao et al., 2021; Chawngthu et al., 2021; Sun et al., 2022; Chen et al., 2021a; Wei et al., 2020; Long et al., 2021). Currently, *Ophiocordyceps* consists of 291 species.

3.6 *Purpureocillium*

The genus *Purpureocillium* was established to accommodate *P. lilacinum* (Thom) Luangsa-ard, Houbraken, Hywel-Jones & Samson (Luangsa-ard et al. 2011). A combined analysis of the morphological and phylogenetic characteristics of *P. lavendulum* Perdomo, Dania Garca, Gené, Cano & Guarro was described by Perdomo et al. (2013). *Nomuraea atypicola* (Yasuda) Samson and *Isaria takamizusanensis* Kobayasi were combined with *Purpureocillium* by Quandt et al. (2014) and Ban et al. (2015), respectively. Two new species, *P. sodanum* Papizadeh, Soudi, Wijayaw., Shahz. Faz. & K.D. Hyde, and *P. roseum* Calvillo and Raymundo, were described by Hyde et al. (2016) and Calvillo-Medina et al. (2021). Currently, *Purpureocillium* consists of 6 species.

3.7 *Samsoniella*

Mongkolsamrit et al. (2018) proposed the genus *Samsoniella* Mongkols., Noisrip., Thanakitp., Spatafora & Luangsa-ard with three new species: *S. alboaurantia* (G. Sm.) Mongkols., Noisrip., Thanakitp., Spatafora & Luangsa-ard, *S. aurantia* Mongkols., et al. and *S. inthanonensis* Mongkols.. Subsequently, three more new species were described by Chen et al. (2020). Wang et al. (2020a) described nine new species and proposed two new combinations. A total of thirteen other new species were reported by Chen et al. (2021b, 2022b, 2023a), Wang et al. (2022b) and Crous et al. (2023). *Samsoniella* currently consists of 31 species.

Morphological characteristics were initially used to delimit *Samsoniella*, and phylogenetic analysis of multiple loci followed. As a new test, Chen et al. (2023a) included a PHI test to differentiate the cryptic *Samsoniella* species. He also recommended that *Samsoniella* cryptic species should be distinguished by the TEF locus, and multiple approaches are recommended to confirm the presence of cryptic species.

3.8 *Tolypocladium*

The genus *Tolypocladium* was proposed by Gams et al. (1971), and three species were described. Barron (1980, 1981, 1983) reported three new species, *T. parasiticum* G.L. Barron, *T. trigonosporum* G.L. Barron and *T. lignicola* G.L. Barron. Bissett (1983) reported two new species, *T. nubicola* Bissett and *T. tundrense* Bissett, and combined three species (*Cephalosporium balanoides* Drechsler, *Verticillium microsporum* Jaap, and *Pachybasium niveum* O. Rostr.) into the genus *Tolypocladium*. He suggested that the combined species *T. niveum* (O. Rostr.) Bissett should be used as the type species in the genus *Tolypocladium* based on the same morphological characteristics. According to Dreyfuss and Gams (1994), *T. inflatum* should be considered as the type species to accommodate for the scientific name of the strain that produces cyclosporin. Eight new species were reported worldwide (Samson and Soares, 1984; Li , 1986; Liu , 1999; Cao, 2014; Gazis et al., 2014).

An analysis of the multi-gene phylogeny of the genus *Elaphocodyceps* by Quandt et al. (2014) suggested it should be merged into *Tolypocladium*. A total of sixteen new species or species combined were reported in the following years (Yamamoto et al., 2022; Wijayawardene et al., 2021; Li et al., 2018; Crous et al., 2020; Crous et al., 2017; Das et al., 2023; Yu et al., 2021; Dong et al., 2022; Soares et al., 2022). Currently, *Tolypocladium* consists of 59 species.

4. Phylogeny of *Cordyceps* Sensu Lato

Gene sequences have gradually been applied to the taxonomy of *Cordyceps* as molecular methodologies advance. For the phylogenetic assessment of *Cordyceps*, the ITS (internal transcribed spacer) sequence is the most popular marker (Liu et al. 2001, 2002). Artjariyasripong et al. (2001) used 18S and 28S sequences to study the relationship between the genus *Cordyceps* and its allies, and the results showed that the subgenus *Neocordyceps* was a monophyletic group, but the subgenus *Ophiocodyceps* and *Eucordyceps* didn't form a monophyletic group. There were some related studies on the genus *Cordyceps*. However, due to the limited taxonomic units and the limited resolution of the gene sequences, the phylogenetic relationships among Cordyceps species have not been clearly resolved (Sung et al. 2001; Stensrud et al. 2005). While, the phylogenetic relationship between the genus *Cordyceps* and the family Clavicipitaceae has not been fully elucidated, resulting in significant limitations in the application of the obtained conclusions in the systematic classification of the genus *Cordyceps* (Artjariyasripong et al., 2001; Sung et al., 2001). With the rapid development of molecular phylogenetic research, more and more scholars are employing multiple gene loci to conduct phylogenetic classification research on the order Hypocreales (Sung et al., 2007b; Chaverri et al. 2008; Kepler et al., 2017).

The multiple loci, including LSU, SSU, TEF, RPB1, RPB2, TUB, and ATP6, were used to analyze 162 taxa to conduct in-depth research on the phylogenetic relationship between the genus *Cordyceps* and the family Clavicipitaceae, and the result showed that both *Cordyceps* and Clavicipitaceae were not monophyletic (Sung et al., 2007a). There are three subfamilies within Clavicipitaceae, Cordycipitaceae, Ophiocordycipitaceae and Clavicipitaceae. The type genus of Cordycipitaceae was *Cordyceps* and the type species was *Cordyceps militaris*, which included most of the previously known species of the genus *Cordyceps*, which were fleshy and brightly colored. The family Ophiocordycipitaceae was mainly based on the subgenus *Ophiocordyceps*, and the majority of species in this family were dark-colored, produced hard or flexible seed bases, and often had sterile tips. The family Clavicepitaceae sensu stricto has been revised to include some species coexisting with herbs and species belonging to the subgenus *Hypocrella* and its related species of insect pathogenic fungi.

4.1 *Clavicipitaceae*

Almost all terrestrial ecosystems contain Clavicipitaceae species. Clavicipitaceous fungi were initially divided into three groups (*Oomyces, Ascopolyporus*), (*Epichloë, Claviceps*) and *Cordyceps* based on their sexual characteristics of reproductive structures. Diehl (1950) divided the family Clavicipitaceae into three subfamilies—Clavicipitoideae, Cordycipitoideae and Oomycetoideae—mainly based on host affiliation. Eriksson (1982) removed *Oomyces* Berk. & Broome from Clavicipitaceae. Sung et al. (2007a) used multi-gene phylogeny to show that the grass-associated subfamily Clavicipitoideae was monophyletic, but not the subfamily Cordycipitoideae. Sung et al. (2007a) studied the *Cordyceps* and clavicipitaceous fungi and found that their phylogenetic analysis contradicts the previous classifications. Then, they amended the morphological characteristics and genera of Clavicipitaceae.

Li et al. (2008) reported a new genus, *Metarhiziopsis* D.W. Li, R.S. Cowles & C.R. Vossbrinck that has characteristics of both *Metarhizium* and *Myrothecium* Tode, and was established based on the analysis of multigene phylogenetics. Chaverri et al. (2008) proposed a new genus, *Samuelsia* P. Chaverri & K.T. Hodge to accommodate species with aschersonia-like anamorphs. Johnson et al. (2009) conducted a systematic and evolutionary reevaluation of *Torrubiella*, and two new genera, *Conoideocrella* and *Orbiocrella* were proposed in the family Clavicipitaceae. Three new genera, *Tyrannicordyceps* Kepler & Spatafora, *Metapochonia* Kepler, S.A. Rehner & Humber, and *Nigelia* Luangsa-ard, Tasan. & Thanakitp. were established by Kepler et al. (2012, 2014) and Luangsa-ard et al. (2017).

Mongkolsamrit et al. (2020) re-evaluated the genus *Metarhizium* and proposed the new genera. *Marquandomyces* Samson, Houbraken & Luangsa-ard, Petchia Thanakitp., Mongkols. & Luangsa-ard; *Purpureomyces* Luangsa-ard, Samson & Thanakitp.; *Sungia* Luangsa-ard, Samson & Thanakitp., and *Yosiokobayasia* Samson, Luangsa-ard & Thanakitp. *Commelinaceomyces* E. Tanaka is a newly established genus to accommodate four clavicipitaceous species that were previously misclassified as *Ustilago* and have the ability to infect plants of the family Commelinaceae (Tanaka et al., 2020). Four new genera: *Allocordyceps* Poinar, *Neoaraneomyces* W.H. Chen, Y.F. Han, J.D. Liang & Z.Q. Liang, *Pseudometarhizium* W.H. Chen, Y.F. Han, J.D. Liang & Z.Q. Liang and *Paraneoaraneomyces* Zhi Y. Zhang & Y.F. Han were reported by Poinar and Maltier (2021), Chen et al. (2022c) and Zhang et al. (2023). Currently, Clavicipitaceae contains 52 genera and over 500 species (Mongkolsamrit et al., 2020; Gao et al., 2021; Chen et al., 2022c; Zhang et al., 2023).

4.2 Cordycipitaceae

The family Cordycipitaceae was formed on the basis of the type genus *Cordyceps* and the type species *Cordyceps militaris*. The family shares a recent ancestor with the order Hypocreales, in which most species have white to brightly colored fleshy stroma (Sung et al. 2007b). However, some species in this family display shank or degradation of the shank on their hosts. Overall the species in this family exhibit the greatest range of morphological features and hosts among any family in the order Hypocreales. Some genera with anamorph and teleomorph species, such as *Akanthomyces*, *Gibellula*, *Beauveria*, *Cordyceps*, *Isaria*, *Lecanicillium* and *Torrubiella*, have many taxonomic issues. Many species of *Cordyceps* are related to the initially described anamorph genus, including *Akanthomyces*, *Beauveria*, *Evlachovaea*, *Isaria*, *Lecanicillium*, *Microhilum* and *Paecilomyces* Bainier. The type species, *Acanthomyces aculeatus* Lebert, was first described by Lebert in 1858. The species *A. pistillariiformis* (Pat.) Samson & H.C. Evans was related to its teleomorph species, *Cordyceps tuberculata* (Lebert) Maire. *Lecanicillium lecanii* (Zimm.) Zare & W. Gams, the anamorph of *Torrubiella confragosa* Mains, was described by Mains (1949). However, the genus *Lecanicillium* was considered as the synonym for the genus *Akanthomyces*.

Kepler et al. (2017) re-analyzed the typical anamorph and teleomorph genera and species to clarify the phylogenetic relationship in the family

Cordycipitaceae based on the five loci, LSU, SSU, TEF, RPB1 and RPB2. The names of sexual and asexual species in the family Cordycipitaceae were revised based on the combined analysis of the latest International Plant Naming Regulations (One Fungi, One Name) and the principle of priority publication, combined with molecular phylogenetic relationships. Nine genera were accepted and eight genera were rejected in the family Cordycipitaceae. Besides, two new genera, *Hevansia* and *Blackwellomyces*, were proposed in this family. Species in the genera *Torrubiella* and *Lecanicillium* were combined into the genus *Akanthomyces*. Species in the genera *Isaria, Microhilm, Phytocordyceps*, and *Evlachovaea* were combined into the genus *Cordyceps*. Species in the genera *Synsterigmatocystis* and *Granulomanus* were combined into the genus *Gibellula*.

In addition, there are also four genera: *Beejasamuha* Subram. & Chandrash., *Coremiopsis* Sizova & Suprun, *Leptobacillium* Zare & W. Gams, and *Rotiferophthora*, which are reported in the family Cordycipitaceae. However, only the genus *Leptobacillium* has available sequences in NCBI, which could be used for phylogenetic analysis. An analysis of Cordycipitaceae morphs with Isaria-like characteristics has been assigned to the new genus *Samsoniella* (Mongkolsamrit et al., 2018). 7 new genera were proposed based on the phylogenetic analysis (Flakus et al., 2019; Wang et al., 2020a; Zhang et al., 2021; Thanakitpipattana et al., 2020, 2022; Kobmoo et al., 2023). Chen et al. (2021c) confirmed that the genus *Pleurodesmospora* Samson, W. Gams & H.C. Evans belongs to the Cordycipitaceae based on multi-gene phylogenetic evidence. Up to date, the family Cordycipitaceae consists of 22 genera.

4.3 *Ophiocordycipitaceae*

To accommodate species of Cordycipitaceae and Clavicipitaceae that differ in their phylogenetic relationships, Sung et al. (2007b) proposed a new family named Ophiocordycipitaceae. The family Ophiocordycipitaceae contains 11 genera, including *Drechmeria* W. Gams & H.-B. Jansson, *Harposporium* Lohde, *Tolypocladium, Polycephalomyces* Kobayasi, *Ophiocordyceps, Hirsutella* Pat., *Hymenostilbe* Petch, *Paraisaria* Samson & B.L. Brady, *Podocrella* Seaver, *Chaunopycnis* W. Gams, and *Sorosporella* Sorokīn (Kirk et al., 2013). Quandt et al. (2014) revised the family Ophiocordycepsitaceae based on multiple gene phylogenetic analysis and divided it into six monophyletic genera: *Ophiocordyceps, Drechmeria, Harposporium, Tolypocladium, Purpureocillium* and *Polycephalomyces*. The genera *Elaphocordyceps* and *Chaunopycnis* were combined into the genus *Tolypocladium*.

Spatafora et al. (2015) suggested the new combinations of the family Ophiocordycipitaceae, following the latest changes in "ICN" and the principles of singularity and priority, using the naming convention of "One Fungus = One Name". A total of 23 new combinations and 15 synonyms were generated by recombining 4 genera of the family Ophiocordycipitaceae (*Ophiocordyceps*, *Drechmeria*, *Harposporium*, and *Purpureocillium*). Additionally, it is recommended to retain the genus *Hirsutella*. Two new genera, *Hantamomyces* Crous and *Pleurocordyceps* Y.J. Yao, Y.H. Wang, S. Ban, W.J. Wang, Yi Li, Ke Wang & P.M. Kirk were reported by Crous et al. (2020) and Wang et al. (2021b), respectively.

Based on phylogenetic analysis, *Polycephalomyces*, *Perennicordyceps*, and *Pleurocordyceps* were separated from Ophiocordycipitaceae by Xiao et al. (2023), and three genera were merged into the new family Polycephalomycetaceae. Currently, the family Ophiocordycipitaceae consists of eight genera.

5. Genetics and Evolution of *Cordyceps*

5.1 Increasing Members of the Cordyceps

With the application of phylogenetic analysis and the implementation of the naming convention "One Fungi, One Name (1 F 1 N)", the membership range of *Cordyceps* is gradually expanding. Kepler et al. (2014) reclassified the members of the genus *Metacordyceps* based on four protein-coding gene sequences and found that all or some members of the genera *Metarhizium*, *Metacordyceps*, *Paecilomyces* Bainier, *Nomuraea* Maubl., and *Chamaelomyces* Sigler belonged to the redefined genus *Metarhizium*.

Quandt et al. (2014) divided the family Ophiocordycipitaceae into six genera, *Ophiocordyceps*, *Tolypocladium*, *Purpureocillium*, *Harposporium*, *Drechmeria*, and *Polycephalomyces*. Chen et al. (2021c) established the genus *Pleurodesmospora* belonging to the Cordycipitaceae based on multi-gene phylogenetic evidence. Xiao et al. (2023) conducted a phylogenetic analysis of the family Ophiocordycipitaceae and revealed that 3 genera, *Polycephalomyces*, *Perennicordyceps* Matočec & I. Kušan, and *Pleurocordyceps* Matočec & I. Kušan, clustered together and differed from other genera of Ophiocordycipitaceae. Thus, a new family, Polycephalomycetaceae, was proposed to accommodate those genera. The relationships among *Cordyceps* species have become diverse with the increasing number of members.

5.2 Increasing Hosts of the Cordyceps

Cordyceps mainly parasitize insects, mites, spiders, and the underground fruit body of the genus *Elaphomycos* Nees & Fr. (Liang et al., 2013). As the membership of *Cordyceps* expands, its relationship with its host also undergoes changes. The *Chamaelomyces* species can infect chameleons and cause granulomatous lesions (Sigler et al., 2010). The species *Purpureocillium lilacinum* can cause infections in immunocompromised mammals and patients (Luangsa-ard et al., 2011). It has been reported that *Polycephalomyces yunnanensis* Hong Yu bis, Y.B. Wang & Y.D. Dai hyperparasite *Ophiocordyceps nutans* (Pat.) G.H. Sung, J.M. Sung, Hywel-Jones & Spatafora (Wang et al., 2015), and *Polycephalomyces tomentosus* (Schrad.) Seifert can parasitize slime molds (Kepler et al., 2013). Besides, *Cordyceps* species can also inhabit the soil as an endophytic fungus of plants and insects (Ownley et al., 2008; Wang et al., 2020b). Currently, the habitats of *Cordyceps* include soil, plants, slime molds, fungi, nematodes, spiders, insects, chameleons, and humans.

5.3 *Evolutionary Relationship of the Cordyceps*

The current phylogenetic analysis methods assume that biological groups evolve in a tree-like bifurcation. However, as research into phylogenetic analysis deepens and expands, phylogenetic trees no longer accurately reflect the entire evolution process, such as species with hybridization origins, gene horizontal transfer, and population-level gene recombination (Cheng and Huang, 2008). Species of the order Hypocreales consume a variety of nutrients, including living organisms and plant residues as well as animals (especially insects) and fungi. In the ancestors of the Hypocreales and Clavicipitaceae, nutrition was primarily derived from plants and their plant residues (Spatafora et al., 2007). Numerous phylogenetic studies have shown that multiple transitions between different lifestyles are crucial in the evolutionary history of fungi in the order Hypocreales. The host jumping phenomenon is common among *Cordyceps* species (Fukatsu and Nikoh, 2003).

Kepler et al. (2012) reported a new plant pathogenic genus, *Tyranicodyceps*, and it's speculated that species within this genus may initially infect Hemiptera insects and subsequently become plant pathogens through host transitions based on ancestral state reconstruction. Lei et al. (2015) reported that *Ophiocordyceps sinensis* exists in the plant *Ranunculus altiplanus*, and its hyphae can be utilized to survive in plant tissue, possibly through herbivorous insects feeding on *Ranunculus altiplanus* and crossing the border based on fluorescence quantitative PCR and fluorescence in situ hybridization technology. A study by Sanjuan et al. (2015) revealed five new Amazonian insect pathogenic fungi, and the formation mechanism of these species was explained based on two hypotheses: host correlation and host habitat. Based on phylogenetic analysis, the results showed that *Ophiocordyceps tiputini* Sanjuan underwent a series of host transformations from ants (Antidae) to fish flies (Pleuroptera) and back to ants, then forming these new taxa.

5.4 *Reticulate Evolution of the Cordyceps*

Phylogenetic networks were proposed to express trait conflicts or network evolution processes and to represent the synthesis of multiple related phylogenetic trees, which can simultaneously represent multiple phylogenetic trees and provide evidence for possible network evolution events (Bryant et al., 2007). Haplotype networks are another useful method for analyzing and displaying differences and evolutionary relationships between DNA sequences (Kang et al., 2015), can accommodate phenomena such as recombination and gene horizontal transfer that occur in sequences, and also allow for circular branching, which not only provides a clear image of the evolutionary branches in the phylogenetic tree but also intuitively reflects the pathways of their developmental differentiation (Posada et al., 2001; Morrison, 2005; Mardulyn, 2012).

Bidochka et al. (2005) reported the recombination phenomenon between *Metarhizium anisopliae* from the same region in Ontario, Canada, active in both hot and cold environments. An AFLP and a single-strand conformation polymorphism were used by Devi et al. (2007) to detect recombination in *Nomuraea rileyi* (Farl.) Samson, and the results revealed that recombination was achieved through the fusion of hyphae from different isolates infected with insects. Ciancio et al. (2013) reported a new taxon, *Hirsutella tunicata* Ciancio, Colagiero & Rosso, based on morphological and phylogenetic analysis, and detected recombination events in the genus *Hirsutella*. In addition, recombination phenomena are often detected in *H. rhossiliansis* (Wang et al. 2016). Kazartsev et al. (2021) found that there was intraspecific recombination of *B. bassiana* and *B. pseudobassiana* in the forests of northern Nordic Russia.

Conclusion

Cordyceps have been integrated into people's lives with multiple identities nowadays. As the number of *Cordyceps* is gradually increasing and its host range is gradually expanding, the relationship between *Cordyceps* and their hosts will also become more complex. The relationship among *Cordyceps* species based on morphological features and phylogenetic analysis will also face challenges. The evolutionary time has successfully been applied to the study of *Lecanicillium* species and explores the attribution of *Lecanicillium* species and their related genera. The phylogenetic networks were used to support morphological and phylogenetic analyses in the study of polymorphism of entomopathogenic fungi in an ant nest. Besides, the phylogenetic networks can aid in confirming interrelationships among *Beauveria* species when studying hidden species within the genus *Beauveria*. In addition, three monophyletic origins within *Isaria fumosorosea* (now combined as *Cordyceps fumosorosea*) were discovered based on a haplotype network. All these results revealed that the elimination of cryptic species needs multiple methods. Furthermore, the analysis of genomic data, phylogenetic networks, and haplotypes is also needed to better understand cryptic species and their taxonomy and phylogenetic relationships, thereby improving the understanding of natural taxonomies.

References

Aebersold, R. (2003). A mass spectrometric journey into protein and proteome research. J. Am. Soc. Mass Spectrom., 14: 685–695. https://doi.org/10.1016/S1044-0305(03)00289-7

Agrawal, Y., Mual, P. and Shenoy, B.D. (2014). Multi-gene genealogies reveal cryptic species *Beauveria rudraprayagi* sp. nov. from India. Mycosphere, 5: 719–736. https://doi.org/10.5943/mycosphere/5/6/3

Araújo, J.P., Evans, H.C., Fernandes, I.O., Ishler, M.J. and Hughes. D.P. (2020). Zombie-ant fungi cross continents: II. Myrmecophilous hymenostilboid species and a novel zombie lineage. Mycologia, 112: 1138–1170. https://doi.org/10.1080/00275514.2020.1822093

Ariyawansa, H.A., Hyde, K.D., Jayasiri, S.C., Buyck, B., Chethana, K.T. et al. (2015). Fungal diversity notes 111–252 - taxonomic and phylogenetic contributions to fungal taxa. Fungal Divers., 75: 27–274. https://doi.org/10.1007/s13225-015-0346-5

Artjariyasripong, S., Mitchell, J.I., Hywel-Jones, N.L. and Jones. E.B.G. (2001). Relationship of the genus *Cordyceps* and related genera, based on parsimony and spectral analysis of partial 18S and 28S ribosomal gene sequences. Mycoscience, 42: 503–517. https://doi.org/10.1007/BF02460949

Ban, S., Azuma, Y., Sato, H., Suzuki, K.I. and Nakagiri. A. (2015). *Isaria takamizusanensis* is the anamorph of *Cordyceps ryogamimontana*, warranting a new combination, *Purpureocillium takamizusanense* comb. nov. Int. J. Syst. Evol. Microbiol., 65: 2459–2465. https://doi.org/10.1099/ijs.0.000284

Barron, G.L. (1980). Fungal parasites of rotifers: a new *Tolypocladium* with underwater conidiation. Can. J. Bot., 58: 439–442. https://doi.org/10.1139/b80-050

Barron, G.L. (1981). Two new fungal parasites of bdelloid rotifers. Can. J. Bot., 59: 1449–1455. https://doi.org/10.1139/b81-198

Barron, G.L. (1983). Structure and biology of a new *Tolypocladium* attacking bdelloid rotifers. Can. J. Bot., 61: 2566–2569. https://doi.org/10.1139/b83-281

Bidochka, M.J., Small, C.L.N. and Spironello, M. (2005). Recombination within sympatric cryptic species of the insect pathogenic fungus *Metarhizium anisopliae*. Environ. Microbiol., 7: 1361–1368. https://doi.org/10.1111/j.1462-5822.2005.00823.x

Bischoff, J.F., Rehner, S.A. and Humber, R.A. (2006). *Metarhizium frigidum* sp. nov.: a cryptic species of M. anisopliae and a member of the *M. flavoviride* complex. Mycologia, 98: 737–745. https://doi.org/10.1080/15572536.2006.11832645

Bischoff, J.F., Rehner, S.A. and Humber, R.A. (2009). A multilocus phylogeny of the *Metarhizium anisopliae* lineage. Mycologia, 101: 512–530. https://doi.org/10.3852/07-202

Bissett, J. (1983). Notes on *Tolypocladium* and related genera. Can. J. Bot., 61: 1311–1329. https://doi.org/10.1139/b83-139

Bryant, D., Moulton, V. and Spillner, A. (2007). Consistency of the neighbor-net algorithm. Algorithms Mol. Biol., 2: 1–11. https://doi.org/10.1186/1748-7188-2-8

Bustamante, D.E., Oliva, M., Leiva, S., Mendoza, J.E., Bobadilla, L. et al. (2019). Phylogeny and species delimitations in the entomopathogenic genus *Beauveria* (Hypocreales, Ascomycota), including the description of *B. peruviensis* sp. nov. MycoKeys, 58: 47–68. https://doi.org/10.3897/mycokeys.58.35764

Calvillo-Medina, R.P., Ponce-Angulo, D.G., Raymundo, T., Müller-Morales, C.A., Escudero-Leyva, E. et al. (2021). *Purpureocillium roseum* sp. nov. A new ocular pathogen for humans and mice resistant to antifungals. Mycoses, 64: 162–173. https://doi.org/10.1111/myc.13198

Cao, Y.P. (2014). Studies on entomogenous fungi in the Anhui Province and antimicrobial activities of the strain RCEF6202D. Anhui Agricultural University, Hefei, Anhui, p. 91.

Chaverri, P., Liu, M. and Hodge, K.T. (2008). A monograph of the entomopathogenic genera *Hypocrella*, *Moelleriella*, and *Samuelsia* gen. nov. (Ascomycota, Hypocreales,

Clavicipitaceae), and their aschersonia-like anamorphs in the Neotropics. Stud. Mycol., 60: 1–66. https://doi.org/10.3114/sim.2008.60.01

Chawngthu, Z., Mc Vabeikhokhei, J.O.S.I.A.H., Zomuanpuii, R., De Mandal, S. and Zothanzama, J. (2021). A new species of *Ophiocordyceps* (Ophiocordycipitaceae) from Mizoram, India. Phytotaxa, 500: 11–20. https://doi.org/10.11646/phytotaxa.500.1.2

Chen W.H., Liang, J.D., Ren, X.X., Zhao, J.H., Han, Y.F. and Liang, Z.Q. (2022c). Phylogenetic, ecological and morphological characteristics reveal two new spider-associated genera in Clavicipitaceae. MycoKeys, 91: 49–66. https://doi.org/10.3897/mycokeys.91.86812

Chen, M.J., Huang, B., Li, Z.Z. and Spatafora, J.W. (2013). Morphological and genetic characterisation of *Beauveria sinensis* sp. nov. from China. Mycotaxon, 124: 301–308. https://doi.org/10.5248/124.301

Chen, M.J., Lin, Y., Wang, T., Zhang, S.L. and Huang, B. (2022a). *Metarhizium macrosemiae* sp. nov. and the anamorph of *M. guniujiangense* on adult cicada from Guniujiang Nature Preserve, southeastern China. Phytotaxa, 575: 68–78. https://doi.org/10.11646/phytotaxa.575.1.4

Chen, S., Wang, Y., Zhu, K. and Yu, H. (2021a). Mitogenomics, phylogeny and morphology reveal *Ophiocordyceps pingbianensis* sp. nov., an entomopathogenic fungus from China. Life (Basel, Switzerland), 11(7): 686. https://doi.org/10.3390/life11070686

Chen, W., Liang, J., Ren, X., Zhao, J., Han, Y. and Liang, Z. (2021b). Cryptic diversity of isaria-like species in Guizhou, China. Life (Basel, Switzerland), 11(10): 1093. https://doi.org/10.3390/life11101093

Chen, W.H., Han, Y.F., Liang, J.D. and Liang, Z.Q. (2019b). Morphological and phylogenetic characterization of novel *Metarhizium* species in Guizhou, China. Phytotaxa, 419: 189–196. https://doi.org/10.11646/phytotaxa.419.2.5

Chen, W.H., Han, Y.F., Liang, J.D., Liang, Z.Q. and Jin, D.C. (2017b). *Metarhizium dendrolimatilis*, a novel *Metarhizium* species parasitic on *Dendrolimus* sp larvae. Mycosphere, 8: 31–37. https://doi.org/10.5943/mycosphere/8/1/4

Chen, W.H., Han, Y.F., Liang, J.D., Tian, W.Y. and Liang, Z.Q. (2020). Morphological and phylogenetic characterisations reveal three new species of *Samsoniella* (Cordycipitaceae, Hypocreales) from Guizhou, China. MycoKeys, 74: 1–15. https://doi.org/10.3897/mycokeys.74.56655

Chen, W.H., Han, Y.F., Liang, J.D., Tian, W.Y. and Liang, Z.Q. (2021c). Multi-gene phylogenetic evidence indicates that *Pleurodesmospora* belongs in Cordycipitaceae (Hypocreales, Hypocreomycetidae) and *Pleurodesmospora lepidopterorum* sp. nov. on pupa from China. MycoKeys, 80: 45–55. https://doi.org/10.3897/mycokeys.80.66794

Chen, W.H., Han, Y.F., Liang, Z.Q. and Jin, D.C. (2017a). A new araneogenous fungus in the genus *Beauveria* from Guizhou, China. Phytotaxa, 302: 57–64. https://doi.org/10.11646/phytotaxa.302.1.5

Chen, W.H., Han, Y.F., Liang, Z.Q., Wang, Y.R. and Zou, X. (2014). Classification of *Gibellula* spp.by DELTA system. Microbiol. China, 41: 399–407. https://doi.org/10.13344/j.microbiol.china.130202

Chen, W.H., Liang, J.D., Ren, X.X., Zhao, J.H. and Han, Y.F. (2023a). Two new species of *Samsoniella* (Cordycipitaceae, Hypocreales) from the Mayao River Valley, Guizhou, China. MycoKeys, 99: 209–226. https://doi.org/10.3897/mycokeys.99.109961

Chen, W.H., Liang, J.D., Ren, X.X., Zhao, J.H. and Han, Y.F. (2023b). Study on species diversity of *Akanthomyces* (Cordycipitaceae, Hypocreales) in the Jinyun Mountains, Chongqing, China. MycoKeys, 98: 299. https://doi.org/10.3897/mycokeys.98.106415

Chen, W.H., Liang, J.D., Ren, X.X., Zhao, J.H., Han, Y.F. and Liang, Z.Q. (2022b). Species diversity of Cordyceps-like fungi in the Tiankeng karst region of China. Microbiol. Spectrum, 10: e01975–22. https://doi.org/10.1128/spectrum.01975-22

Chen, W.H., Liu, M., Huang, Z.X., Yang, G.M., Han, Y.F., Liang, J.D. and Liang, Z.Q. (2018a). *Beauveria majiangensis*, a new entomopathogenic fungus from Guizhou, China. Phytotaxa, 333: 243–250. https://doi.org/10.11646/phytotaxa.333.2.8

Chen, Z.H., Chen, K., Dai, Y.D., Zheng, Y., Wang, Y.B. et al. (2019a). *Beauveria* species diversity in the Gaoligong Mountains of China. Mycol. Prog., 18: 933–943. https://doi.org/10.1007/s11557-019-01497-z

Chen, Z.H., Dai, Y.D., Chen, K., Zhang, Y.F., Xu, L. and Wang, Y.B. (2023c). *Papiliomyces puniceum* and *Metarhizium lymantriidae*: two new species from the Gaoligong Mountains in southwestern China. Phytotaxa, 594: 53–63. https://doi.org/10.11646/phytotaxa.594.1.3

Chen, Z.H., Zhang, Y.G., Yang, X.N., Chen, K., Liu, Q. and Xu, L. (2018b). A new fungus *Metarhizium gaoligongense* from China. Int. J. Agric. Biol., 20: 2271–2276. https://doi.org/10.17957/IJAB/15.0777

Cheng, C.H. and Huang, Y. (2008). Construction and application of phylogenetic network. Entomotaxonomia, 3: 215–221.

Chu, H.L., Chen, W.H., Wen, T.C., Liang, Z.Q., Zheng, F.C. et al. (2016). Delimitation of a novel member of genus *Metarhizium* (Clavicipitaceae) by phylogenetic and network analysis. Phytotaxa, 288: 51–60. https://doi.org/10.11646/phytotaxa.288.1.5

Ciancio, A., Colagiero, M., Rosso, L.C., Gutierrez, S.N.M. and Grasso, G. (2013). Phylogeny and morphology of *Hirsutella tunicata* sp. nov. (Ophiocordycipitaceae), a novel mite parasite from Peru. Mycoscience, 54: 378–386. https://doi.org/10.1016/j.myc.2013.01.002

Clifton, E.H., Castrillo, L.A. and Hajek, A.E. (2021). Discovery of two hypocrealean fungi infecting spotted lanternflies, *Lycorma delicatula*: *Metarhizium pemphigi* and a novel species, *Ophiocordyceps delicatula*. J. Invertebr. Pathol., 186: 107689. https://doi.org/10.1016/j.jip.2021.107689

Crous, P.W., Wingfield, M.J., Burgess, T.I., Hardy, G.E.ST.J., Barber, P.A. et al. (2017). Fungal Planet description sheets: 558–624. Persoonia, 38: 240–384. https://doi.org/10.3767/003158517X69894

Crous, P.W., Cowan, D.A., Maggs-Kölling, G., Yilmaz, N., Larsson, E. et al. (2020). Fungal Planet description sheets: 1112–1181. Persoonia, 45: 251–409(159). https://doi.org/10.3767/persoonia.2020.45.10

Crous, P.W., Osieck, E.R., Shivas, R.G., Tan, Y.P., Bishop-Hurley, S.L. et al. (2023). Fungal Planet description sheets: 1478–1549. Persoonia, 50: 158–310(153). https://doi.org/10.3767/persoonia.2023.50.05

Curran, J., Driver, F., Ballard, J.W.O. and Milner, R.J. (1994). Phylogeny of *Metarhizium*: analysis of ribosomal DNA sequence data. Mycol. Res., 98: 547–552. https://doi.org/10.1016/S0953-7562(09)80478-4

Das, K., Ryu, J.J., Hong, S.M., Lim, S.K., Lee, S.Y. and Jung, H.Y. (2023). Molecular phylogeny and morphology of *Tolypocladium globosum* sp. nov. isolated from soil in Korea. Mycobiology, 51: 79–86. https://doi.org/10.1080/12298093.2023.2192614

Devi, U.K., Reineke, A., Rao, U.C.M., Reddy, N.R.N. and Khan, A.P.A. (2007). AFLP and single strand conformation polymorphism studies of recombination in the entomopathogenic fungus *Nomuraea rileyi*. Mycol. Res., 111: 716–725. https://doi.org/10.1016/j.mycres.2007.03.003

Diehl, W.W. (1950). *Balansia* and the Balansiae in America. Agricultural Monograph No. 4., Unted States Bureau of Plant Industry, Soils and Agricultural Engineering, US Department of Agriculture, Washington D.C., p. 794.

Domon, B. and Aebersold, R. (2006). Mass spectrometry and protein analysis. Science, 312: 212–217. https://doi.org/10.1126/science.112461

Dong, Q.Y., Wang, Y., Wang, Z.Q., Liu, Y.F. and Yu, H. (2022). Phylogeny and systematics of the genus *Tolypocladium* (Ophiocordycipitaceae, Hypocreales). J. Fungi, 8(11): 1158. https://doi.org/10.3390/jof8111158

Dreyfuss, M.M. and Gams, W. (1994). Proposal to reject *Pachybasium niveum* Rostr. in order to retain the name *Tolypocladium inflatum* W. Gams for the fungus that produces cyclosporin. Taxon, 43: 660–661. https://doi.org/10.2307/1223557

Driver, F., Milner, R.J. and Trueman. J.W. (2000). A taxonomic revision of *Metarhizium* based on a phylogenetic analysis of rDNA sequence data. Mycol. Res., 104: 134–150. https://doi.org/10.1017/S0953756299001756

Eriksson, O. (1982). Outline of the ascomycetes-1982. Mycotaxon, 15: 203–248.

Fan, Q., Wang, Y.B., Zhang, G.D., Tang, D.X. and Yu, H. (2021). Multigene phylogeny and morphology of *Ophiocordyceps alboperitheciata* sp. nov., a new entomopathogenic fungus attacking Lepidopteran larva from Yunnan, China. Mycobiology, 49: 133–141. https://doi.org/10.1080/12 298093.2021.1903130

Flakus, A., Etayo, J., Miadlikowska, J., Lutzoni, F., Kukwa, M. et al. (2019). Biodiversity assessment of ascomycetes inhabiting Lobariella lichens in Andean cloud forests led to one new family, three new genera and 13 new species of lichenicolous fungi. Pl. Fungal Syst., 64: 283–344. https://doi. org/10.2478/pfs-2019-0022

Fukatsu, T. and Nikoh, N. (2003). Interkingdom host shift in the *Cordyceps* fungi. pp. 301–317. *In*: Clavacipitalean Fungi. White Jr., J.F., Bacon, C.W., Hywel-Jones and Spatafora, J.W., CRC Press, Boca Raton, USA. https://doi.org/10.1201/9780203912706.ch10

Gams, W. (1971). *Tolypocladium*, eine Hyphomycetengattung mit geschwollenen Phialiden. Persoonia, 6: 185–191.

Gao, S., Meng, W., Zhang, L., Yue, Q., Zheng, X. and Xu, L. (2021). *Parametarhizium* (Clavicipitaceae) gen. nov. with two new species as a potential biocontrol agent isolated from forest litters in Northeast China. Front Microbiol., 12: 131. https://doi.org/10.3389/fmicb.2021.627744

Gazis, R., Skaltsas, D. and Chaverri, P. (2014). Novel endophytic lineages of *Tolypocladium* provide new insights into the ecology and evolution of *Cordyceps*-like fungi. Mycologia, 106: 1090–1105. https://doi.org/10.3852/13-346

Gutierrez, A.C., Leclerque, A., Manfrino, R.G., Luz, C., Ferrari, W.A. et al. (2019). Natural occurrence in Argentina of a new fungal pathogen of cockroaches, *Metarhizium argentinense* sp. nov. Fungal Biol., 123: 364–372. https://doi.org/10.1016/j.funbio.2019.02.005

Han, Y.F. (2007). Studies on the systematics of the genus *Paecilomyces* Bainier in China and the pathogenicity of some isolates. Guizhou University, Guiyang, Guizhou, PR China, p. 165.

Han, Y.F., Liang, J.D., Liang, Z.Q., Zou, X. and Dong, X. (2010). Two new *Taifanglania* species identified through DELTA-assisted phonetic analysis. Mycotaxon, 112: 325–333. https:// doi.org/10.5248/112.325

Han, Y.F., Zhang, Y.W., Liang, J.D. and Liang, Z.Q. (2009). Studies on the genus *Paecilomyces* in China. Application of DELTA expert system on the entomopathogenic *Paecilomyces* sensu lato. Mycotaxon, 109: 75–84. https://doi.org/10.5248/109.75

Hricakova, N., Medo, J., Hleba, L., Barta, M. and Makova, J. (2018). Use of MALDI-TOF mass spectrometry in rapid identification of *Beauveria bassiana* and *Beauveria pseudobassiana*. J. Cent. Eur. Agric., 19: 394–407. https://doi.org/10.5513/JCEA01/19.2.2151

Hyde, K.D., Hongsanan, S., Jeewon, R., Bhat, D.J., Mckenzie, E.H.C. et al. (2016). Fungal diversity notes 367–490: taxonomic and phylogenetic contributions to fungal taxa. Fungal Divers., 80: 1–270. https://doi.org/10.1007/s13225-016-0373-x

Imoulan, A., Wu, H.J., Lu, W.L., Li, Y., Li, B.B. et al. (2016). *Beauveria medogensis* sp. nov., a new fungus of the entomopathogenic genus from China. J. Invertebr. Pathol., 139: 74–81. https://doi.org/10.1016/j.jip.2016.07.006

Isaka, M., Sappan, M., Luangsa-ard, J.J., Hywel-Jones, N.L., Mongkolsamrit, S. and Chunhametha, S. (2011). Chemical taxonomy of *Torrubiella* s. lat.: Zeorin as a marker of *Conoideocrella*. Fungal Biol., 115: 401–405. https://doi.org/10.1016/j.funbio.2011.02.007

Johnson, D., Sung, G.H., Hywel-Jones, N.L., Luangsa-ard, J.J., Bischoff, J.F. et al. (2009). Systematics and evolution of the genus *Torrubiella* (Hypocreales, Ascomycota). Mycol. Res., 113: 279–289. https://doi.org/10.1016/j.mycres.2008.09.008

Kang, B., Cheng, S.M., Yan, J.J., Peng, J.B. and Zhang, S.Y. (2015). Constructing haplotype median-joining network with highly divergent intraspecific and interspecific sequences. Acta Biophys. Sin., 2: 165–171.

Kazartsev, I.A. and Lednev, G.R. (2021). Distribution and diversity of *Beauveria* in boreal forests of northern European Russia. Microorganisms, 9(7): 1409. https://doi.org/10.3390/ microorganisms9071409

Kepler, R., Ban, S., Nakagiri, A., Bischoff, J., Hywel-Jones, N. et al. (2013). The phylogenetic placement of hypocrealean insect pathogens in the genus *Polycephalomyces*: An application of One Fungus One Name. Fungal Biol., 117: 611–622. https://doi.org/10.1016/j. funbio.2013.06.002

Kepler, R.M., Humber, R.A., Bischoff, J.F. and Rehner, S.A. (2014). Clarification of generic and species boundaries for *Metarhizium* and related fungi through multigene phylogenetics. Mycologia, 106: 811–829. https://doi.org/10.3852/13-319

Kepler, R.M., Luangsa-ard, J.J., Hywel-Jones, N.L., Quandt, C.A., Sung, G.H. et al. (2017). A phylogenetically-based nomenclature for Cordycipitaceae (Hypocreales). IMA Fungus, 8: 335–353. https://doi.org/10.5598/imafungus.2017.08.02.08

Kepler, R.M., Sung, G.H., Harada, Y., Tanaka, K., Tanaka, E. et al. (2012). Host jumping onto close relatives and across kingdoms by *Tyrannicordyceps* (Clavicipitaceae) gen. nov. and *Ustilaginoidea_* (Clavicipitaceae). Am. Bot., 99: 552–561. https://doi.org/10.3732/ ajb.1100124

Khao-ngam, S., Mongkolsamrit, S., Rungjindamai, N., Noisripoom, W., Pooissarakul, W. et al. (2021). *Ophiocordyceps asiana* and *Ophiocordyceps tessaratomidarum* (Ophiocordycipitaceae, Hypocreales), two new species on stink bugs from Thailand. Mycol. Prog., 20: 341–353. https://doi.org/10.1007/s11557-021-01684-x

Khonsanit, A., Luangsa-ard, J.J., Thanakitpipattana, D., Kobmoo, N. and Piasai, O. (2019). Cryptic species within *Ophiocordyceps myrmecophila* complex on formicine ants from Thailand. Mycol. Prog., 18: 147–161. https://doi.org/10.1007/s11557-018-1412-7

Khonsanit, A., Luangsa-ard, J.J., Thanakitpipattana, D., Noisripoom, W., Chaitika, T. and Kobmoo, N. (2020). Cryptic diversity of the genus *Beauveria* with a new species from Thailand. Mycol. Prog., 19: 291–315. https://doi.org/10.1007/s11557-020-01557-9

Kirk, P.M., Cannon, P.F., David, J.C. and Stalpers, J.A. (2001). Dictionary of the Fungi (9th ed.). CAB International, Wallingford, UK, p. 655.

Kirk, P.M., Stalpers, J.A., Braun, U., Crous, P.W., Hansen, K. et al. (2013). A without-prejudice list of generic names of fungi for protection under the International Code of Nomenclature for algae, fungi, and plants. IMA Fungus, 4: 381–443. https://doi. org/10.5598/imafungus.2013.04.02.1

Kobayasi, Y. (1941). The genus *Cordyceps* and its allies. Sci. Rep. Tokyo Bunrika Daigaku, Section B, 5: 53–260.

Kobayasi, Y. (1983). *Cordyceps* species from Japan 6. Bull. Natl. Sci. Mus. Tokyo, series B, 9: 1–21.

Kobmoo, N., Arnamnart, N., Pootakham, W., Sonthirod, C., Khonsanit, A. et al. (2021). The integrative taxonomy of *Beauveria asiatica* and *B. bassiana* species complexes with whole-genome sequencing, morphometric and chemical analyses. Persoonia, 47: 136–150. https://doi.org/10.3767/persoonia.2021.47.04

Kobmoo, N., Tasanathai, K., Araújo, J.P.M., Noisripoom, W., Thanakitpipattana, D. et al. (2023). New mycoparasitic species in the genera *Niveomyces* and *Pseudoniveomyces* gen. nov. (Hypocreales: Cordycipitaceae), with sporothrix-like asexual morphs, from Thailand. Fungal Syst. Evol., 12: 91–110. https://doi.org/10.3114/fuse.2023.12.07

Lao, T.D., Le, T.A.H. and Truong, N.B. (2021). Morphological and genetic characteristics of the novel entomopathogenic fungus *Ophiocordyceps langbianensis* (Ophiocordycipitaceae, Hypocreales) from Lang Biang Biosphere Reserve, Vietnam. Sci. Rep., 11(1): 1412(2021). https://doi.org/10.1038/s41598-020-78265-7

Latch, G.C.M. (1965). *Metarhizium anisopliae* (Metsch.) Sorok. strains in New Zealand and the possible use for controlling pasture inhabiting insects. N. Z. J. Agric. Res., 8: 384–396. https://doi.org/10.1080/00288233.1965.10422370

Lei, W., Zhang, G., Peng, Q. and Liu, X. (2015). Development of *Ophiocordyceps sinensis* through plant-mediated interkingdom host colonization. Int. J. Mol. Sci., 16: 17482–17493. https:// doi.org/10.3390/ijms160817482

Li, C., Hywel-Jones, N., Cao, Y., Nam, S. and Li, Z. (2018). *Tolypocladium dujiaolongae* sp. nov. and its allies. Mycotaxon, 133: 229–241. https://doi.org/10.5248/133.229

Li, C.L. (1986). Study of a new species of *Tolypocladium*. J. Nanjing Univ., 22: 401–407.

Li, D.W., Cowles, R.S. and Vossbrinck, C.R. (2008). *Metarhiziopsis microspora* gen. et sp. nov. associated with the elongate hemlock scale. Mycologia, 100: 460–466. https://doi.org/10.3852/07-078R1

Liang, Z.Q., Han, Y.F., Liang, J.D. and Zou, X. (2013). Artificial culture of *Cordyceps*. Guizhou Science and Technology Press, Guiyang, Guizhou, PR China, p. 162.

Liang, Z.Q., Liu, A.Y. and Liu, J.L. (1991). A new species of the genus *Cordyceps* and its Metarhizium anamorph. Acta Mycol. Sin., 10: 257–262.

Liu, Z.Y. (1999). Studies on relationship between *Cordyceps* spp. and their anamorphs. Huazhong Agricultural University, Wuhan, Huibei, PR China, p. 188.

Liu, Z.Y., Liang, Z.Q., Liu, A.Y., Yao, Y.J. and Yu, Z.N. (2002). Molecular evidence for teleomorph–anamorph connections in *Cordyceps* based on ITS-5.8S rDNA sequences. Mycol. Res., 106: 1100–1108. https://doi.org/10.1017/S0953756202006378

Liu, Z.Y., Yao, Y.J., Liang, Z.Q., Liu, A.Y., Pegler, D.N. and Chase, M.W. (2001). Molecular evidence for the anamorph–teleomorph connection in *Cordyceps sinensis*. Mycol. Res., 105: 827–832. https://doi.org/10.1017/S095375620100377X

Long, F.Y., Qin, L.W., Xiao, Y.P., Hyde, K.D., Wang, S.X. and Wen, T.C. (2021). Multigene phylogeny and morphology reveal a new species, *Ophiocordyceps vespulae*, from Jilin Province, China. Phytotaxa, 478: 33–48. https://doi.org/10.11646/phytotaxa.478.1.2

Lopes, R.B., Faria, M., Souza, D.A., Bloch Jr, C., Silva, L.P. and Humber, R.A. (2014). MALDI-TOF mass spectrometry applied to identifying species of insect-pathogenic fungi from the *Metarhizium anisopliae* complex. Mycologia, 106: 865–878. https://doi.org/10.3852/13-401

Luangsa-ard, J., Houbraken, J., van Doorn, T., Hong, S.B., Borman, A.M. et al. (2011). *Purpureocillium*, a new genus for the medically important *Paecilomyces lilacinus*. FEMS Microbiol. Lett., 321: 141–149. https://doi.org/10.1111/j.1574-6968.2011.02322.x

Luangsa-ard, J., Tasanathai, K., Thanakitpipattana, D., Khonsanit, A. and Stadler, M. (2018). Novel and interesting *Ophiocordyceps* spp. (Ophiocordycipitaceae, Hypocreales) with superficial perithecia from Thailand. Stud. Mycol., 89: 125–142. https://doi.org/10.1016/j.simyco.2018.02.001

Luangsa-ard, J.J., Berkaew, P., Ridkaew, R., Hywel-Jones, N.L. and Isaka, M. (2009). A beauvericin hot spot in the genus *Isaria*. Mycol. Res., 113: 1389–1395. https://doi.org/10.1016/j.mycres.2009.08.017

Luangsa-ard, J.J., Mongkolsamrit, S., Thanakitpipattana, D., Khonsanit, A., Tasanathai, K. et al. (2017). Clavicipitaceous entomopathogens: new species in *Metarhizium* and a new genus *Nigelia*. Mycol. Prog., 16: 369–391. https://doi.org/10.1007/s11557-017-1277-1

Luz, C., Rocha, L.F., Montalva, C., Souza, D.A., Botelho, A.B.R. et al. (2019). Metarhizium humberi sp. nov. (Hypocreales: Clavicipitaceae), a new member of the PARB clade in the Metarhizium anisopliae complex from Latin America. J. Invertebr. Pathol., 166: 107216. https://doi.org/10.1016/j.jip.2019.107216

Mains, E.B. (1949). New species of *Torrubiella*, *Hirsutella* and *Gibellula*. Mycologia, 41: 303–310.

Mains, E.B. (1950). Entomogenous species of *Akanthomyces*, *Hymenostilbe* and *Insecticola* in north America. Mycologia, 42: 566–589.

Mardulyn, P. (2012). Trees and/or networks to display intraspecific DNA sequence variation? Mol. Ecol., 21: 3385–3390. https://doi.org/10.1111/j.1365-294X.2012.05622.x

Mayr, E. (1942). Systematics and the origin of species. Columbia University Press, New York, USA, p. 344.

Mongkolsamrit, S., Khonsanit, A., Thanakitpipattana, D., Tasanathai, K., Noisripoom, W. et al. (2020). Revisiting *Metarhizium* and the description of new species from Thailand. Stud. Mycol., 95: 171–251. https://doi.org/10.1016/j.simyco.2020.04.001

Mongkolsamrit, S., Noisripoom, W., Pumiputikul, S., Boonlarppradab, C., Samson, R.A. et al. (2021). *Ophiocordyceps flavida* sp. nov. (Ophiocordycipitaceae), a new species from Thailand associated with *Pseudogibellula formicarum* (Cordycipitaceae), and their bioactive secondary metabolites. Mycol. Prog., 20: 477–492. https://doi.org/10.1007/s11557-021-01683-y

Mongkolsamrit, S., Noisripoom, W., Thanakitpipattana, D., Wuthikun, T., Spatafora, J.W. and Luangsa-ard, J.J. (2018). Disentangling cryptic species with Isaria-like morphs in Cordycipitaceae. Mycologia, 110: 230–257. https://doi.org/10.1080/00275514.2018.1446651

Montalva, C., Collier, K., Rocha, L.F.N., Inglis, P.W., Lopes, R.B. et al. (2016). A natural fungal infection of a sylvatic cockroach with *Metarhizium blattodeae* sp. nov., a member of the *M. flavoviride* species complex. Fungal Biol., 120: 655–665. https://doi.org/10.1016/j.funbio.2016.03.004

Moonjely, S., Barelli, L. and Bidochka, M.J. (2016). Insect pathogenic fungi as endophytes. Adv. Genet., 94: 107–135. https://doi.org/10.1016/bs.adgen.2015.12.004

Morrison, D.A. (2005). Networks in phylogenetic analysis: new tools for population biology. Int. J. Parasitol., 35: 567–582. https://doi.org/10.1016/j.ijpara.2005.02.007

Nishi, O., Shimizu, S. and Sato, H. (2017). *Metarhizium bibionidarum* and *M. purpureogenum*: new species from Japan. Mycol. Prog., 16: 987–998. https://doi.org/10.1007/s11557-017-1333-x

Ownley, B.H., Griffin, M.R., Klingeman, W.E., Gwinn, K.D., Moulton, J.K. and Pereira, R.M. (2008). *Beauveria bassiana*: endophytic colonization and plant disease control. J. Invertebr. Pathol., 98: 267–270. https://doi.org/10.1016/j.jip.2008.01.010

Perdomo, H., Cano, J., Gené, J., García, D., Hernández, M. and Guarro, J. (2013). Polyphasic analysis of *Purpureocillium lilacinum* isolates from different origins and proposal of the new species *Purpureocillium lavendulum*. Mycologia, 105: 151–161. https://doi.org/10.3852/11-190

Petch, T. (1931). Notes on entomogenous fungi. Trans. Br. Mycol. Soc., 16: 55–75.

Petch, T. (1935). Notes on entomogenous fungi. Trans. Br. Mycol. Soc., 19: 161–194.

Pinel, C., Arlotto, M., Issartel, J.P., Berger, F., Pelloux, H. et al. (2011). Comparative proteomic profiles of *Aspergillus fumigatus* and *Aspergillus lentulus* strains by surface-enhanced laser desorption ionization time-of-flight mass spectrometry (SELDI-TOF-MS). BMC Microbiol., 11: 1–10. https://doi.org/10.1186/1471-2180-11-172

Poinar, G. and Maltier, Y.M. (2021). *Allocordyceps baltica* gen. et sp. nov. (Hypocreales: Clavicipitaceae), an ancient fungal parasite of an ant in Baltic amber. Fungal Biol., 125: 886–890. https://doi.org/10.1016/j.funbio.2021.06.002

Posada, D. and Crandall, K.A. (2001). Intraspecific gene genealogies: trees grafting into networks. Trends Ecol. Evol., 16: 37–45. https://doi.org/10.1016/S0169-5347(00)02026-7

Quandt, C.A., Kepler, R.M., Gams, W., Araújo, J.P., Ban, S. et al. (2014). Phylogenetic-based nomenclatural proposals for Ophiocordycipitaceae (Hypocreales) with new combinations in *Tolypocladium*. IMA Fungus, 5: 121–134. https://doi.org/10.5598/imafungus.2014.05.01.12

Rakotonirainy, M.S., Cariou, M.L., Brygoo, Y. and Riba, G. (1994). Phylogenetic relationships within the genus *Metarhizium* based on 28S rRNA sequences and isozyme comparison. Mycol. Res., 98: 225–230. https://doi.org/10.1016/S0953-7562(09)80190-1

Rehner, S.A., Minnis, A.M., Sung, G.H., Luangsa-ard, J.J., Devotto, L. and Humber, R.A. (2011). Phylogeny and systematics of the anamorphic, entomopathogenic genus *Beauveria*. Mycologia, 103: 1055–1073. https://doi.org/10.3852/10-302

Robène-Soustrade, I., Jouen, E., Pastou, D., Payet-Hoarau, M., Goble, T. et al. (2015). Description and phylogenetic placement of *Beauveria hoplocheli* sp. nov. used in the biological control of the sugarcane white grub, *Hoplochelus marginalis*, on Reunion Island. Mycologia, 107: 1221–1232. https://doi.org/10.3852/14-344

Rombach, M.C., Humber, R.A. and Roberts, D.W. (1986). *Metarhizium flavoviride* var. *minus*, var. nov., a pathogen of plant-and leafhoppers on rice in the Philippines and Solomon Islands. Mycotaxon, 27: 87–92.

Samson, R.A. and Soares Jr, G.G. (1984). Entomopathogenic species of the hyphomycete genus *Tolypocladium*. J. Invertebr. Pathol., 43: 133–139. https://doi.org/10.1016/0022-2011(84)90130-7

Sanjuan, T.I., Franco-Molano, A.E., Kepler, R.M., Spatafora, J.W., Tabima, J. et al. (2015). Five new species of entomopathogenic fungi from the Amazon and evolution of neotropical *Ophiocordyceps*. Fungal Biol., 119: 901–916. https://doi.org/10.1016/j.funbio.2015.06.010

Sato, H., Ban, S. and Hosoya, T. (2022). Reassessment of type specimens of *Cordyceps* and its allies, described by Dr. Yosio Kobayasi and preserved in the mycological herbarium of the National Museum of Nature and Science. Part 4. *Cordyceps* sl on Lepidoptera. Mycoscience, 64: 40–46. https://doi.org/10.1007/S10267-009-0015-1

Schroeter, J. and von Schlesien Halfte, K.F. (1894). In: Cohn, F. (ed.). Kryptogamen-Flora von Schlesien. Volume 3(2), Breslau, J.U. Kern's Verlag, p. 597.

Sharma, A., Kaur, E., Joshi, R., Kumari, P., Khatri, A. et al. (2023). Systematic analyses with genomic and metabolomic insights reveal a new species, *Ophiocordyceps indica* sp. nov. from treeline area of Indian Western Himalayan region. Front. Microbiol., 14: 1188649. https://doi.org/10.3389/fmicb.2023.1188649

Sigler, L., Gibas, C.F.C., Kokotovic, B. and Bertelsen, M.F. (2010). Disseminated mycosis in veiled chameleons (Chamaeleo calyptratus) caused by *Chamaeleomyces granulomatis*, a new fungus related to *Paecilomyces viridis*. J. Clin. Microbiol., 48: 3182–3192. https://doi.org/10.1128/jcm.01079-10

Soares, J.M., Karlsen-Ayala, E., Salvador-Montoya, C.A. and Gazis, R. (2022). Two novel endophytic *Tolypocladium* species identified from native pines in south Florida. Fungal Syst. Evol., 11: 51–61. https://doi.org/10.3114/fuse.2023.11.04

Sorokin, N. (1883). Plant Parasites of Man and Animals as causes of infectious diseases. J. Mil. Med., 2: 268–291.

Spatafora, J.W., Quandt, C.A., Kepler, R.M., Sung, G.H., Shrestha, B. et al. (2015). New 1F1N species combinations in Ophiocordycipitaceae (Hypocreales). IMA Fungus, 6: 357–362. https://doi.org/10.5598/imafungus.2015.06.02.07

Spatafora, J.W., Sung, G.H., Sung, J.M., Hywel-Jones, N.L. and White Jr, J.F. (2007). Phylogenetic evidence for an animal pathogen origin of ergot and the grass endophytes. Mol. Ecol., 16: 1701–1711. https://doi.org/10.1111/j.1365-294X.2007.03225.x

Stadler, M., Tichy, H.V., Katsiou, E. and Hellwig, V. (2003). Chemotaxonomy of *Pochonia* and other conidial fungi with *Verticillium*-like anamorphs. Mycol. Prog., 2: 95–122. https://doi.org/10.1007/s11557-006-0048-1

Stensrud, Ø., Hywel-Jones, N.L. and Schumacher, T. (2005). Towards a phylogenetic classification of *Cordyceps*: ITS nrDNA sequence data confirm divergent lineages and paraphyly. Mycol. Res., 109: 41–56. https://doi.org/10.1017/S095375620400139X

Sun, T., Zou, W., Dong, Q., Huang, O., Tang, D. and Yu, H. (2022). Morphology, phylogeny, mitogenomics and metagenomics reveal a new entomopathogenic fungus *Ophiocordyceps nujiangensis* (Hypocreales, Ophiocordycipitaceae) from Southwestern China. MycoKeys, 94: 91–108. https://doi.org/10.3897/mycokeys.94.89425

Sung, G.H., Hywel-Jones, N.L., Sung, J.M., Luangsa-ard, J.J., Shrestha, B. and Spatafora, J.W. (2007b). Phylogenetic classification of *Cordyceps* and the clavicipitaceous fungi. Stud. Mycol., 57: 5–59. https://doi.org/10.3114/sim.2007.57.01

Sung, G.H., Spatafora, J.W., Zare, R., Hodge, K.T. and Gams, W. (2001). A revision of *Verticillium* sect. Prostrata. II. Phylogenetic analyses of SSU and LSU nuclear rDNA sequences from anamorphs and teleomorphs of the Clavicipitaceae. Nova Hedwigia, 72: 311–328. https://doi.org/10.1127/nova.hedwigia/72/2001/311

Sung, G.H., Sung, J.M., Hywel-Jones, N.L. and Spatafora, J.W. (2007a). A multi-gene phylogeny of Clavicipitaceae (Ascomycota, Fungi): Identification of localized incongruence using a combinational bootstrap approach. Mol. Phylogenet. Evol., 44: 1204–1223. https://doi.org/10.1016/j.ympev.2007.03.011

Tanaka, E., Shrestha, B. and Shivas, R.G. (2020). *Commelinaceomyces*, gen. nov., for four clavicipitaceous species misplaced in *Ustilago* that infect Commelinaceae. Mycologia, 112: 649–660. https://doi.org/10.1080/00275514.2020.1745524

Tang, D., Huang, O., Zou, W., Wang, Y., Wang, Y. et al. (2023). Six new species of zombie-ant fungi from Yunnan in China. IMA Fungus, 14: 9. https://doi.org/10.1186/s43008-023-00114-9

Tasanathai, K., Noisripoom, W., Chaitika, T., Khonsanit, A., Hasin, S. and Luangsa-ard, J. (2019). Phylogenetic and morphological classification of *Ophiocordyceps* species on termites from Thailand. MycoKeys, 56: 101–129. https://doi.org/10.3897/mycokeys.56.37636

Tasanathai, K., Thanakitpipattana, D., Himaman, W., Phommavong, K., Dengkhhamounh, N. and Luangsa-ard, J. (2020). Three new *Ophiocordyceps* species in the *Ophiocordyceps pseudoacicularis* species complex on Lepidoptera larvae in Southeast Asia. Mycol. Prog., 19: 1043–1056. https://doi.org/10.1007/s11557-020-01611-6

Taylor, J.W. (2011). One fungus = One name: DNA and fungal nomenclature twenty years after PCR. IMA Fungus, 2: 113–120. https://doi.org/10.5598/imafungus.2011.02.02.01

Thanakitpipattana, D., Mongkolsamrit, S., Khonsanit, A., Himaman, W., Luangsa-ard, J.J. and Pornputtapong, N. (2022). Is Hyperdermium congeneric with Ascopolyporus? Phylogenetic relationships of *Ascopolyporus* spp. (Cordycipitaceae, Hypocreales) and a new genus *Neohyperdermium* on scale insects in Thailand. J. Fungi, 8(5): 516. https://doi.org/10.3390/jof8050516

Thanakitpipattana, D., Tasanathai, K., Mongkolsamrit, S., Khonsanit, A., Lamlertthon, S. and Luangsa-ard, J.J. (2020). Fungal pathogens occurring on Orthopterida in Thailand. Persoonia, 44: 140–160. https://doi.org/10.3767/persoonia.2020.44.06

Tulloch, M. (1976). The genus *Metarhizium*. Trans. Br. Mycol. Soc., 66: 407–411. https://doi.org/10.1016/S0007-1536(76)80209-4

Wang, N., Zhang, Y., Jiang, X., Shu, C., Hamid, M.I. et al. (2016). Population genetics of *Hirsutella rhossiliensis*, a dominant parasite of cyst nematode juveniles on a continental scale. Appl. Environ. Microbiol., 82: 6317–6325. https://doi.org/10.1128/AEM.01708-16

Wang, Y., Fan, Q., Wang, D., Zou, W.Q., Tang, D.X. et al. (2022a). Species diversity and virulence potential of the *Beauveria bassiana* complex and *Beauveria scarabaeidicola* complex. Front. Microbiol., 13: 841604. https://doi.org/10.3389/fmicb.2022.841604

Wang, Y., Wu, H.J., Tran, N.L., Zhang, G.D., Souvanhnachit, S. et al. (2021a). *Ophiocordyceps furcatosubulata*, a new entomopathogenic fungus parasitizing beetle larvae (Coleoptera: Elateridae). Phytotaxa, 482: 268–278. https://doi.org/10.11646/phytotaxa.482.3.5

Wang, Y.B., Wang, Y., Fan, Q., Duan, D.E., Zhang, G.D. et al. (2020a). Multigene phylogeny of the family Cordycipitaceae (Hypocreales): new taxa and the new systematic position of the Chinese cordycipitoid fungus *Paecilomyces hepiali*. Fungal Divers., 103: 1–46. https://doi.org/10.1007/s13225-020-00457-3

Wang, Y.B., Yu, H., Dai, Y.D., Chen, Z.H., Zeng, W.B. et al. (2015) *Polycephalomyces yunnanensis* (Hypocreales), a new species of *Polycephalomyces* parasitizing *Ophiocordyceps nutans* and stink bugs (hemipteran adults). Phytotaxa, 208: 34–44. https://doi.org/10.11646/phytotaxa.208.1.3

Wang, Y.H., Ban, S., Wang, W.J., Li, Y., Wang, K. et al. (2021b). *Pleurocordyceps* gen. nov. for a clade of fungi previously included in *Polycephalomyces* based on molecular phylogeny and morphology. J. Syst. Evol., 59: 1065–1080. https://doi.org/10.1111/jse.12705

Wang, Z., Li, M., Ju, W., Ye, W., Xue, L. et al. (2020b). The entomophagous caterpillar fungus *Ophiocordyceps sinensis* is consumed by its lepidopteran host as a plant endophyte. Fungal Ecol., 47: 100989. https://doi.org/10.1016/j.funeco.2020.100989

Wang, Z., Wang, Y., Dong, Q., Fan, Q., Dao, V.M. and Yu, H. (2022b). Morphological and phylogenetic characterization reveals five new species of *Samsoniella* (Cordycipitaceae, Hypocreales). J. Fungi, 8(7): 747. https://doi.org/10.3390/jof8070747

Watanabe, N., Hattori, M., Yokoyama, E., Isomura, S., Ujita, M. and Hara, A. (2006). Entomogenous fungi that produce 2, 6-pyridine dicarboxylic acid (dipicolinic acid). J. Biosci. Bioeng., 102: 365–368. https://doi.org/10.1263/jbb.102.365

Wei, D.P., Wanasinghe, D.N., Hyde, K.D., Mortimer, P.E., Xu, J.C. et al. (2020). *Ophiocordyceps tianshanensis* sp. nov. on ants from Tianshan mountains, PR China. Phytotaxa, 464: 277–292. https://doi.org/10.11646/phytotaxa.464.4.2

Wijayawardene, N.N., Dissanayake, L.S., Li, Q.R., Dai, D.Q., Xiao, Y. et al. (2021). Yunnan–Guizhou Plateau: a mycological hotspot. Phytotaxa, 523: 1–31. https://doi.org/10.11646/phytotaxa.523.1.1

Xiao, Y.P., Hongsanan, S., Hyde, K.D., Brooks, S., Xie, N. et al. (2019). Two new entomopathogenic species of *Ophiocordyceps* in Thailand. MycoKeys, 47: 53–74. https://doi.org/10.3897/mycokeys.47.29898

Xiao, Y.P., Wang, Y.B., Hyde, K.D., Eleni, G., Sun, J.Z. et al. (2023). Polycephalomycetaceae, a new family of clavicipitoid fungi segregates from Ophiocordycipitaceae. Fungal Divers., 120: 1–76. https://doi.org/10.1007/s13225-023-00517-4

Yamamoto, K., Ohmae, M. and Orihara, T. (2020). *Metarhizium brachyspermum* sp. nov. (Clavicipitaceae), a new species parasitic on Elateridae from Japan. Mycoscience, 61: 37–42. https://doi.org/10.1016/j.myc.2019.09.001

Yamamoto, K., Sugawa, G., Takeda, K. and Degawa Y. (2022). *Tolypocladium bacillisporum* (Ophiocordycipitaceae): A new parasite of Elaphomyces from Japan. Truffology, 5: 15–21.

Yang, Y., Xiao, Y., Yu, G., Wen, T., Deng, C. et al. (2021). *Ophiocordyceps aphrophoridarum* sp. nov., a new entomopathogenic species from Guizhou, China. Biodivers. Data J., 9: e66115. https://doi.org/10.3897/BDJ.9.e66115

Yu, F.M., Chethana, K.W.T., Wei, D.P., Liu, J.W., Zhao, Q. et al. (2021). Comprehensive review of *Tolypocladium* and description of a novel lineage from Southwest China. Pathogens, 10(11): 1389. https://doi.org/10.3390/pathogens10111389

Zhang, M.L. (2009). DELTA system, a recommendable information processing tool for taxonomic description. J. Plant Resour. Environ., 18: 87–90.

Zhang, S.L., He, L.M., Chen, X. and Huang, B. (2013). *Beauveria lii* sp. nov. isolated from *Henosepilachna vigintioctopunctata*. Mycotaxon, 121: 199–206.

Zhang, Z.F., Zhou, S.Y., Eurwilaichitr, L., Ingsriswang, S., Raza, M. et al. (2021). Culturable mycobiota from Karst caves in China II, with descriptions of 33 new species. Fungal Divers., 106: 29–136. https://doi.org/10.1007/s13225-020-00453-7

Zhang, Z.Y., Feng, Y., Tong, S.Q., Ding, C.Y., Tao, G. and Han, Y.F. (2023). Morphological and phylogenetic characterisation of two new soil-borne fungal taxa belonging to Clavicipitaceae (Hypocreales, Ascomycota). MycoKeys, 98: 113–132. https://doi.org/10.3897/mycokeys.98.106240

Zou, W., Tang, D., Xu, Z., Huang, O., Wang, Y. et al. (2022). Multigene phylogeny and morphology reveal *Ophiocordyceps hydrangea* sp. nov. and *Ophiocordyceps bidoupensis* sp. nov.(Ophiocordycipitaceae). MycoKeys, 92: 109. https://doi.org/10.3897/mycokeys.92.86160

3

An Overview of the Phylogeny of *Cordyceps*

Shivannegowda Mahadevakumar[1],* and *Kandikere R. Sridhar*[2]

1. Introduction

Within the Ascomycota class of parasitic fungi, the Cordycipitaceae belong to the Hypocreales order and class of Sordariomycetes. The stromata of Cordycipitaceae members can be fleshy, highly pigmented, or pale. Their perithecia are placed perpendicular to the exterior of the stroma. The cylindrical asci have thickened asci tips. Ascospores are often cylindrical, have several septa, and may disarticulate into part-spores or stay whole when they mature. The majority of species with fleshy, pale, or vividly pigmented stromata belong to the Cordycipitaceae, which is also related to the Hypocreaceae (Sung et al., 2007; Maharachchikumbura et al., 2015). Instead, several species are distinguished by having smaller stipes. With its diverse physical traits and large range of hosts, it is the most complicated group in the Hypocreales (Wang et al., 2020a).

Cordyceps was first recognized as a genus in Pyrenomycetes by Fries in 1818, and it was later classified as a tribe within Sphaeria (Shrestha et al., 2014). So far, various sources have identified up to 20 distinct taxa of *Cordyceps* as synonyms (Shrestha et al., 2014). A historical account

[1] Botanical Survey of India, Andaman and Nicobar Regional Center, Haddo 744102, Port Blair, India.

[2] Department of Biosciences, Mangalore University, Mangalagangotri, Mangalore, Karnataka, India.

* Corresponding author: mahadevakumars@gmail.com

of *Cordyceps* taxonomy was comprehensively reviewed by Shrestha et al. (2014), who concluded that this genus is the oldest legitimate genus in the family Cordycipitaceae and is distinguished by teleomorph. The type species of *Cordyceps*, *C. militaris*, was previously characterized in the literature in the 17th and early 18th centuries (Shrestha et al., 2014). The insect associated with *C. militaris* demonstrates an intriguing life cycle. This fungus has evolved the ability to penetrate and take over the body of an insect host in order to complete the sexual phase of its life cycle. The fungus sticks to its host and germinates, releasing enzymes that allow it to pierce the exoskeleton of the insect. Once inside the host, the fungus undergoes an initial stage of growth that resembles that of yeast, circulating poisons that ultimately cause the host to die. After feeding on the poor insect, fungus hyphae spread throughout the body and devovour every internal organ. Eventually, the insect's inner tissue is replaced by a mass of fungal mycelium, leaving the host reduced to an exoskeleton that holds the fungal stroma. The mushroom is frequently discovered exploding from the victim's head (Jenkinson, 2010).

Some of the species in the genus *Cordyceps* are significant for the economy, medicine, and ecology, making it a valuable source of fungal resources. Because of its numerous health advantages, *C. chanhua* has long been used in traditional Chinese medicine as a tonic and nutraceutical (Li et al., 2021). Since, *C. tenuipes* has significant analgesic, antibacterial, anticancer, antimalarial, immunoregulatory, and other pharmacological activities, it is widely employed as a raw ingredient in functional foods in Japan as well as South Korea (Wang et al., 2020b). *Cordyceps fumosorosea* is an important and ecologically safe substitute for chemical pesticides (Mascarin et al., 2013; Rojas et al., 2023). *Cordyceps militaris* is one of the most well-known edibles and therapeutic entomopathogens, and it has been used since time immemorial in nations in East Asia (China, Japan, and Korea) and is currently popular in Western nations (Cui et al., 2015; Lou et al., 2019). Recent investigations have demonstrated that *C. militaris* consists of numerous physiologically active compounds that are advantageous to human health (e.g., carotenoids, cordycepic acid, cordycepin, pentastatin, and polysaccharides) (Xia et al., 2017; Nurmamat et al., 2018; Kunhorm et al., 2019). According to Lou et al. (2019), currently there is a sizable *C. militaris* business in China that generates a projected 10 billion Chinese Yuan in economic value annually.

2. Fungal Phylogeny (1990–2005)

The swift development in the application of molecular tools, diagnostic methods, and advanced sequencing platforms has changed the face of fungal systematics. As a result of the development of the Sanger sequencing method and the polymerase chain reaction (PCR), molecular phylogeny has been the most effective and widely used method since the early 1990s. As a result, the characteristics used to categorize fungi have changed over time, starting with morphology and development and moving on to biochemistry and physiology, and more recently, DNA sequences and proteins. It has been demonstrated that molecular phylogeny is essentially better than traditional techniques in fungal systematics owing to the fact that phenotypic features are frequently subject to convergent evolution, reduction, or disappearance.

Phylogenetic research has significantly revised fungal systematics and improved our understanding of fungal evolution on a regular basis. According to morphological, physiological, and biochemical characteristics, the first phylogeny reflects the conventional systematics of fungi (Alexopoulos, 1962). According to this system, fungi were classified into two subdivisions and ten classes under the kingdom Plantae, or Mycota. In 2000, fungi were classified as a monophyletic group and grouped into nine subgroups depending on the ITS-rDNA (SSU) phylogeny (Tehler et al., 2000). Based on multilocus DNA sequences, the phylogeny of the fungal kingdom was rebuilt in 2007. Single-gene sequences were the only ones utilized to construct the phylogeny at the beginning of molecular phylogenetic investigations of fungi (Liu et al., 1999; Tehler et al., 2000). Multilocus phylogenies, which typically involve fewer than six loci, were first reported in the literature in the following years (Blackwell et al., 2006; Spatafora et al., 2006; Zhang et al., 2006). Compared to single-gene trees, the latter possessed a comparatively better resolution power. The LSU and SSU of rDNA, including a number of genes coding proteins such as RPB1, RPB2, TEF1, mtSSU, and ATP6, were selected in Assembling the Fungal Tree of Life (AFTOL) project to investigate the fungal phylogeny (http:// aftol.org/about.php), which served as the foundation for the kingdom fungi's classification (James et al., 2006; Hibbett et al., 2007). Many recently published fungal monographs, including those on *Cladosporium, Phoma, Alternaria, Massarineae,* and families of Sordariomycetes, are among those that have also relied on multilocus phylogenies (Maharachchikumbura et al., 2016).

The evolutionary connections among fungal taxa at all taxonomic levels have been investigated through a multitude of molecular investigations based on one or a few genes (Schoch et al., 2009). As the ribosomal RNA genes (rDNA) are multiple copies that are isolated and cloned easily, and because different sections of rDNA evolve at different rates, making them valuable for taxonomic resolution of lineages at diverse taxonomic levels, rDNA is the most studied and frequently used locus in fungal phylogeny (Bruns et al., 1991; Hibbett, 1992). Since the beginning of molecular phylogenetics, extensive research on fungi has been performed on the internal transcribed spacers (ITS), intergenic spacers, large subunits (LSU), and small subunits (SSU) of rDNA (White et al., 1990). The SSU and LSU rDNA genes were commonly used to delineate higher taxonomic levels, whereas ITS was employed to resolve specific species or genera. The ITS of rDNA was designated as a universal DNA barcode for fungi due to its high-level discrimination rates of species and success rate of PCR (Schoch et al., 2012). In response to this issue, a group of mycologists and curators of the NCBI's fungal RefSeq have begun curating a set of common fungal ITS sequences fetched from the type materials (Schoch et al., 2014).

Other than rDNA, single-copy as well as housekeeping protein genes make up the vast bulk of the cellular genome. They are vital to biological processes, and serve as useful markers for tracking the evolution of organisms. Actin (ACT), calmodulin (CAL), DNA replication licensing factor (MCM7), glycerol-3-phosphate dehydrogenase (GAPDH), histone (H3, H4), RNA polymerase II largest subunit (RPB1), RNA polymerase II 2nd largest subunit (RPB2), and translation elongation factor 1 (TEF1) are among the genes that are significant in fungal systematics (Templeton et al., 1992; Glass and Donaldson, 1995; Liu et al., 1999; Matheny et al., 2002; Guerber et al., 2003; Helgason et al., 2003; Hong et al., 2005; Rehner and Buckley, 2005; Schmitt et al., 2009). Additionally, mitochondrial genes like ATPase subunit 6 (ATP6), cytochrome c oxidase I (COI), and mitochondrial LSU and SSU of rDNA have been employed in fungal phylogenetics (Kretzer and Bruns, 1999; Zoller et al., 1999; Seif et al., 2005; Seifert et al., 2007).

3. Classical Genomics of *Cordyceps*

Fungal taxonomy is mainly based on morphology and other phenotypic traits; nevertheless, several features are unstable and offer only limited systematic details. More illustrative features are required for fungus categorization that is more trustworthy. Taxonomy of fungi and fungus-like taxa goes back to the 17th century (i.e., after the invention of the compound microscope). In classical taxonomy, fungi were identified based on morphological grounds. There are plentiful monographs, identification manuals, and other literature to categorize fungi based on morphology. Further, the development of electron microscopy facilitated understanding the fungal ultrastructure, which has been followed in fungal systematics (Guarro et al., 1999). Data generated through SEM has shed light on the evolution of morphological characters and contributed to the investigation of fungal systematics (Bracker, 1967; Beckett et al., 1974; Kimbrough, 1994; Lutzoni et al., 2004).

The next phase of development is the advancement and utilization of biochemical and physiological traits. Colour reactions to various stains to distinguish lichens and ascomycetes were established in the early 19th century (Nylander, 1866), now regarded as the spot test, a mandatory protocol to be followed by lichenologists (Elix, 1992). This is mainly because morphologically identifiable species typically have the same or comparable chemical components, causing continuous colour reactions in lichens. The fungal systematics are equipped by the extensive use of physiological and biochemical techniques brought about by technological advancements.

4. Molecular Analysis of *Cordyceps*

With more than 600 species, *Cordyceps* was formerly categorized as belonging to the Clavicipitaceae family based on features such as filiform ascospores, thickened ascus apices, and cylindrical asci. Producing well-developed, frequently stipitate stromata, *Cordyceps* was identified by its ecology as an arthropod and Elaphomyces pathogen, with infrageneric arrangements highlighting the structure of the perithecia, morphological variations in ascospores, and its associated host details. In order to advance the arrangement of *Cordyceps* and the family Clavicipitaceae, Sung et al. (2007) conducted an analysis using phylogenetic relationships involving 162 taxa that were calculated using analyses involving five to seven loci (nrSSU, nrLSU, tef1, rpb1, rpb2, tub2 and atp6). They rejected the monophyly of *Cordyceps* as well as Clavicipitaceae by providing compelling evidence for the presence of three clavicipitaceous clades in their first comprehensive morphological and multilocus phylogenetic analysis. The morphological characteristics, such as ascospore fragmentation and perithecia arrangement, that are taken into account for the classification of *Cordyceps* are not very useful and frequently cause either confusion or overlap of morphological features across many of the recognized taxa. The form, texture, and pigmentation of stromata are among the morphological characteristics that align with the molecular phylogeny. The work of Sung et al. (2007) is significant in this regard. Based on the nature of *C. militaris*, the family Cordycipitaceae has been approved and comprises the majority of *Cordyceps* species with fleshy, vividly coloured stromata.

The Ophiocordycipitaceae family was created to accommodate the species that are known to produce stromata that are tough to pliable, darkly pigmented, and frequently have perithecial apices. The proposal for the family was based on *Ophiocordyceps* Petch. Additionally, two new genera were created. For closely related *Cordyceps* to the grass symbionts in the Clavicipitaceae, a second new genus, the *Metacordyceps*, was proposed, while a subclade of the Ophiocordycipitaceae, comprising all *Cordyceps* species parasitizing the fungus *Elaphomyces* as well as a few closely related spp. that parasitize arthropods, is suggested to be home to the new genus *Elaphocordyceps*. They include the teleomorph connected to *Metarhizium* and other related anamorphs.

In order to incorporate individuals of the unaffiliated clade "Polycephalomyces" with the clavicipitaceous fungal clades Clavicipitaceae, Ophiocordycipitaceae, and Cordycipitaceae, Wang et al. (2021) created a new genus *Pleurocordyceps* within the *Cordyceps* members. Wang and colleagues gave a multigene (nrSSU, nrLSU, tef1, rpb1, and rpb2) phylogenetic support and created the new genus *Pleurocordyceps* due to the significant increase in the taxa documented in this clade. Furthermore, the research by Wang et al. (2021) revealed that the *Polycephalomyces formosus*-like clade described from the Japan group is sister to the *Pleurocordyceps* clade, supporting a major *Pleurocordyceps* clade.

According to Mongkolsamrit et al. (2020), phylogenetic investigations considering ITS, nrLSU, rpb1, rpb2, and tef-1α and showed a close link between *C. neopruinosa*, *C. pruinose*, and *C. ninchukispora*. Close connections existed amongst *C. cocoonihabita*, *C. neopruinosa*, *C. ninchukispora*, and *C. pruinosa*. They shared several morphological traits, such as superficial perithecia and orange-to-red-coloured stromata. These features were also shared by *C. bullispora*. The phylogenetic study revealed that *C. bullispora* was distinct from *C. cocoonihabita*, *C. neopruinosa* and closely related to a taxon that had not yet been identified, *C.* cf. *pruinosa*. Within this subclade are *C. pruinosa* and *C. ninchukispora*. Contrary to what was observed in *C. pruinosa* and *C. ninchukispora*, the *C. neopruinosa* perithecia were longer and wider. The stroma of *C. cocoonihabita* was longer. Conidia exhibited an *Isaria*-like micromorphological arrangement that distinguished it from *C. ninchukispora* and *C. pruinosa*, which had anamorphs of *Mariannaea* and *Acremonium*, respectively (Liang, 1983, 1991; Su and Wang, 1986; Mongkolsamrit et al., 2020).

In Vietnam, Wang (2023) conducted a comprehensive morphological as well as phylogenetic study of the *C. militaris* complex. Macroscopic as well as microscopic studies revealed some physical characteristics overlapped with some distinct differences when the specimens were compared to other distinguished species in *C. militaris*, illustrating the cryptic behaviour of species in a species complex. All of the *Cordyceps* species that Wang et al. (2023) described were distinct from one another and from other closely resembling species. For instance, *C. militaris* produces a profusion of stromata (orange to reddish-orange) emerging from the entire body of the lepidopteran larvae, cylindrical to clavate fertile parts with a spinous surface, short part-spores, superficial ovoid perithecia, and asexual conidiogenous features that resemble *Paecilomyces*. Contrarily, *C. sapaensis* possessed longer superficial perithecia, banana-shaped fertile sections, and a single yellowish to orange stroma. Molecular phylogenetic investigations (data set of nrSSU, nrLSU, TEF, RPB1, and RPB2) supported the presence of the two distinct species and the known species in the complex of *C. militaris*, highlighting the importance of molecular and morphological identification (Wang et al., 2023).

5. Multigene Analysis of *Cordyceps*

Since large-scale cultivation has not been accomplished and natural fungal populations are dwindling, harvesting wild specimens has become more difficult. Globally, there has been a rise in demand recently for cultivable fungus species with potential medical uses. A wide variety of entomogenous fungi can be found in the Yunnan Province as well as the Tibetan Plateau (Liang, 1981; Chen et al., 2019). *Cordyceps chaetoclavata, C. cocoonihabita, C. shuifuensis,* and *C. subtenuipes.*

The majority of diagnostic traits applied in the current classifications of cordycepitoid fungi, such as perithecia arrangement, conidiogenous structures, conidial shape, conidial size, and ascospore fragmentation, are not phylogenetically useful, according to entomopathogenic fungal phylogenetic classifications (Sung et al., 2007; Kepler et al., 2012, 2017; Mongkolsamrit et al., 2018). But the morphology, texture, and coloration of the stromata and synnemata are the traits that most closely match the phylogeny. Nevertheless, these macro- and micromorphological traits may help distinguish between fungi that resemble *Cordyceps, Isaria,* and *Lecanicillium.* Fleshy stromata with red to orange hues, superficial perithecia, cylindrical asci with apex-thickened ascus, and multiseptate cylindrical ascospores are characteristics of *Cordyceps'* sexual morphs. These bear a striking resemblance to *Samsoniella,* whose primary hosts are lepidopterans. Research on cordycipitoid fungi demonstrates that *Samsoniella* and *Cordyceps* generate comparable anamorphic conidiogenous structures (Samson, 1974; Mongkolsamrit et al., 2018). Similar asexual morphs generate branched, white to orange synnemata, white to cream conidial dry mass on the synnemata, phialides in the shape of a flask that are formed in whorls, and conidia with diverging chains that are shared by *Samsoniella* as well as *Cordyceps.* The isolates designated as *S. alboaurantium* serve as examples of the polyphyletic species complex that the morphology of *C. farinosa* represents, thus; it is not an authentic diagnosis (Kepler et al., 2017; Mongkolsamrit et al., 2018). Recent inventions and descriptions of 17 species of *Cordyceps* and allied species are well supported by multigene sequence data as well as concatenated phylogenetic analysis. The list of recently described new species of *Cordyceps,* along with gene and locus sequence data, is provided in Table 1 followed by phylogenetic tree in Fig. 1.

6. DNA Barcoding of *Cordyceps*

It is argued that in some rare situations where the holotype material is very limited and no other fruiting structures or live collections are available, DNA barcoding will help in resolving the identity crisis because the requirement of material for DNA barcoding is very little. On the contrary, if

Table 1. List of new species of *Cordyceps* described recently with multigene sequence data.

	Multilocus sequence data					Reference
	ITS	LSU	TEF	RPB1	RPB2	
Cordyceps bullispora Hong Yu bis, Q.Y. Dong & Zhi Yuan Zhao	+	+	+	+	+	Dong et al., 2022
Cordyceps chaetoclavata H. Yu, Y.B. Wang, Y. Wang, Q. Fan & Zhu L. Yang	+	+	+	+	+	Wang et al., 2020a
Cordyceps cocoonihabita H. Yu, Y.B. Wang, Y. Wang, Q. Fan & Zhu L. Yang	+	+	+	+	+	Wang et al., 2020a
Cordyceps kuiburiensis Himaman, Mongkols., Noisrip. & Luangsa-ard	+	+	+	+	-	Crous et al., 2019
Cordyceps longiphialis Hong Yu bis, Q.Y. Dong & D.X. Tang	+	+	+	+	+	Dong et al., 2022
Cordyceps nabanheensis Hong Yu bis, Q.Y. Dong	+	+	+	+	+	Dong et al., 2022
Cordyceps polystromata Hong Yu bis, Y. Wang & Q.Y. Dong	+	+	+	+	+	Wang et al., 2023
Cordyceps pseudotenuipes Hong Yu bis, Q.Y. Dong & Yao Wang	+	+	+	+	+	Dong et al., 2022
Cordyceps sandindaengensis Mongkols., Noisrip. & Luangsa-ard	+	+	+	+	+	Crous et al., 2023
Cordyceps sapaensis Hong Yu bis, Y. Wang & Q.Y. Dong	+	+	+	+	+	Wang et al., 2023
Cordyceps shuifuensis H. Yu, Y.B. Wang, Y. Wang & Zhu L. Yang	+	+	+	+	+	Wang et al., 2020a
Cordyceps simaoensis Hong Yu bis, Q.Y. Dong & Z.Q. Wang	+	+	+	+	+	Dong et al., 2022
Cordyceps subtenuipes H. Yu, Y.B. Wang, D.E. Duan & Zhu L. Yang	+	+	+	+	+	Wang et al., 2020a
Metacordyceps neogunnii T.C. Wen & K.D. Hyde	+	+	+	+	-	Wen et al., 2017
Ophiocordyceps indica Girish Nadda & Aakriti Sharma	+	+	+	+	+	Sharma et al., 2023
Ophiocordyceps hydrangea H. Yu, W.Q. Zou & D.X. Tang	+	-	+	+	+	Zou et al., 2022
Ophiocordyceps bidoupensis H. Yu, W.Q. Zou & D.X. Tang	+	+	+	+	+	Zou et al., 2022

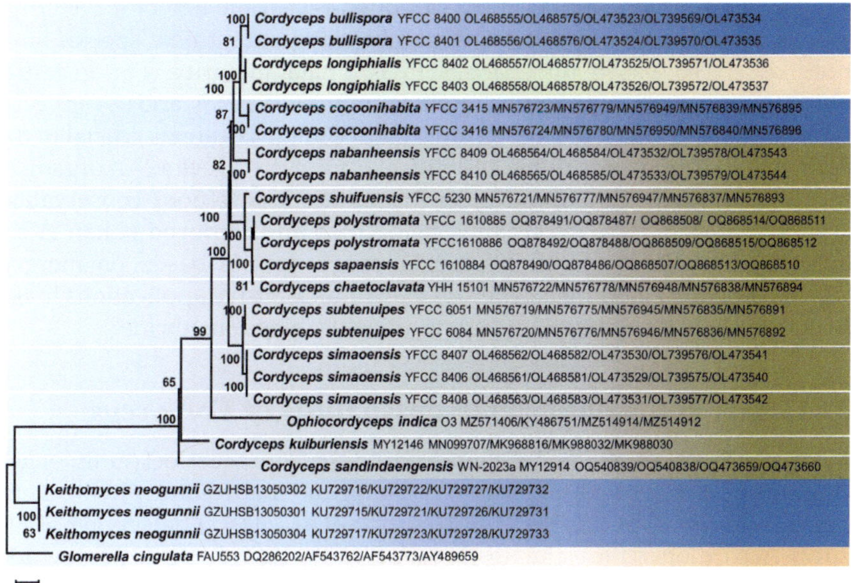

Fig. 1. Phylogenetic tree of recently described *Cordyceps* spp. (represented in Table 1) constructed using ITS-LSU-TEF-RPB1-RPB2 gene sequence data in MEGA11. The evolutionary history was inferred using the Neighbor-Joining Method with 1000 bootstrap replicates. The tree was computed using the Kimura 2-Parameter Model, which involved 25 nucleotide sequence data (concatenated) and there were a total of 2095 positions in the final dataset. The tree is rooted in *Glomerella cingulata* (FAU553).

any well-known fungus faces contrast in taxonomical terms, such examples can be resolved with the help of barcoding. Here, an example is illustrated where, by utilizing DNA barcoding, the identity crisis of *C. gunnii* was resolved. The collections of *C. gunnii* from Tasmania and China have not been thoroughly examined (Stensrud et al., 2005; Sung et al., 2007a, b). Liu et al. (2002) presented the only article to propose that 5.8S-ITS rDNA sequences indicate conspecificity between Chinese *C. hawkesii* and *C. gunnii*. Many aspects regarding these two species are still unknown due to a lack of sufficient DNA data and a lack of comparisons in morphology. Therefore, it was imperative that these significant fungi have their taxonomic status reviewed (Chan et al., 2011). This taxonomic ambiguity was cleared up when Wen et al. (2017) provided clear evidence that the Chinese collections that were mistakenly identified as *C. gunnii* are not the actual species. Quandt et al. (2015) and Spatafora et al. (2015) also confirmed that Tasmanian *Cordyceps gunnii* belongs to the family Ophiocordycipitaceae. After being misidentified for thirty years, the Chinese species "*Cordyceps gunnii*" is now known as *Metacordyceps neogunnii* (Liang, 1983). According to Li et al. (1999, 2010), *Cordyceps gunnii* var. *minor* from Anhui Province is also a member of *M. neogunnii*, based on phylogeny and morphological study.

The varying form of stromata can often reveal the age of a *Cordyceps* species, as seen in *M. neogunnii*. A number of apparent new species have been described based on changes in stromata maturity. For instance, *O. sinensis* is synonymized with *C. gansuensis*, *C. multiaxialis*, and *O. nepalensis* (Liu et al., 2001; Liang, 2007). Accurate source identification is crucial in the realms of traditional medicine and the pharmaceutical sector. An organism named incorrectly can result in legal scientific issues, poor conservation efforts, and even false information being filed and retrieved (Chan et al., 2011). The methods for classifying *Cordyceps* species based on merged sequence data from several gene loci may find extensive applications in the fields of fungal biotechnology and traditional Chinese medicine.

7. Trends in the Phylogenomic Analysis of *Cordyceps*

Many poorly supported nodes come from one or from a few loci (multilocus), producing incongruent phylogenies owing to inadequate phylogenetic details and gene-specific noise (Fitzpatrick et al., 2006; Ebersberger et al., 2012). According to the estimates, to rebuild a strong systematic framework, a minimum of 20 unlinked genes or 8,000 randomly chosen orthologous nucleotides on a genomic scale are necessary (Rokas et al., 2003). Large-scale gene sets from different autonomously evolving regions across the entire transcriptome or genome will be employed in phylogenomic analyses to increase informativeness, minimize stochastic error, and enhance phylogenetic correctness. This strategy is based on genomic data. 1996 saw the publication of the first fungal genome, that of *Saccharomyces cerevisiae* (Goffeau et al., 1996). *Neurospora crassa* was sequenced (Galagan et al., 2003) and *Schizosaccharomyces pombe* was sequenced in 2002 (Wood et al., 2002). The white-rot *Phanerochaete chrysosporium*'s first Basidiomycota genome was subsequently published in 2004 (Martinez et al., 2004). Fungal genome sequencing has made significant strides thanks to developments in the technology of next-generation sequencing (NGS). Several sequencing systems, including AB SOLiD, Illumina GA/HiSeq System, PacBio RS, and Roche 454, are examples of NGS, which is fast with a high-throughput (Goodwin et al., 2016).

A large-scale genomic study is now faster and more reasonably priced than NGS technology, which excels in immense parallel processing with a high throughput and low cost compared to Sanger sequencing, which requires separate processes for DNA synthesis and detection. This is due to the efforts of genome consortia like the 1000 Fungal Genomes Project at the Joint Genome Institute, Broad Institute Fungal Genome Initiatives, Genoscope sequencing projects, TIGR, and others. The number of fungal genomes available has increased exponentially, especially in recent years (Grigoriev et al., 2011). According to the Fungal Genome Project (http://

genome.jgi.doe.gov/programs/fungi/1000fungalgenomes.jsf), 23 entries of *Cordyceps* and related species genomes were sequenced and made available in the public domain. The details of the species available on the JGI platform are presented in Table 2. Additionally, in the NCBI genome database repository, there are several entries for organisms with whole genome sequence data (*Beauveria bassiana*, *B. brongniartii*, *Cordyceps cateniannulata*, *C. cicadae*, *C. farinose*, *C. fumosorosea*, *C. javanica*, *C. militaris*, *C. pruinosa*, *C. tenuipes* and *Ophiocordyceps sinensis*) (https://www.ncbi.nlm.nih.gov/genome).

De Bekker et al. (2017) have compared the genomic sequence data of five *Ophiocordyceps* species that infect ants from three species complexes. These species, which incite varying degrees of manipulation, were gathered from five distinct ant species spread over three continents. These ant-manipulating *Ophiocordyceps* species were the only ones to have a significant number of secreted and pathogenicity-related small proteins that were not present in other ascomycetous fungi. Although a few of those were shared by all of them, indicating that various approaches to behaviour modulation had developed throughout time. An additional piece of evidence for this is the comparatively quick emergence of putative manipulation genes linked to biting behaviour that have been previously documented. Zheng et al. (2011) reported the genome sequence of the type species *Cordyceps militaris*. According to phylogenomic studies, distinct species within the *Cordyceps/Metarhizium* genus have undergone independent development to become insect pathogens. Convergent evolution is believed to be responsible for the enormous secretomes and gene family expansions that these species share. There are no genes for recognized human mycotoxins in the DNA of *Cordyceps*. Xiao et al. (2012) have reported the whole genome sequence and assembly of *Beauveria bassiana*.

Phylogenomic studies revealed that *B. bassiana* ascomycete entomopathogenicity is polyphyletic but shows convergent evolution to insect pathogenicity. The genome also shows that, despite the infrequent observation of its sexual stage, *B. bassiana* has heterothallic characteristics with the closely related *Cordyceps militaris*. According to a high-throughput RNA-seq transcriptomic analysis, by turning on specific gene sets, *B. bassiana* could detect as well as adapt to various environmental niches. Gao et al. (2011) reported the whole genome sequencing of *Metarhizium acridum* and *M. anisopliae*. Based on whole-genome analysis, the genome architectures of these two species are quite syntenic, which implies that the genus *Metarhizium* originated from plant endophytes or pathogens. Compared to other fungi, *M. anisopliae* and *M. acridum* both have a remarkably higher percentage of genes expressing secreted proteins; also, around thirty percent of these genes lack functionally defined homologs, indicating hitherto unknown interactions between fungal diseases and insects. *Ophiocordyceps sinensis*, an endemic entomopathogen to the Qinghai-Tibet

Table 2. List of *Cordyceps* species with whole genome sequence data available from a public database (https://mycocosm.jgi.doe.gov/mycocosm/home/releases?flt=Cordyceps).

NCBI taxon	Isolate	Assembly length	Number of genes	Reference
Beauveria bassiana (Bals.-Criv.) Vuill.	ARSEF 2860	33,693,936	10,364	Xiao et al., 2012
Beauveria brongniartii (Sacc.) Petch	RCEF3172	32,515,992	9,595	Shang, et al., 2016
Cordyceps militaris (L.) Fr.	CM01	32,268,578	9,651	Zheng et al., 2011
Cordyceps sp.	RAO 2017	32,302,071	11,255	de Bekker et al., 2017
Isaria fumosorosea Wize	ARSEF 2679	33,485,962	10,061	Shang, et al., 2016
Lecanicillium lecanii (Zimm.) Zare & W. Gams	RCEF 1005	35,594,087	11,030	Shang et al., 2016
Metarhizium acridum (Driver & Milner) J.F. Bisch., S.A. Rehner & Humber	CQMa 102	39,422,329	9,849	Gao et al., 2011
Metarhizium album Petch	ARSEF 1941	30,361,122	8,472	Hu et al., 2014
Metarhizium anisopliae (Metschn.) Sorokin	ARSEF 549	38,504,274	10,891	Hu et al., 2014
Metarhizium brunneum Petch	ARSEF3297	37,016,100	10,689	Hu et al., 2014
Metarhizium guizhouense Q.T. Chen & H.L. Guo	ARSEF977	43,436,776	11,787	Hu et al., 2014
Metarhizium majus (J.R. Johnst.) J.F. Bisch., S.A. Rehner & Humber	ARSEF 297	41,974,566	11,535	Hu et al., 2014
Metarhizium rileyi (Farl.) Kepler, S.A. Rehner & Humber	RCEF4871	31,982,430	8,764	Shang et al., 2016
Metarhizium robertsii J.F. Bisch., S.A. Rehner & Humber	ARSEF 23	41,656,800	11,688	Hu et al., 2014
Ophiocordyceps australis (Speg.) G.H. Sung, J.M. Sung, Hywel-Jones & Spatafora	Map64	23,324,075	8,171	de Bekker et al., 2017
Ophiocordyceps camponoti-rufipedis H.C. Evans & D.P. Hughes	Map16	21,899,844	7,618	de Bekker et al., 2017

Species	Strain	Genome size (bp)	Gene count	Reference
Ophiocordyceps kimflemingiae Araújo, H.C. Evans & D.P. Hughes	SC16a	23,913,698	8,577	de Bekker et al., 2017
Ophiocordyceps sinensis (Berk.) G.H. Sung, J.M. Sung, Hywel-Jones & Spatafora	IOZ07	110,880,992	9,961	Shu et al., 2020
Pochonia chlamydosporia (Gaddard) Zare & W. Gams	170	44,215,803	14,204	Wang et al., 2016
Tolypocladium capitatum (Holmsk.) C.A. Quandt, Kepler & Spatafora	CBS 113982	22,985,095	7,646	Quandt et al., 2018
Tolypocladium inflatum W. Gams	NRRL 8044	30,348,711	9,998	Bushley et al., 2013
Tolypocladium ophioglossoides (J.F. Gmel.) C.A. Quandt, Kepler & Spatafora	CBS 100239	31,240,685	9,317	Quandt et al., 2015
Tolypocladium paradoxum (Kobayasi) C.A. Quandt, Kepler & Spatafora	NRBC 100945	27,610,157	8,983	Quandt et al., 2018

Plateau, is valued in traditional Chinese medicine as well as a main source of revenue for plentiful Himalayan tenants. Through single-molecule sequencing technology, Shu et al. (2020) reported the genome assembly of *O. sinensis*. The RNA-seq and homologous protein sequence identified up to 8,916 protein-coding genes in the IOZ07 assembly, and 63 gene clusters of secondary metabolites in the upgraded assembly. Their research will further enhance the evolutionary investigations and utilization of Chinese *Cordyceps* resources.

8. Current Status of *Cordyceps* Research in India

Research activities and discoveries of a significant number of *Cordyceps* and their allies are predominatly from China and Thailand. In India, there is hardly any research on the *Cordyceps* fungus. Earlier, it was reported from Himalayan regions only. However, recent developments have incited the assumption that *Cordyceps* and allies are also occurring in the Western Ghats. There are 641 records of *Cordyceps* fungi in MycoBank (https://www.mycobank.org/); of these, there is relatively lower species representation from India. So far from the Indian scenario, a total of eight *Cordyceps* species and seven *Ophiocordyceps* species representations have been recorded. They include: *Cordyceps canadensis* from Uttar Pradesh; *C. cicadae* from Maharashtra and Himachal Pradesh; *C. forquignoni* from Maharashtra; *C. gracilis* from Himachal Pradesh; *C. militaris* from Maharashtra, Karnataka, and Nagaland; *C. sinclairii* from Himachal Pradesh; *C. superficialis* from Maharashtra; *C. tuberculata*; *Ophiocordyceps blattae* from Maharashtra; *O. indica* from Himachal Pradesh; *O. mizoramensis* from Mizoram; *O. neovolkiana* from Kerala; *O. nutans* from Chhattisgarh, Karnataka, and Maharashtra; *O. sinensis* from Uttarakhand; *O. unilateralis* from Karnataka. Among them, only *O. indica* was recently described with complete multilocus barcoding as well as genomic information (Sharma et al., 2023).

Since many of the *Cordyceps* species and their allies described across the globe are supported by multilocus sequence data, it will support the identity of the *Cordyceps* and their allies in the Indian scenario. Therefore, it is the need of the hour to reinvestigate many of the previously recorded *Cordyceps* and their allies for molecular barcoding, since there might be a difference between the isolates. One of the main areas of their occurrence is the Indian subcontinent's Himalayan range. Nonetheless, it is clear from people and experts that *Cordyceps* are accessible and that harvesting occurs frequently in the Himalayan region. Reports regarding the presence of *Cordyceps* and allied species from India's southwest and Western Ghats have been reported recently. But because of its diverse flora and fauna, experts have also looked into its occurrence in the Western Ghats. Despite the fact that the species composition of *Cordyceps* varies between various places, there is a great deal of room to focus and discover the variety of undiscovered gems found in the Western Ghats and Himalayan regions. If we analyse the situation of Vietnam in Southeast Asia, located in a tropical area, it has an extraordinarily high biodiversity, where, more than 100 *Cordyceps* species have been recorded (Xu et al., 2022).

Lao et al. (2021) attributed the amazing diversification of flora and fauna observed in these forests to the high temperatures and ample rainfall of the tropical monsoon climate. These conditions have produced an environment that is favourable to the growth of fungi that are toxic to arthropods. Situations are not different in the biodiversity hotspots in India. Since India represents four biodiversity hotspots, each may represent a diversified and varied species composition of *Cordyceps* and allies. Therefore, it is encouraging to anticipate that the next decade will involve exploration and commercialization, paving the way for strengthening a better understanding of the *Cordyceps* and their allies.This will also contribute to taxonomic support for basic research, repositories, and establishment of commercial ventures in our country. There are several success stories associated with the commercial exploitation of *Cordyceps* and their allies. There is huge scope for strengthening the basic research in taxonomy, relationships with associated insects, and environmental factors influencing colonization and fruiting body development. If a consolidated list of repositories in culture collections is established for the *Cordyceps* of India, then there is a boon for commercial exploitation and their conservation as well. Considering the great economic values associated with the *Cordyceps* and allies, mycologists strive hard to cover larger areas, with a special focus on documenting the diversity of the *Cordyceps* and their allies. Since the ecology, geography, and a great wealth of flora and fauna are abundant in the Indian subcontinent, there is a high likelihood of enumerating a large number of hidden treasures of *Cordyceps* and allies. Meanwhile, researchers are exploring how the bioprospects will position India in the market for the producing and supplying of novel raw materials for the pharmaceutical and functional food industries..

Conclusion

Cordyceps and allied species are highly valuable ascomycetes with pharmaceutical, nutritional, and agricultural significance. *Cordyceps* taxonomy is challenging, as it has evolved from phenotypic morphology to insect-associated characters to multigene phylogeny to provide a comprehensive approach for clear delineation of lineages. The traditional taxonomy-based illustration of identification, supported by multigene sequence data and genomic resources in the recent past have paved the way for an appreciation of the diversity of *Cordyceps* and allied species and also offered mycologists ways to exploit the available resources. The phylogenetic analysis of *Cordyceps* and allied genera has undergone multiple changes. In view of the applications of *Cordyceps* and allies in the food, pharmaceutical, and agricultural industries, the authentication of the strains of *Cordyceps* is an essential task using classical genomics, molecular analysis, multigene approaches, DNA barcoding, and phylogenetic analysis. Advances have taken place in metagenomics and multi-omics for better utilization of traits of *Cordyceps* in the future. sPhylogenetic studies based on molecular data, such as DNA sequencing, have been used to understand the evolutionary relationships within the *Cordyceps* genus and its placement within the fungal kingdom. It's important to note that the taxonomy and phylogeny of fungi, including *Cordyceps*, can be complex and subject to ongoing research and revision. Advances in molecular techniques and genomic studies continue to contribute to our understanding of the evolutionary relationships within the fungal kingdom.

References

Alexopoulos, C.J. (1962). Introductory Mycology. John Wiley & Sons, New York, p. 482.

Beckett, A., Heath, I.B. and Mclaughlin, D.J. (1974). An Atlas of Fungal Ultrastructure. Longman Group, London, p. 221.

Blackwell, M., Hibbett, D.S., Taylor, J.W. and Spatafora, J.W. (2006). Research coordination networks: A phylogeny for kingdom fungi (Deep Hypha). Mycologia, 98: 829–837.

Bracker, C.E. (1967). Ultrastructure of fungi. Ann. Rev. Phytopathol., 5: 343–374.

Bruns, T.D., White, T.J. and Taylor, J. W. (1991). Fungal molecular systematics. Ann. Rev. Ecol. Evol. Syst., 22: 525–564.

Bushley, K.E., Raja, R., Jaiswal, P., Cumbie, J.S., Nonogaki, M. et al. (2013). The Genome of Tolypocladium inflatum: Evolution, Organization, and Expression of the Cyclosporin Biosynthetic Gene Cluster. PLoS Genet., 9(6): e1003496. 10.1371/journal.pgen.1003496

Chan, W.H., Ling, K.H., Chiu, S.W., Shaw, P.C. and But, P.P.H et al. (2011). Molecular analyses of *Cordyceps gunnii* in China. J. Food Drug Anal., 19(1): 18–25.

Chen, Z.H., Wang, Y.B., Dai, Y.D., Chen, K., Xu, L et al. (2019). Species diversity and seasonal fluctuation of entomogenous fungi of Ascomycota in Taibaoshan Forest Park in western Yunnan. Biodivers. Sci., 27: 993–1001. 10.17520/biods.2019135

Crous, P.W., Osieck, E.R., Shivas, R.G., Tan, Y.P., Bishop-Hurley, S.L. et al. (2023). Fungal Planet description sheets: 1478–1549. Persoonia - Mol. Phylog. Evol. Fungi, 50: 158–310. 10.3767/persoonia.2023.50.05

Crous, P.W., Wingfield, M.J., Lombard, L., Roets, F., Swart, W.J. et al. (2019). Fungal Planet description sheets: 951–1041. Persoonia - Mol. Phylog. Evol. Fungi, 43: 223–425. 10.3767/persoonia.2019.43.06

Cui, J.D. (2015). Biotechnological production and applications of *Cordyceps militaris*, a valued traditional Chinese medicine. Crit. Rev. Biotechnol., 35: 475–484.

de Bekker, C., Ohm, R.A., Evans, H.C., Brachmann, A. and Hughes, D.P et al. (2017). Ant-infecting *Ophiocordyceps* genomes reveal a high diversity of potential behavioral manipulation genes and a possible major role for enterotoxins. Sci. Reports, 7(1): 12508. 10.1038/s41598-017-12863-w.

Dong, Q.-Y., Wang, Y., Wang, Z.-Q., Tang, D.-X., Zhao, Z.-Y. et al. (2022). Morphology and phylogeny reveal five novel species in the genus *Cordyceps* (Cordycipitaceae, Hypocreales) from Yunnan, China. Front. Microbiol., 13: 846909. 10.3389/fmicb.2022.846909

Ebersberger, I., Simoes, R.M, Kupczok, A., Gube, M., Kothe, E. et al. (2012). A consistent phylogenetic backbone for the fungi. Mol. Biol. Evol. 29: 1319–1334.

Elix, J.A. (1992). Lichen chemistry. Flora of Australia, 54: 23–29.

Fitzpatrick, D.A., Logue, M.E., Stajich, J.E. and Butler, G. (2006). A fungal phylogeny based on 42 complete genomes derived from supertree and combined gene analysis. BMC Evol. Biol., 6, 99. 10.1186/1471-2148-6-99

Fries, E.M. (1818) Observationes mycologicae Praecipue ad Illustrandam Floram Suecicam. Pars secunda (Cancellans issue). G. Bonnieri, Copenhagen, Denmark. https://books.google.com/books/about/Observationes_mycologicae_praecipue_ad_i.html?id=GOIVAAAAYAAJ

Galagan, J.E., Calvo, S.E., Borkovich, K.A., Selker, E.U., Read, N.D. et al. (2003). The genome sequence of the filamentous fungus *Neurospora crassa*. Nature, 422: 859–868.

Gao, Q., Jin, K., Ying, S.H., Zhang, Y., Xiao, G. et al. (2011). Genome sequencing and comparative transcriptomics of the model entomopathogenic fungi *Metarhizium anisopliae* and *M. acridum*. PLoS Genet., 7(1): e1001264. 10.1371/journal.pgen.1001264.

Glass, N.L. and Donaldson, G.C. (1995). Development of primer sets designed for use with the PCR to amplify conserved genes from filamentous ascomycetes. Appl. Environ. Microbiol., 61: 1323–1330.

Goffeau, A., Barrell, B.G., Bussey, H., Davis, R.W., Dujon, B. et al. (1996). Life with 6000 genes. Science, 274: 546–567.

Goodwin, S., McPherson, J.D. and McCombie, W.R. (2016). Coming of age: Ten years of next-generation sequencing technologies. Nature Rev. Genet., 17: 333–351.

Grigoriev, I.V., Cullen, D., Goodwin, S.B., Hibbett, D., Jeffries, T.W. et al. (2011). Fueling the future with fungal genomics. Mycol., 2: 192–209.

Guarro, J., Gene, J. and Stchigel, A.M. (1999). Developments in fungal taxonomy. Clin. Microbiol. Rev., 12: 454–500.

Guerber, J.C., Liu, B., Correll, J.C. and Johnston, P.R. (2003). Characterization of diversity in *Colletotrichum acutatum* sensu lato by sequence analysis of two gene introns, mtDNA and intron RFLPs, and mating compatibility. Mycologia, 95: 872–895.

Helgason, T., Watson, I.J. and Young, J.P.W. (2003). Phylogeny of the glomerales and diversisporales (Fungi: Glomeromycota) from actin and elongation factor 1-alpha sequences. FEMS Microbiol. Lett., 229: 127–132.

Hibbett, D. (1992). Ribosomal RNA and fungal systematics. Trans. Mycol. Soc. Japan, 33: 533–556.

Hibbett, D.S., Binder, M., Bischoff, J.F., Blackwell, M., Cannon, P.F. et al. (2007). A higher-level phylogenetic classification of the Fungi. Mycol. Res., 111: 509–547.

Hong, S.B., Go, S.J., Shin, H.D., Frisvad, J.C. and Samson, R.A et al. (2005). Polyphasic taxonomy of *Aspergillus fumigatus* and related species. Mycologia, 97: 1316–1329.

Hu, X., Xiao, G., Zheng, P., Shang, Y. and Su, Y. (2014). Trajectory and genomic determinants of fungal-pathogen speciation and host adaptation. Proc. Natl. Acad. Sci. Unit. States Am., 111(47): 16796–16801. 10.1073/pnas.14126621

James, T.Y., Kauff, F., Schoch, C.L., Matheny, P.B., Hofstetter, V. et al. (2006). Reconstructing the early evolution of Fungi using a six-gene phylogeny. Nature, 443: 818–822.

Jenkinson, S.T. (2010). MycoDigest: The evolutionary story of *Cordyceps* and its allies. Mycena News, 61(01): 1 and 7.

Kepler, R.M., Luangsa-Ard, J.J., Hywel-Jones, N.L., Quandt, C.A., Sung, G.H. et al. (2017). A phylogenetically-based nomenclature for Cordycipitaceae (Hypocreales). IMA Fungus **8**: 335–353. 10.5598/imafungus.2017.08.02.08

Kepler, R.M., Sung, G.H., Ban, S., Nakagiri, A., Chen, M.J. et al. (2012). New teleomorph combinations in the entomopathogenic genus *Metacordyceps*. Mycologia, 104: 182–197. 10.3852/11-070

Kimbrough, J.W. (1994). Septal ultrastructure and ascomycete systematics. *In*: Hawksworth, D. (ed.). Ascomycete Systematics: Problems and Perspectives in the Nineties. pp. 127–141. Plenum Press, New York.

Kretzer, A.M. and Bruns, T.D. (1999). Use of atp6 in fungal phylogenetics: An example from the boletales. Mol. Phylogenet. Evol., 13: 483–492.

Kunhorm, P., Chaicharoenaudomrung, N. and Noisa, P. (2019). Enrichment of cordycepin for cosmeceutical applications: Culture systems and strategies. Appl. Microbiol. Biotechnol., 103: 1681–1691.

Lao, T.D., Le, T.A.H. and Truong, N.B. (2021). Morphological and genetic characteristics of the novel entomopathogenic fungus *Ophiocordyceps langbianensis* (Ophiocordycipitaceae, Hypocreales) from Lang Biang Biosphere Reserve, Vietnam. Sci. Rep., 11: 1412. 10.1038/s41598-020-78265-7

Li, C.R., Huang, B., Fan, M.Z. and Lin, Y.R. (2010). *Metacordyceps guniujiangensis* and its *Metarhizium* anamorph: A new pathogen on cicada nymphs. Mycotaxon, 111: 221–231. 10.5248/111.221

Li, Z.Z., Li, C.R., Huang, B., Fan M.Z. and Lee, M.W et al. (1999). New variety of *Cordyceps gunnii* (Berk.) Berk. and its *Paecilomyces* anamorph. J. Mycol. Infec., 27(3): 231–233.

Li, Z.-Z., Luan, F.-G., Hywel-Jones, N.L., Zhang, S.-L., Chen, M.-J. et al. (2021). Biodiversity of cordycipitoid fungi associated with *Isaria cicadae* Miquel II: Teleomorph discovery and nomenclature of chanhua, an important medicinal fungus in China. Mycosystema, 40: 95–107. 10.13346/j.mycosystema.200119

Liang, Z.Q. (1981). Entomogenous fungi from the pests of tea-plants. Acta Phytopathol. Sin., 11: 9–16.

Liang, Z.Q. (1983). A record and description on *Cordyceps pruinosa* Petch and its conidial state. J. Guizhou Agric. Coll., 2: 72–80.

Liang, Z.Q. (1991). Verification and identification of the anamorph of *Cordyceps pruinosa* Petch. Acta Mycol. Sin., 10: 104–107.

Liang, Z.Q. (2007). Flora Fungorum Sinicorum, Volume 32 - *Cordyceps*. Science Press, Beijing, p. 190.

Liu, Y.J., Whelen, S. and Hall, B.D. (1999). Phylogenetic relationships among ascomycetes: Evidence from an RNA polymerase II subunit. Mol. Biol. Evol., 16: 1799–1808.

Liu, Z.Y., Liang, Z.Q., Liu, A.Y., Yao, Y.J., Hyde, K.D et al. (2002). Molecular evidence for teleomorph-anamorph connections in *Cordyceps* based on ITS-5.8S rDNA sequences. Mycol. Res., 106(9): 1100–1108. 10.1017/S09537756202006378

Liu, Z.Y., Yao, Y.J., Liang, Z.Q., Liu, A.Y., Pegler, D.N et al. (2001). Molecular evidence for the anamorph-teleomorph connection in *Cordyceps sinensis*. Mycol. Res., 105(7): 827–832. 10.1017/S095375620100377X

Lou, H.W., Lin, J.F., Guo, L.Q., Wang, X.W., Tian, S.Q. et al. (2019). Advances in research on *Cordyceps militaris* degeneration. Appl. Microbiol. Biotechnol., 103: 7835–7841.

Lu, H.S. and Mclaughlin, D.J. (1991). Ultrastructure of the septal pore apparatus and early septum initiation in *Auricularia auricula-judae*. Mycologia, 83: 322–334.

Lutzoni, F., Kauff, F., Cox, C.J., McLaughlin, D., Celio, G. et al. (2004). Assembling the fungal tree of life: Progress, classification, and evolution of subcellular traits. Am. J. Bot., 91: 1446–1480.

Maharachchikumbura, S.S.N., Hyde, K.D., Jones, E.B.G., McKenzie, E.H.C., Huang, S.K. et al. (2015). Towards a natural classification and backbone tree for Sordariomycetes. Fungal Divers., 72: 199–301.

Maharachchikumbura, S.S.N., Hyde, K.D., Jones, E.B.G., McKenzie, E.H.C., Bhat, J.D. et al. (2016). Families of sordariomycetes. Fungal Divers., 79: 1–317.

Martinez, D., Larrondo, L.F., Putnam, N., Gelpke, M.D., Huang, K. et al. (2004). Genome sequence of the lignocellulose degrading fungus *Phanerochaete chrysosporium* strain RP78. Nat. Biotechnol., 22: 695–700.

Mascarin, G.M., Kobori, N.N., Quintela, E.D. and Delalibera, I.J. (2013). The virulence of entomopathogenic fungi against *Bemisia tabaci* biotype B (Hemiptera: Aleyrodidae) and their conidial production using solid substrate fermentation. Biol. Cont., 66: 209–218.

Matheny, P.B., Liu, Y.J., Ammirati, J.F. and Hall, B.D. (2002). Using RPB1 sequences to improve phylogenetic inference among mushrooms (*Inocybe*, Agaricales). Am. J. Bot., 89: 688–698.

Mongkolsamrit, S., Noisripoom, W., Tasanathai, K., Khonsanit, A., Thanakitpipattana, D. et al. (2020). Molecular phylogeny and morphology reveal cryptic species in *Blackwellomyces* and *Cordyceps* (Cordycipitaceae) from Thailand. Mycol. Prog., 19: 957–983. 10.1007/s11557-020-01615-2

Mongkolsamrit, S., Noisripoom, W., Thanakitpipattana, D., Wutikhun, T., Spatafora, J.W et al. (2018). Disentangling cryptic species with *Isaria*-like morphs in Cordycipitaceae. Mycologia, 110: 230–257. 10.1080/00275514.2018.1446651

Nurmamat, E., Xiao, H., Zhang, Y. and Jiao, Z. (2018). Effects of different temperatures on the chemical structure and antitumor activities of polysaccharides from *Cordyceps militaris*. Polymers (Basel), 10(4): 430. 10.3390/polym10040430

Nylander, W. (1866). Hypochlorite of lime and hydrate of potash, two new criteria in the study of Lichens. Bot. J. Linn. Soc., 9: 358–365.

Quandt, C.A., Bushley, K.E. and Spatafora, J.W. (2015). The genome of the truffle-parasite *Tolypocladium ophioglossoides* and the evolution of antifungal peptaibiotics. BMC Genomics, 16(1): 553. 10.1186/s12864-015-1777-9.

Quandt, C.A., Patterson, W., and Spatafora, J.W. (2018). Harnessing the power of phylogenomics to disentangle the directionality and signatures of interkingdom host jumping in the parasitic fungal genus *Tolypocladium*. Mycologia, 110(1): 104–117. 10.1080/00275514.2018.1442618

Rehner, S.A. and Buckley, E. (2005). A *Beauveria* phylogeny inferred from nuclear ITS and EF1-alpha sequences: Evidence for cryptic diversification and links to *Cordyceps* teleomorphs. Mycologia, 97: 84–98.

Rojas, V.M.A., Iwanicki, N.S.A., Alessandro, C.P., Fatoretto, M.B., Demétrio, C.G.B et al. (2023). Characterization of Brazilian *Cordyceps fumosorosea* isolates: Conidial production, tolerance to ultraviolet-B radiation, and elevated temperature. J. Invertebr. Pathol., 197: 1–9. 107888. 10.1016/j.jip.2023.107888

Rokas, A., Williams, B.L., King, N. and Carroll, S.B. (2003). Genome-scale approaches to resolving incongruence in molecular phylogenies. Nature, 425: 798–804.

Samson, R.A. (1974). *Paecilomyces* and some allied Hyphomycetes. Stud. Mycol., 6: 1–119. 10.1186/1471-2180-12-3

Schmitt, I., Crespo, A., Divakar, P.K., Fankhauser, J.D., Herman-Sackett, E. et al. (2009). New primers for promising single-copy genes in fungal phylogenetics and systematics. Persoonia, 23: 35–40.

Schoch, C.L., Robbertse, B., Robert, V., Vu, D. and Cardinali, G. (2014). Finding needles in haystacks: linking scientific names, reference specimens and molecular data for Fungi. Database, 2014: bau061. 10.1093/database/bau061

Schoch, C.L., Seifert, K.A., Huhndorf, S., Robert, V. and Spouge, J.L. (2012). Fungal Barcoding Consortium; Fungal Barcoding Consortium Author List. Nuclear ribosomal internal transcribed spacer (ITS) region as a universal DNA barcode marker for Fungi. Proc. Natl. Acad. Sci. USA. 109(16): 6241-6. 10.1073/pnas.1117018109

Schoch, C.L., Sung, G.H., Lopez-Giraldez, F., Townsend, J.P., Miadlikowska, J. et al. (2009). The Ascomycota tree of life: A phylum-wide phylogeny clarifies the origin and evolution of fundamental reproductive and ecological traits. Syst. Biol., 58: 224–239.

Seif, E., Leigh, J., Liu, Y., Roewer, I., Forget, L. et al. (2005). Comparative mitochondrial genomics in zygomycetes: Bacteria-like RNase P RNAs, mobile elements and a close source of the group I intron invasion in angiosperms. Nucleic Acids Res., 33: 734–744.

Seifert, K.A., Samson, R.A., Dewaard, J.R., Houbraken, J., Levesque, C.A. et al. (2007). Prospects for fungus identification using CO1 DNA barcodes, with *Penicillium* as a test case. Proc. Nat. Acad. Sci., 104: 3901–3906.

Shang, Y., Xiao, G., Zheng, P., Cen, K., Zhan, S. and Wang, C. (2016). Divergent and Convergent Evolution of Fungal Pathogenicity. Genome Biol Evol., 8(5): 1374–87. 10.1093/gbe/evw082

Sharma, A., Kaur, E., Joshi, R., Kumari, P., Khatri, A. et al. (2023). Systematic analyses with genomic and metabolomic insights reveal a new species, *Ophiocordyceps indica* sp. nov.

from treeline area of Indian Western Himalayan region. Front. Microbiol. 14: 1188649. 10.3389/fmicb.2023.1188649

Shrestha, B., Tanaka, E., Han, J.G., Oh, J.S., Han, S.K. et al. (2014). A brief chronicle of the genus *Cordyceps* Fr., the oldest valid genus in Cordycipitaceae (Hypocreales, Ascomycota). Mycobiology, 42: 93–99. 10.5941/MYCO.2014.42.2.93

Shu, R., Zhang, J., Meng, Q., Zhang, H., Zhou, G. et al. (2020). A New High-Quality Draft Genome Assembly of the Chinese *Cordyceps Ophiocordyceps* sinensis. Genome Biol. Evol., 12(7): 1074–1079. 10.1093/gbe/evaa112.

Spatafora, J.W., Quandt, C.A., Kepler, R.M., Sung, G.H., Shrestha, B. et al. (2015). New 1F1N species combinations in Ophiocordycipitaceae (Hypocreales). IMA Fungus 6: 357–362

Spatafora, J.W., Sung, G.H., Johnson, D., Hesse, C., O'Rourke, B. et al. (2006). A five-gene phylogeny of Pezizomycotina. Mycologia, 98: 1018–1028.

Stensrud, Ø., Hywel-Jones, N.L. and Schumacher, T. (2005). Towards a phylogenetic classification of *Cordyceps*: ITS nrDNA sequence data confirm divergent lineages and paraphyly. Mycol. Res., 109: 41–56.

Su, C.H. and Wang, H.H. (1986). *Phytocordyceps*, a new genus of the Clavicipitaceae. Mycotaxon, 26: 337–344.

Sung, G.H., Hywel-Jones, N.L., Sung, J.M., Luangsa-ard, J.J., Shrestha, B et al. (2007). Phylogenetic classification of *Cordyceps* and the clavicipitaceous fungi. Stud. Mycol., 57: 5–59. 10.3114/sim.2007.57.01

Sung, G.H., Sung, J.M., Hywel-Jones, N.L. and Spatafora, J.W. (2007). A multi-gene phylogeny of Clavicipitaceae (Ascomycota, Fungi): identification of localized incongruence using a combinational bootstrap approach. Mol. Phylogenet. Evol., 443): 1204–1223. 10.1016/j.ympev.2007.03.011

Tehler, A., Farris, J.S., Lipscomb, D.L. and Kallersjo, M. (2000). Phylogenetic analyses of fungi based on large rDNA data sets. Mycologia, 92: 459–474.

Templeton, M.D., Rikkerink, E.H.A., Solon, S.L. and Crowhurst, R.N. (1992). Cloning and molecular characterization of the glyceraldehyde-3-phosphate dehydrogenase encoding gene and cDNA from the plant pathogenic fungus *Glomerella cingulata*. Gene, 122: 225–230.

Wang, G., Liu, Z., Lin, R., Li, E., Mao, Z. et al. (2016). Biosynthesis of Antibiotic Leucinostatins in Bio-control Fungus Purpureocillium lilacinum and Their Inhibition on Phytophthora Revealed by Genome Mining. PLoS Pathog., 12(7):e1005685. 10.1371/journal.ppat.1005685

Wang, Y., Dong, Q.-Y., Luo, R., Fan, Q., Duan, D.-E. et al. (2023). Molecular Phylogeny and Morphology Reveal Cryptic Species in the *Cordyceps militaris* Complex from Vietnam. J. Fungi, 9: 676. 10.3390/jof9060676

Wang, Y., Liu, Y.F., Fan, Q., Wang, Y.B., Yu, H et al., 2020b. Cordycipitoid Fungi Powders Promote Mycelial Growth and Bioactive-Metabolite Production in Liquid Cultures of the Stout Camphor Medicinal Mushroom *Taiwanofungus camphoratus* (Agaricomycetes). Int. J. Med. Mushrooms, 22: 615–626. 10.1615/IntJMedMushrooms.2020035440

Wang, Y.B., Wang, Y., Fan, Q., Duan, D.E., Zhang, G.D. et al. (2020a). Multigene phylogeny of the family Cordycipitaceae (Hypocreales): new taxa and the new systematic position of the Chinese cordycipitoid fungus *Paecilomyces hepialid*. Fungal Divers., 103: 1–46. 10.1007/s13225-020-00457-3

Wang, Y.H., Ban, S., Wang, W.G., Li, Y., Wang, K. et al. (2021). *Pleurocordyceps* gen. nov. for a clade of fungi previously included in *Polycephalomyces* based on molecular phylogeny and morphology. J. Systemat. Evol., 59(5): 1065–1080. 10.1111/jse.12705

Wen, T.C., Xiao, Y.P., Han, Y.F., Huang, S.K., Zha, L.S. et al. (2017). Multigene phylogeny and morphology reveal that the Chinese medicinal mushroom 'Cordyceps gunnii' is *Metacordyceps neogunnii* sp. nov. Phytotaxa, 302(1): 27–39. 10.11646/phytotaxa.302.1.2

White, T.J., Bruns, T., Lee, S. and Taylor, J.W. (1990). Amplification and direct sequencing of fungal ribosomal RNA genes for phylogenetics. pp. 315–322. *In*: Innis, M.A., Gelfand, D.H., Sninsky, J.J. and White, T.J. (eds.). PCR Protocols: A Guide to Methods and Applications. Academic Press, New York.

Wood, V., Gwilliam, R., Rajandream, M.A., Lyne, M., Lyne, R. et al. (2002). The genome sequence of *Schizosaccharomyces pombe*. Nature, 415: 871–880.

Xia, Y., Luo, F., Shang, Y., Chen, P., Lu, Y. et al. (2017). Fungal cordycepin biosynthesis is coupled with the production of the safeguard molecule pentostatin. Cell Chem. Biol., 24: 1479–1489.

Xiao, G., Ying, S.H., Zheng, P., Wang, Z.L., Zhang, S. et al. (2012). Genomic perspectives on the evolution of fungal entomopathogenicity in *Beauveria bassiana*. Sci. Rep., 2: 483(2012). 10.1038/srep00483.

Xu, Z.H., Tran, N.L., Wang, Y., Zhang, G.D., Dao, V.M. et al. (2022). Phylogeny and morphology of *Ophiocordyceps puluongensis* sp. nov. (Ophiocordycipitaceae, Hypocreales), a new fungal pathogen on termites from Vietnam. J. Invertebr. Pathol., 192: 107771. 10.1016/j.jip.2022.107771

Zhang, N., Castlebury, L.A., Miller, A.N., Huhndorf, S.M., Schoch, C.L. et al. (2006). An overview of the systematics of the Sordariomycetes based on a four-gene phylogeny. Mycologia, 98: 1076–1087.

Zheng, P., Xia, Y., Xiao, G., Xiong, C., Hu, X. et al. (2011). Genome sequence of the insect pathogenic fungus Cordyceps militaris, a valued traditional Chinese medicine. Genome Biol., 12(11): R116. 10.1186/gb-2011-12-11-r116

Zoller, S., Scheidegger, C. and Sperisen, C. (1999). PCR primers for the amplification of mitochondrial small subunit ribosomal DNA of lichen-forming ascomycetes. The Lichenologist, 31: 511–516.

Zou, W.Q., Tang, D.X., Xu, Z.H., Huang, O., Wang, Y.B. et al. (2022). Multigene phylogeny and morphology reveal *Ophiocordyceps hydrangea* sp. nov. and *Ophiocordyceps bidoupensis* sp. nov. (Ophiocordycipitaceae). MycoKeys, 92: 109–130.

Diversity and Distribution

4

Diversity and Distribution of *Cordyceps* in the Indian Subcontinent

Kandikere R. Sridhar[1],* and *Shivannegowda Mahadevakumar*[2]

1. Introduction

Macrofungi are well known for their diversity, distribution, and ecological as well as economic significance (Hyde et al., 2019; Pérez-Moreno, 2021; Zhou et al., 2023). Although some of them have been identified as toxic, several others possess useful traits such as ectomycorrhizal (mutualistic) relationships, nutritional value, and versatility in metabolites with applications to treat human diseases.

Interestingly, all three traits are found in several mushrooms (e.g., some *Amanita* spp., *Astraeus* spp., and truffles) (Pavithra et al., 2015, 2016; Greeshma et al., 2018a, 2018b; Graziosi et al., 2022; Sridhar et al., 2024). Ethnic knowledge of tribals on edible mushrooms also underscores the importance of mushrooms in human nutrition and health (Karun and Sridhar, 2017). As an alternative source of nutrition and health, mushrooms have gained increased awareness worldwide for their utilization and cultivation.

Among the medicinally valued mushrooms, the entomopathogenic fungi belonging to *Cordyceps* and allies have high prominence (in oriental and modern medicines) owing to their wide nutraceutical and therapeutic applications. They have been in use for over 2000 years in oriental Chinese

[1] Department of Biosciences, Mangalore University, Mangalagangotri, Mangalore 574 199, Karnataka, India.
[2] Botanical Survey of India, Andaman and Nicobar Regional Center, Haddo 744 102, Port Blair, Andaman, India.
* Corresponding author: kandikeremanasa@gmail.com

medicine to cure multiple infectious diseases (Zhou et al., 2009). The majority of studies on the distribution and descriptions of *Cordyceps* are from the Asian continent (Boesi and Francesca, 2011). About 750 species of *Cordyceps* have been identified, largely from the regions of South Asia, Europe, and North America (Kała et al., 2024). Tingo Maria Valley in Peru houses over 1000 species of *Cordyceps*, and many of them need precise descriptions (Holliday, 2018). *Ophiocordyceps sinensis* is one of the authorized food ingredients in the European Union; *Cordyceps militaris* is used as a dietary supplement in the European Union; and in Asia, it serves as a valuable food source (Fenglin et al., 2020; Kała et al., 2024). According to Lo et al. (2013) and Holliday (2018), one kilogram of *Cordyceps* costs more than US$100,000. Owing to modern technological developments, several *Cordyceps* products in nutrition and medicine have been introduced into the markets. Several ethnic communities in the Himalayan range depend on marketing *Cordyceps* and allies for their livelihood; thus, Wang et al. (2022) proposed a model for transboundary conservation to encourage fair trade.

Recent developments prompted the conclusion that *Cordyceps* and allies are not only restricted to the Himalayas but also to other parts of the Indian subcontinent, especially the Western Ghats and the west coast of India (Dattaraj et al., 2018). Most of the studies on *Cordyceps* and allies in southwest India revealed the occurrence of a few species, while a large number of varied species on arthropod hosts were reported for the first time in the scrub jungles southwest of India (Dattaraj et al., 2018). Along with some recently found *Cordyceps* and related species, this brief review extends our knowledge on their occurrence, distribution, and diversity mainly in the Western Ghats and southwest coast of India.

2. A Brief Historical Account

The generic name *Cordyceps* is derived from the Latin *kordyle* (club) and *ceps* (head). The oldest report on *Cordyceps* goes back to 1439–1475 (Lu, 2017). The documented record (possibly *Ophicordyceps sinensis*) in China dates back to AD 620 during the Tang Dynasty (Bensky et al., 2004). *Ophiocordyceps sinensis* was employed in medicine by Tibetans in the 15th and 18th centuries as it was integrated into Chinese Materia Medica (CMM) (Lu, 2017). The CMM's significance shifted dramatically between the early and mid-twentieth centuries. *Cordyceps* have been used for over 2000 years in Chinese medicine to cure numerous infectious diseases (Zhou et al., 2009). The basis for considering *Cordyceps* and allies in oriental medicine is the belief, as well as clinical trials, that *O. sinensis* enhances sexual function in men and women (Zhu et al., 1998). From 1980 onwards, scientific perceptions were applied with advanced methodology to validate the traditional claims on *Cordyceps* (Holliday and Cleaver, 2008).

3. Diversity and Distribution

There are several hotspots of *Cordyceps* in the Indian subcontinent. Diverse species of *Cordyceps* are inhabitants of grasslands located in the highlands of alpine meadows (~3000–5000 m asl) of 27 mountain ranges of the trans-Himalayan ecoregion, infecting mainly caterpillars (*Thitarodes*) in Bhutan, China, India, Nepal, Sikkim, and Tibet (Li et al., 2011; Yan, 2014; Baral et al., 2015). In northern India, *Cordyceps* and allies are well known in Arunachal Pradesh, Chhattisgarh, Himachal Pradesh, Mizoram, Nagaland, Uttrakhand, and Uttar Pradesh (Table 1). Parihar (2021) has given more detailed information on the specific locations of availability of *Cordyceps* and allied species in the Indian subcontinent. Usually, in the Himalayas, after the winter season (October–April), the rainy season prevails (June-September) (50–90% annual rainfall), thus intensive collections of *Cordyceps* happen during May and June (Caplins and Halvorson, 2017).

Nag Raj (1962) first reported the parasitizing ants by *Ophiocordyceps unilataralis* in the Western Ghats region of Karnataka (Table 1). After a gap of 22 years, *Cordyceps forquignoni* and *Cordyceps militaris* were reported from the Western Ghats region of Maharashtra (Jadale and Patil, 1983). In 2002, two more species (*Cordyceps superficialis* and *Ophiocardyceps blattae*) were reported from the Western Ghats of Maharashtra (Nanaware, 2002), followed by three additional species of *Cordyceps* in 2005 by Bhagwat et al. (2005). Afterward, a steady increase was seen in reports of *Cordyceps* and allies in the Western Ghats (Table 1). *Cordyceps* spp. in the Western Ghats is usually found during the mid-rainy season of the southwest monsoon (July-September).

Surprisingly, many *Cordyceps* spp. have also been reported from the lateritic scrub jungles and coconut plantations of the southwest coast of India (Kumar and Aparna, 2014; Prathibha, 2015; Dattaraj et al., 2018; Laya et al., 2019). The present study added four more *Cordyceps* from the scrub jungles in and around Mangalore University Campus. The main reason for the occurrence of varied *Cordyceps* in scrub jungles of the University Campus is the presence of rare, endemic, and endangered species of medicinal plants and forest tree species cultivated in the botanical gardens (medicinal garden and arboretum). Such phytogeographic conditions attract a variety of insect populations, which leads to bipartite and tripartite associations among plants, insects, and fungi. As seen in the Western Ghats, the usual season of occurrence of *Cordyceps* on the west coast will is from July to September of the southwest monsoon season. Figure 1 provides a few known and unknown species of *Cordyceps* in the Western Ghats and west coast of India.

Table 1. *Cordyceps* spp. reported from the Himalayas and the Western Ghats of India (-, not defined).

Species	Host	Location	Reference
C. canadensis Ellis & Everh.	Decaying wood or cow dung	Gorakhpur, Uttar Pradesh	Singh et al., 2019
C. cicadae (Miq,) Massee	Cicada (*Cryptotympana atrata*)	Kollhapur, Maharashtra	Bagam et al., 2022
C. forquignoni Quél.	Hexapoda fly	Panhala forest, Maharashtra	Jagdale and Patil, 1983
C. militaris (L.) Fr.	Adult Lepedoptera	Kolhapur, Maharashtra	Jagdale and Patil, 1983
C. militaris	Adult insect	Tungareshwar Sanctuary, Maharashtra	Kamble and Agre, 2012
C. militaris	Insect larvae and pupae	Tar Devi, Himachal Pradesh	Pathania and Sagar, 2014
C. militaris	Oval-shaped cocoons	Mangalore, Karnataka	Dattaraj et al., 2018
C. militaris	Larvae or pupae of butterflies and moths	Nagaland	Ao and Deb, 2019
C. superficialis (Peck) G.H. Sung, J.M.Sung, Hywel-Jones & Spatafora	Larvae of *Torrubia superficialis*	Malkapur, Maharashtra	Nanaware, 2002
Cordyceps sp.	-	Kodagu, Karnataka	Bhagwat et al., 2005
Cordyceps sp.	-	Kodagu, Karnataka	Bhagwat et al., 2005
Cordyceps sp.	-	Kodagu, Karnataka	Bhagwat et al., 2005
Cordyceps sp.	Insect (Hymenoptera)	Northern Kerala	Julia et al., 2012
Cordyceps sp.	Coconut root grub (*Leucopholis coneophora*)	Kasaragod, Kerala	Kumar and Aparna, 2014
Cordyceps sp.	Coconut root grub (*Leucopholis coneophora*)	Sringeri, Karnataka	Prathibha, 2015
Cordyceps sp.	Coconut root grub (*Leucopholis coneophora*)	Kasaragod, Kerala	Prathibha, 2015
Cordyceps sp.	Grass hopper	Mangalore, Karnataka	Dattaraj et al., 2018
Cordyceps sp.	Larvae	Mangalore, Karnataka	Dattaraj et al., 2018
Cordyceps sp.	Millipede	Mangalore, Karnataka	Dattaraj et al., 2018
Cordyceps sp.	Spherical cocoons	Mangalore, Karnataka	Dattaraj et al., 2018

Table 1. contd. ...

Table 1. contd. …

Species	Host	Location	Reference
Cordyceps sp.	Adult insect	Mangalore, Karnataka	Dattaraj et al., 2018
Cordyceps sp. (Fig. 1d and e)	Oval cocoon	Mangalore, Karnataka	This study, 2022
Cordyceps sp. (Fig. 1f and g)	Millipede	Mangalore, Karnataka	This study, 2022
Cordyceps sp. (Fig. 1f-k)	Insect	Mangalore, Karnataka	This study, 2022
Cordyceps sp. (Fig. 1l)	Caterpillar	Mangalore, Karnataka	This study, 2022
Cordyceps sp. (Fig. 1m and n)	Oval cocoon	Virajpet, Karnataka	This study, 2022
O. blattae (Petch) Petch	Adult insect and dead larvae	Panhala, Maharashtra	Nanaware, 2002
O. indica Girish Nanda & Aakriti Sharma	Lepidopteran caterpillar	Kullu, Himachal Pradesh	Sharma et al., 2023
Ophiocordyceps mizoramensis Zohmangaiha, Zothanzama & Vabeikhokhei	Paper wasp (*Polistes olivaceous*)	Zawlnuam, Mizoram	Zohmangaiha et al., 2021
O. neovolkiana (Kobayasi) G.H. Sung, J.M. Sung, Hywel-Jones & Spatafora	coconut root grub (*Leucopholis coneophora*)	Kasaragod, Kerala	Laya et al., 2019
O. nutans (Pat.) G.H. Sung, J.M. Sung, Hywel-Jones & Spatafora	Pentatomid stink bug (*Halyomorpha halys*)	Makutta, Karnataka	Sridhar and Karun, 2017
O. nutans (Fig.1a–c)	Pentatomid stink bug (*Halyomorpha halys*)	B'Shettigeri, Karnataka	Sridhar and Karun, 2017
O. nutans	Pentatomid stink bug (*Tessaratoma javanca*)	Sindudurg, Maharashtra	Patil and Biranje, 2018
O. nutans	Pentatomid stink bug (*Halyomorpha halys*)	Kanger Valley, Chhattisgarh	Paul et al., 2020
Ophiocordyceps sinensis (Berk.) G.H. Sung, J.M. Sung, Hywel-Jones & Spatafora	Moth larvae	Gori Ganga River valley, Uttrakhand	Sharma, 2004
O. sinensis	Caterpillars (*Hepialus*)	Pithoragarh, Uttrakhand; Arunachal Pradesh	Negi et al., 2006
O. sinensis	Insect larvae	Dharchula and Munsyari, Uttrakhand	Arora et al., 2013
O. sinensis	Caterpillar	Chamoli, Uttrakhand	Kuniyal and Sundriyal, 2013

… Table 1. contd.

Table 1. contd. ...

Species	Host	Location	Reference
O. sinensis	Cjhaudabise	Munsyari Tehsil, UIttrakhand	Chandra and Tewari, 2014
O. sinensis	Thitarodes	Pithoragarh, Uttrakhand	Negi et al., 2015
O. sinensis	Caterpillar	Chamoli, Uttarakhand	Caplins and Halvorson, 2017
O. sinensis	Caterpillar	Gori Valley, Pithoragarh, Uttrakhand	Laha et al., 2018
O. sinensis	Caterpillar	Garhwal, Uttrakhand	Yadav et al., 2019
O. sinensis	Larvae	Gori Ganga watershed, Uttrakhand	Parihar, 2021
O. unilateralis (Tul. & C. Tul.) Petch	Ants	Balehonnur, Karnataka	Nag Raj, 1962

4. Biological Control

Several entomopathogenic fungi are useful as bioindicators (e.g., *Beauvaria, Cordyceops, Entomophora, Hirsutella, Metarhizium, Paecilomyces,* and others) (Ali, 2012). They infect different insects, especially ants, beetles, butterflies, grasshoppers, spiders, wasps, and so on. *Cordyceps* are known to infect mainly lepidopteran insects. *Cordyceps* spp. were known to attack caterpillars of 57 different insects (Wang and Yao, 2011). In fact, *O. sinensis* is consumed by the ethnic population along with its host (the ghost moth, *Thitarodes armoricanus*). Shreshta et al. (2017) recorded more than 50 *Cordyceps* spp. infecting the suborder Apocrita (ants, bees, and wasps). *Ophiocordyceps nutans* is known to parasitize up to 30 species of stinkbugs (Shreshtra et al., 2017). Singh et al. (2019) reported the occurrence of *Cordyceps canadensis* on rotting wood or cow dung; it is likely this species infects insects or larvae that feed on wood or dung (Table 1). *Ophiocordyceps neovolkiana* has been reported from the coastal region of Kasaragod (Kerala) during 2017 and 2018 on the coconut root grub (*Leucopholis coneophora*) (Laya et al., 2019). Similar observations have been made on the infection of coconut root grub in coconut plantations in the Western Ghats and the west coast (Kumar and Aparna, 2014; Prathibha, 2015). Such *Cordyceps* spp. are highly useful in controlling the coconut root grub, which is a perineal problem in coconut plantations in southwest India.

Complex bipartite and tripartite associations or interactions among plants, insects, and fungi are fascinating subject matter in biology and lead to confusion in understanding mutualism, antagonism and parasitism (Dattaraj et al., 2018). Many entomopathogenic fungi are known to

Fig. 1. *Cordyceps* found in the Western Ghats (a-c, m and n) and scrub jungles (d-l) of southwest India: a, *Ophiocordyceps nutans* grown on stink bugs; b, branched tip of the sporocarp and two sporocarps *O. nutans* emerging from stink bug; c, infected dead stink bugs; d, habitat of *Cordyceps* sp. 1; e, *Cordyceps* sp. 1 emerged from an oval cocoon; f, *Cordyceps* sp. 2; g, millipede host of *Cordyceps* sp. 2; h, habitat of *Cordyceps* sp. 3; i, closeup view of sporocarp of *Cordyceps* sp. 3; j, cross section of sporocarp of *Cordyceps* sp. 3; k, dead host insect of Cordyceps sp. 3; l, *Cordyceps* sp. 4 growing on a bluish caterpillar; m, *Cordyceps* sp. 5 grown on a spherical cocoon; n, closeup view of *Cordyceps* sp. 5 (Scale bar, a-i, k-m, 1 cm; j, 500 μm).

colonize live plants as endophytes (Suryanarayanan, 2023). *Cordyceps* are no exception, as *Hymenostilbe nutans* (anamorph of *Ophiocordyceps nutans*) harbour the bark of *Cassina glauca* in the Western Ghats (Sasaki et al., 2008; Sridhar and Karun, 2017). Similarly, *Beauvaria* and *Isaria* (Cordycepitaceae) also exist as endophytes in non-crop as well as crop plant species (Vega, 2008; Vidal and Jaber, 2015). Knowledge on the bipartite (plant-fungus) and tripartite (plant-fungus-insect) associations needs more attention to harness the biopesticide potential (Raman and Suryanarayanan, 2007; Suryanarayanan, 2023).

5. Convergent Evolution

A detailed account on the occurrence of *Cordyceps* and allies showed the occurrence of six species in Himalayas, 13 species each in Western Ghats, and the west coast of India. A wide variation could be seen in the ecoregions of India possessing *Cordyceps* and allied species (Himalayas, Western Ghats, and West coast). They differ in altitudes, climatic conditions, endemism, geology, insect population, phytogeography, soil nutrients, vegetation, and wildlife (Dattaraj et al., 2018). There are numerous grounds to follow the evolutionary lines of *Cordyceps* in the Himalayas and the Western Ghats. Sridhar and Karun (2017) argued that the Himalayan high-altitude cold-adapted *Ophiocordyceps sinensis* and the Western Ghats high-altitude warm-adapted *Ophiocordyceps nutans* might have evolved in parallel. Such convergent evolution was possible owing to the climatic conditions, vegetation, insect populations, and fungal inhabitants of these ecoregions.

6. Conservation

According to Xu et al. (2003), the growth of *Cordyceps* depends on the soil pH with the optimum pH required being 6.0. In addition, soil nutrients and combinations of vegetation will influence the distribution of *Cordyceps* (Wu et al., 2009). Entomopathogenic fungi (e.g., *Cordyceps* and allies) are in association with specific plants or tree species (as endophytes or saprophytes in anamorphic or teleomorphic stages) and pathogens of specific herbivore insects, which need further investigation to harness the benefits (e.g., medicinally valuable metabolites, biocontrol of insects, and conservation of plants). There seem to be bipartite and tripartite associations that are mainly controlled by environmental conditions. Conservation of economically important fungi is possible through precise attempts at cultivation by simulating the natural conditions to harness benefits similar to naturally harvested fungi. In this context, Shreedevasena et al. (2022) provided information on various aspects of the cultivation of *Ophiocordyceps sinensis* (isolation, mass multiplication, inoculation of the host, mechanism of infection, extraction, and purification methods). Similarly, the importance of such artificial culture techniques has been highlighted for *Cordyceps militaris* by Sharma et al. (2023). Yadav and Badola (2019) addressed the challenges in conservation and sustainable trade of caterpillar fungi in the Himalayan ecoregion.

According to Caro and O'Doherty (1999), the 'flagship' species are defined as those species having high-profile noteworthy ecological roles in a specific ecosystem. Karun and Sridhar (2013) and Sridhar and Karun (2017) urged to consider *Ophiocordyceps nutans* as the flagship species of the Western Ghats of India, similar to the global flagship species conferred to *Ophiocordyceps sinensis* (Cannon, 2011; Zhang et al., 2012). Similarly, *Ophiocardyceps unilateralis* has been considered a keystone species and *Ophiocardyceps* as the flagship genus in the tropical forests of Brazil (Evans et al., 2011). Such demarcations of species and their ecological status will lead to global concern for proper conservation measures.

Conclusion

The Indian subcontinent is endowed with as many as 29 species of *Cordyceps* and allies. There is a wide scope for following their diversity, distribution, and economic significance towards nutraceutical and biopesticide values. In addition, *Cordyceps* and allied species serve as biocontrol agents against several insect pests. Their tripartite association (teleomorph- or anamorph-plant-insect) and interactions need precise studies. The occurrence of *Cordyceps* in southwest India has gained momentum more recently. A wide variety of adults, cocoons, and caterpillars are attacked by the *Cordyceps* in southwest India, which needs further exploration. *Cordyceps* and allies found in the Western Ghats and west coast should not dwindle like the Himalayas due to human interference. Their conservation (in type localities), utilization and commercial exploitation should be given utmost importance. There is a need to follow the impact of climate change and the parallel evolution of *Cordyceps* in the Himalayas and the Western Ghats. A deeper investigation into the uniqueness of *Cordyceps* and related species is warranted to overcome several challenges, especially to tackle lifestyle diseases. *Cordyceps* spp., reported from the Western Ghats and west coast of India, is morphologically versatile, but scientific validation is necessary based on molecular phylogenetic approaches.

Acknowledgements

The authors are grateful for the helpful discussion on *Cordyceps* studies by N.C. Karun, H.R. Dattaraj, B.R. Jagadish, and K. Sharathchandra. Support from the Mangalore University and Department of Biosciences to conduct studies on *Cordyceps* is highly acknowledged.

References

Ali, M.L. (2012). The caterpillar fungus *Cordyceps sinensis* as a natural source of bioactive compounds. J. Pharma. Bio. Sci., 1: 41–43.

Ao, T. and Deb, C.R. (2019). Wild mushrooms of Nagaland, India—an important bioresource. Stud. Fungi, 4: 54–71.

Arora, R.K., Singh, N. and Singh, R.P. (2013). Characterization of an entomophagous medicinal fungus *Cordyceps sinensis* (Berk.) Sacc. of Uttarakhand, India. The Bioscan, 8: 195–200.

Bagam, P.H., Bhosale, V.V., Kamble, S.S., Magdum, A.B., Momin, S.A. et al. (2022). A new report on *Cordyceps ciadae* from Western Ghats, India. ANVESAK, 82: 1–4.

Baral, B., Shrestha, B. and Silva, J.A.T. (2015). A review of Chinese *Cordyceps* with special reference to Nepal, focusing on conservation. Environ. Exp. Biol., 13: 61–73.

Bensky, D., Clavey, S. and Stöger, E. (2004). Chinese Herbal Medicine: Materia Medica, 3rd Edition. Eastland Press, Seattle, p. 770.

Bhagwat, S.A., Kushalappa, C.G., Williams, P.H. and Brown, N.D. (2005). The Role of Informal Protected Areas in Maintaining Biodiversity in the Western Ghats of India. Ecol. Soc., 10(1): 8. http://www.ecologyandsociety.org/vol10/iss1/art8/

Boesi, A. and Francesca, C. (2011). *Cordyceps sinensis* medicinal fungus: traditional use among Tibetan people, harvesting techniques, and modern uses. Med. Environ. Sci. https://api.semanticscholar.org/CorpusID:202734805

Cannon, P.F. (2011). The caterpillar fungus, a flagship species for conservation of fungi. Fungal Conser., 1: 35-39.

Caplins, L. and Halvorson, S.J. (2017). Collecting *Ophiocordyceps sinensis*: an emerging livelihood strategy in the Garhwal, Indian Himalaya. J. Mountain Sci., 14: 390-402. https://doi.org/10.1007/s11629-016-3892-8

Caro, T.M. and O'Doherty, G. (1999). On the use of surrogate species in conservation biology. Conser. Biol., 13: 805-814.

Chandra, V.V. and Tewari, A. (2014). Contribution of (Berk.) Sung et al. (Yarsa Gumba) in the livelihood of rural communities in Kumaun Himalaya: Management and conservation issues. Ind. Forester, 140(4): 384-388.

Dattaraj, H.R., Jagadish, B.R., Sridhar, K.R. and Ghate, S.D. 2018. Are the scrub jungles of southwest India potential habitats of *Cordyceps*? KAVAKA – Trans. Mycol. Soc. India,51: 20-22.

Evans, H.C., Elliot, S.I. and Hughes, D.P. (2011). *Ophiocordyceps unilateralis*. Commun. Integr. Biol., 45: 598-602.

Fenglin, L., Li, L., Xiaokun, J., Yan, L. and Hengwei, L. (2020). Analysis of seven mineral elements in cultivated fruiting bodies of *Cordyceps militaris*. International Conference on Energy, Environment and Bioengineering, E3S Web Conf., 185: 04019. https://doi.org/10.1051/e3sconf/202018504019

Graziosi, S., Hall, I.R. and Zambonelli, A. 2022. The mysteries of the white truffle: Its biology, ecology and cultivation. Encyclopedia 2022, 2(4): 1959–1971. https://doi.org/10.3390/encyclopedia2040135

Greeshma, A.A., Sridhar, K.R. and Pavithra, M. (2018a). Nutritional perspectives of an ectomycorrhizal edible mushroom *Amanita* of the southwestern India. Curr. Res. Environ. Appl. Mycol., 8: 54-68.

Greeshma, A.A., Sridhar, K.R., Pavithra, M. and Tomita-Yokotani, K. (2018b). Bioactive potential of nonconventional edible wild mushroom *Amanita*. pp. 719-738. *In*: Gehlot, P. and Singh, J. (eds.). Fungi and their Role in Sustainable Development: Current Perspectives. Springer Verlag, Singapore.

Holliday, J.C. and Cleaver, M.P. (2008). Medicinal value of the caterpillar fungi species of the genus *Cordyceps* (Fr.) link (Ascomycetes). A review. Int. J. Med. Mushr., 10(3): 219–234.

Holliday, J. (2018). *Cordyceps*: A highly coveted medicinal mushroom. pp. 59–91. *In*: Agrawal, D.C., Tsay, H.-S., Shyur, L.F., Wu, Y.C. and Wang, S.-Y. (eds.). Medicinal Plants and Fungi: Recent Advances in Research and Development. Springer Verlag, Singapore.

Hyde, K.D., Xu, J., Rapior, S., Jeewon, R., Lumyong, S. et al. (2019). The amazing potential of fungi: 50 ways we can exploit fungi industrially. Fung. Divers., 97: 1–136.

Jagdale, S.V. and Patil, M.S. (1983). Studies in Pyrenomycetes of Maharastra—Family Clavicipitaceae. Indian J. Mycol. Pl. Pathol., 13: 150–153.

Juliya, R.F., Sankaran, K.V. and Varma, R.V. (2012). Diversity of entomopathogenic fungi in the Kerala part of the Western Ghats. Ind. Forester, 138: 182–188.

Kała, K., Jędrejko, K., Sułkowska-Ziaja, K. and Muszyńska, B. (2024). *Cordyceps militaris* and its applications. pp. 345–368. *In*: Deshmukh, S.K., Sridhar, K.R. and Enshasy, H.A.E. (eds.). Bioprospects of Macrofungi, CRC Press, Boca Raton, USA.

Kamble, V.R. and Agre, D.G. (2012). Reinvestigation of *Cordyceps militaris* (L. ex. St. Amaans) Link from Maharashtra. Bionano Forester, 5: 224–225.

Karun, N.C. and Sridhar, K.R. (2013). The stink bug fungus *Ophiocordyceps nutans*—a proposal for conservation and flagship status in the Western Ghats of India. Fungal Conserv., 3: 43–49.

Karun, N.C. and Sridhar, K.R. (2017). Edible wild mushrooms in the Western Ghats: Data on the ethnic knowledge. Data in Brief, 14: 320–328.

Kumar, T.S. and Aparna, N.S. (2014). *Cordyceps* species as a bio-control agent against coconut root grub, *Leuopholis coneophora* Burm. J. Environ. Res. Dev., 8: 614–618.

Kuniyal, C.P. and Sundriyal, R.C. (2013). Conservation salvage of *Cordyceps sinensis* collection in the Himalayan mountains is neglected. Ecosyst. Ser., 3: 40–43.

Laha, A., Badola, R. and Hussain, S.A. (2018). Earning a Livelihood from Himalayan Caterpillar Fungus in Kumaon Himalaya: Opportunities, Uncertainties, and Implications. Mountain Res. Develop., 38(4): 323–331.

Laya, P.K., Verma, C.K.Y., Cherian, A.K., Anees, M.M. and Rashmi, R. (2019). Observations on entomopathogenic fungus *Ophiocordyceps neovolkiana* on coconut root grub *Leucopholis coneophora*. KAVAKA—Trans. Mycol. Soc. India, 53: 55–60.

Li, Y., Wang, X. L., Jiao, L., Jiang, Y., Li, H. et al. (2011). A survey of the geographic distribution of *Ophiocordyceps sinensis*. J. Microbiol., 49: 913–919. https://doi.org/10.1007/s12275-011-1193-z

Lo, H.C., Hsieh, C., Lin, F.Y. and Hsu, T.H. (2013). A systematic review of the mysterious caterpillar fungus *Ophiocordyceps sinensis* in Dong Chong Xia Cao and related bioactive ingredients. J. Trad. Comp. Med., 3(1): 16–32. https://doi.org/10.1016/S2225-4110(16)30164-X

Lu, D. 2017. Transnational Travels of the Caterpillar Fungus in the Fifteenth through Nineteenth Centuries. Asian Med., 12: 7-55.

Nag Raj, T.R. (1962). Addition to the Indian species of *Cordyceps*. Curr. Sci., 7: 301–302.

Nanaware, S.D. (2002). Taxonomical Studies in the Fungi from Western Ghats of the Maharashtra. Ph.D. Thesis, Department of Botany, Shivaji University, Kolhapur, Maharashtra, India, p. 235. http://hdl.handle.net/10603/139134

Negi, C.S., Joshi, P. and Bohra, S. (2015). Rapid Vulnerability Assessment of Yartsa Gunbu (*Ophiocordyceps sinensis* [Berk.] G.H. Sung et al.) in Pithoragarh District, Uttarakhand State, India. Mountain Res. Develop., 35(4): 382–391.

Negi, C.S., Koranga, P.R. and Ghinga, H.S. (2006). Yartsa gumba (*Cordyceps sinensis*): a call for its sustainable exploitation. Int. J. Sust. Develop. World Ecol., 13: 1–8.

Parihar, D.S. (2021). IUCN red listed caterpillar fungus effected by overharvesting on Gori Ganga watershed. Int. J. Creative Res. Thoughts, 9(1): 2320–1882.

Pathania, P. and Sagar, A. (2014). Studies on the biology of *Cordyceps militaris*: A medicinal mushroom from North West Himalaya. KAVAKA—Trans. Mycol. Soc. India, 43: 35–40.

Patil, A. and Biranje, S. (2018). Report on *Ophiocordyceps nutans* (Pat.) G.H. Sung and Spataphora on *Tessartoma javania* (Thunberg) from Western Ghats, Maharashtra. Aayushi Int. Interdis. Res. J. (Special issue): 636–638.

Paul, J.S., Jadhav, S.K., Auraishi, A. and Naik, M.L. (2020). Ferret out a natural bio-pesticide: *Ophicordyceps nutans* in central India and its interaction analysis with tree stink bug. Proc. Zool. Soc., 73: 316–219.

Pavithra, M., Greeshma, A.A., Karun, N.C. and Sridhar, K.R. (2015). Observations on the *Astraeus* spp. of Southwestern India. Mycosphere, 6: 421–432.

Pavithra, M., Sridhar, K.R., Greeshma, A.A. and Tomita-Yokotani, K. (2016). Bioactive potential of the wild mushroom *Astraeus hygrometricus* in the southwest India. Mycology, 7: 191–202.

Pérez-Moreno, J., Mortimer, P.E., Xu, J., Karunarathna, S.C. and Li, H. (2021). Global perspectives on the ecological, cultural, and socioeconomic relevance of wild edible fungi. Stud. Fungi, 6(1): 408–424. https://doi.org/10.5943/sif/6/1/31

Prathibha, P.S. (2015). Behavioural Studies of Palm White Grubs *Leufopholis* spp. (Coleoptera: Scarabaeidae) and Evaluation of New Insecticides for their Management. Ph.D. Thesis, Agricultural Entomology, University of Agricultural Sciences, Bangalore, India, p.119.

Raman, A. and Suryanarayanan, T.S. (2017). Fungus-plant interaction influences plant-feeding insects. Fungal Ecol., 29: 123–132.

Sasaki, F., Miyamoto, T., Yamamoto, A., Tamai, Y. and Yajima, T. (2008). Morphological and genetic characteristics of the entomopathogenic fungus *Ophiocordyceps nutans* and its host insects. Mycol. Res., 112: 1241–1244.

Sharma R., Kumar, A. and Shukla, A.C. (2023). *Cordyceps militaris*: Artificial Culture Technique: A Soft Gold Medicinal Fungus. National Seminar cum Workshop on Mushroom Cultivation and Agri Business Management, May 8–9, pp. 1–3.

Sharma, A., Kaur, E., Joshi, R., Kumari, P., Khatri, A. et al. (2023). Systematic analyses with genomic and metabolomic insights reveal a new species, *Ophiocordyceps indica* sp. nov. from treeline area of Indian Western Himalayan region. Front. Microbiol., 14: 1188649. https://doi.org/10.3389/fmicb.2023.1188649

Sharma, S. (2004). Trade of *Cordyceps sinensis* from high altitudes of Indian Himalaya: conservation and biotechnological properties. Curr. Sci., 86: 1614–1619.

Shreedevasena, S., Karthiba, L., Raveena, R., Ramyabharathi, S.A., Salama, E.A. et al. (2022). Mass production and marketing of compost caterpillar fungus *Cordyceps sinensis*. pp. 239–263. In: Amaresan, N., Dharumadurai, D. and Cundell, D.R. (eds.). Industrial Microbiology Based Entrepreneurship. Springer Nature Singapore. https://doi.org/10.1007/978-981-19-6664-4_16

Shrestha, B., Tanaka, E., Hyun, M.W., Han, J.G., Kim, C.S. et al. (2017). *Cordyceps* species parasitizing hymenopteran and hemipteran insects. Mycosphere, 8(9): 1424–1442. https://doi.org/10.5943/mycosphere/8/9/8

Singh, R.P., Kashyap, A.S., Pal, A., Singh, P. and Tripathi, N.N. (2019). Macrofungal diversity of North-Eastern part of Uttar Pradesh (India). Int. J. Curr. Microbiol. Appl. Sci., 8: 823–838.

Sridhar, K.R. and Karun, N.C. (2017). Observations on *Ophiocordyceps nutans* in the Western Ghats. J. New Biol. Rep., 6: 104–111.

Sridhar, K.R., Mahadevakumar, S. and Karun, N.C. (2024). On the *Amanita* in southwest India. pp. 159–176. In: Ecology of Macrofungi: An Overview. Sridhar, K.R. and Deshmukh, S.K. (eds.). CRC Press, Boca Raton, USA.

Suryanarayanan, T.S. (2023). Fungal endophyte-plant-insect interaction: a tripartite association needing attention. Ind. J. Entomol., e23262. 10.55446/IJE.2023.1262

Vega, F.E. (2008). Insect pathology and fungal endophytes. J. Invert. Pathol., 98: 277–279.

Vidal, S. and Jaber, L.R. (2015). Entomopathogenic fungi as endophytes: plant-endo phytedherbivore interactions and prospects for use in biological control. Curr. Sci., 109: 46–54.

Wang, X.-L. and Yao, Y.-J. (2011). Host insect species of *Ophiocordyceps sinensis*: A review. ZooKeys, 127: 43–59. https://doi.org/10.3897/zookeys.127.802

Wang, Z., Da, W., Negi, C.S., Ghimire, P.L., Wangdi, K. et al. (2022). Profiling, monitoring and conserving caterpillar fungus in the Himalayan region using anchored hybrid enrichment markers. Proc. R. Soc. B, 289: 2021–2650. 10.1098/rspb.2021.2650

Wu, Q.G., Su, Z.X., Su, R.J., Hu, J.Y. and Wang, H. (2009). The dominant factors of habitat selection of *Cordyceps sinensis*. Guangxi Zhiwu/Guihaia, 29: 331–336.

Xu, C.-P., Kim, S.-W., Hwang, H.-J., Choi, J.-W. and Yun, J.-W. (2003). Optimization of submerged culture conditions for mycelial growth and exo-biopolymer production by *Paecilomyces tenuipes* C240. Proc. Biochem., 38: 1025–1030.

Yadav, P.K. and Badola, S. (2019). Challenges in conservation and sustainable trade of Caterpillar Fungus in India. Newsletter on Wildlife Trade in India (Special Issue on Medicinal Plants), Issue, 31, pp.18–26.

Yadav, P.K., Saha, S., Mishra, A.K., Kapoor, M., Kaneria, M. et al. (2019). Yartsagunbu: transforming people's livelihoods in the Western Himalaya. Oryx, 53(2): 247–255.

Yan, J.-K., Wang, W.-Q. and Wu, J-Y. (2014). Recent advances in *Cordyceps sinensis* polysaccharides: Mycelial fermentation, isolation, structure and bioactivities: A review. J. Func. Foods, 6: 33–47.

Zhang, Y., Li, E., Wang, C., Li, Y. and Liu, X. (2012). *Ophiocordyceps sinensis*, the flagship of China: terminology, life strategy and ecology. Mycology, 3: 2–10.

Zhou, X., Gong, Z., Su, Y., Lin, J. and Tang, K. (2009). *Cordyceps* fungi: natural products, pharmacological functions and developmental products. J. Pharm. Pharmacol., 61(3): 279–291.

Zhou, Y., Chu, M., Ahmadi, M., Agar, O.T., Barrow, C.J. et al. (2023). A comprehensive review on phytochemical profiling in mushrooms: Occurrence, biological activities, applications and future prospective. Food Rev. Int. https://doi.org/10.1080/87559129.2023.2202738

Zhu, J.S., Halpern, G. and Jones, K. (1998). The scientific rediscovery of an ancient Chinese herbal medicine: *Cordyceps sinensis*: part I. J. Alt. Complem. Med., 4(3): 289–303.

Zohmangaiha, C., Vabeikhokhei, J.M.C., Zomuanpuii, R., De Mandal, S. and Zothanzama, J. (2021). A new species of *Ophiocordyceps* (Ophiocordycipitaceae) from Mizoram, India. Phytotaxa, 500: 11–20.

5

Diversity of Cordycipitaceae in the Sahyadri Mountain Ranges of the Western Ghats

Jithu Unni Krishnan,,# Devika Lal Mudavarthottiyil Raju,#
Anju Mathiyil Subramanian, Murali Revanasiddappa.,
Shambu Kumar* and *Rakhi Kunnath Radhakrishnan*

1. Introduction

Entomopathogenic fungi, as their name implies, are a functional class that kills or seriously disables the mightiest animal class, the 'Insects'. These fungal parasites have a global presence, inhabiting nearly all terrestrial ecosystems. Their highest diversity is found in tropical forests (Vega and Kaya, 2011). However, their distribution is not limited to hot and humid regions. Eilenberg (2002) identified potential extremophiles in the temperate wilderness, such as the High Arctic Tundra and in the frigid environment of Antarctica (Bridge and Worland, 2004; Tosi et al., 2004). Hence, these special fungi can be considered to enjoy a cosmopolitan distribution. Within the Hypocreales order, Cordycipitaceae family members primarily inhabit either an anamorphic state, characterized by an asexual mode of reproduction, or a teleomorphic state, signifying sexual reproduction. However, they commonly display both states throughout their life

Forest Health Division, KSCSTE-Kerala Forest Research Institute, Peechi 680653, Kerala, India.
* Corresponding author: jithuukrishnan@gmail.com
# Note: First and second authors have contributed equally to this work.

cycle, affording them adaptive advantages across diverse environmental conditions and host organisms.

Trends in global distribution patterns show teleomorphs are mostly found in tropical habitats, while anamorphs are found in both tropical and temperate habitats (Hajek, 1997). Owing to their extensive geographic range and remarkable adaptive mechanisms, research endeavours are not constrained by quantity; instead, three prominent research spheres emerge: (i) taxonomy and diversity; (ii) ecological roles and functions; (iii) both direct and indirect applications. The primary focus of our knowledge is predominantly directed towards the latter category, with taxonomy serving as an essential adjunct. Consequently, there is a notable dearth of comprehensive data about the distinct taxonomic groups of entomopathogenic fungi associated with insects in tropical forest ecosystems (Varma et al., 2006; Prathiba, 2015; Dattaraj et al., 2018).

Detecting fungal entomopathogens in the environment can present a formidable challenge due to their often-inconspicuous characteristics, which may indeed account for the limited availability of relevant information. These pathogens are not always immediately visible, necessitating meticulous scrutiny for their identification. A reliable indicator of their presence is the emergence of fungal structures from a deceased host, a tell-tale sign of infection. However, many infected hosts are diminutive and tend to perish in concealed locations such as subterranean or partially aquatic habitats, highly humid macro-environments (e.g., tropical rainforests) and microenvironments (e.g., beneath tree bark), further complicating their detection. Field studies aimed at gathering and estimating fungal diversity within their natural habitat require dual expertise, combining entomological and pathological knowledge. Notably, some members of the Entomophthorales order cause their host to perish at an elevated point on the nearest available vegetation, which is very much publicized as 'Summit disease'. This strategy promotes the dispersal of fungal conidia and ensures the cadaver remains attached to the substrate through the production of rhizoids (Roy et al., 2006).

Certain ascomycete taxa, particularly within the Hypocreales (family, Cordycipitaceae), display conspicuous stroma (both in colour and structure) when in their teleomorphic stages, making them more noticeable. However, the anamorphic stages of several fungal entomopathogens are widespread in terrestrial ecosystems, with subtle and often minute fungal structures. This further imposes additional constraints on research studies within this category. Additionally, these microscopic fungal spores can be found outside their hosts in various ecosystem partitions, primarily in the soil environment. The soil serves as a stable habitat for these fungi, buffering population variations and protecting them from adverse abiotic conditions (Van der Putten et al., 2001). Chen et al. (2021) worked out an interest in estimating the species diversity of potential soil-inhabitant entomopathogenic fungi as insect pest control agents. From 212 soil samples from three different provinces in China, the study observed around 10% potential entomopathogenic fungi from 490 different cultures.

There is also the chance of various climatic conditions that can adversely affect the population of these interesting groups of fungi. At least in the case of specialist entomopathogenic fungus, the host population can be vital in determining its occurrence. Research conducted on the diversity of entomopathogenic fungi within the tropical rainforests of Africa and South America has revealed a relatively lower diversity compared to other ecosystems (Samson et al., 1988). Several factors may contribute to this, including the absence of specific host organisms, suboptimal conditions for fungal infection, or a combination of both. Notably, Aung et al. (2008) observed a higher diversity of entomopathogenic fungi in 'disturbed forests' compared to 'conserved forests' and 'agricultural land'. This may be because disturbed rainforests encompass a combination of forest and agricultural habitats, facilitating the coexistence of both specialized and generalist entomopathogens within this environment. *Cordyceps* species, one of the well-known members of the Cordycipitaceae tends to thrive in undisturbed habitats characterized by clean air, high humidity and ample shade (Aung et al., 2008). Though diversity studies of entomopathogenic fungi across different habitats are limited, from the available information, tropical rain forests filter out the majority of other entomopathogenic fungi under their specific micro-climatic conditions, potentially favouring the dominance of *Cordyceps* spp.

The Western Ghats, also known as the Sahyadri mountain ranges, are biodiversity hotspots that represent niche complexity at their peak, supporting commendable species richness and endemism from the tropical belt. Its gene pool cannot compromise on the demand for its vital microbial diversity and hence for *Cordyceps* spp. as well. Throughout the Western Ghats region, one can anticipate a significant array of teleomorphic forms, owing to the favourable fulfilment of microhabitat prerequisites. However, entomo-pathological investigations concerning taxonomy and a broader systematic understanding have yet to reach their full potential in this biodiversity hotspot. In addition to the imperative need for expanded species inventories, comprehensive research is essential to ascertain the prospective utility of indigenous entomopathogenic fungi within forest ecosystems. This includes their potential application as biological pest control agents for combating significant insect pests across a spectrum of forest management practices, encompassing silviculture and general forestry. This chapter is dedicated to providing a comprehensive overview of the diversity among some notable genera of Cordycipitaceae in Sahyadri mountain ranges of the Western Ghats.

2. Diversity and Morphology

The family Cordycipitaceae comprises a wide range of species, not only those that parasitize insects but also those that live as endophytes in plants or other substrates, showcasing various ecological roles and adaptations. The family includes some well-publicised genera, each with its unique characteristics, economic importance, life cycle and host preference. Some of the most notable genera within the Cordycipitaceae include *Cordyceps*, known for their parasitic relationships with insects; *Beauveria* and *Lecanicillium*, which are highly valued for their biocontrol potential against a broad range of insect pests; and *Akanthomyces*, frequently employed in biocontrol programs. Other significant genera like *Isaria, Engyodontium, Hirsutella, Simplicilium, Gibellula, Torrubiella, Microhilum Ascopolyporus* and *Hyperdermium* also play essential roles in regulating insect populations or infecting other fungi. Wijayawardene et al. (2020) check-listed global distribution of the genera under the family Cordycipitaceae (Table 1). Morphological features of anamorphs and teleomorphs of Cordicipitiaceae have been summarized in Table 2.

2.1 *Akanthomyces*

This genus is reported to be a generalist on invertebrates as a pathogenic fungus. *Akanthomyces* spp. is known for its capacity to biocontrol certain insects. It can be found in both spiders and insects. *Akanthomyces* spp. was known to attack some insect orders (e.g., Coleoptera, Hemiptera, Lepidoptera and Orthoptera) (Chen et al., 2021; Mongkolsamrit et al., 2018; Hodge, 2003). Generally, the host range of *Akanthomyces* for both teleomorphs and anamorphs is similar. Detailed information on the morphological characteristics of *Akanthomyces* in the anamorphic state is presented in Table 2.

2.2 *Beauveria*

This most publicised group is characterised by a complex life cycle that involves both saprophytic and entomopathogenic phases. In the saprophytic phase, the fungi live as soil-dwelling organisms, recycling organic matter and nutrients. However, when the opportunity arises, many species of *Beauveria* can switch to their entomopathogenic phase, infecting and ultimately killing a wide range of arthropod hosts, including insects and mites. This dual lifestyle is a hallmark of *Beauveria*'s diversity, as it allows the fungi to thrive in various habitats and exploit a multitude of ecological niches. The diversity within the genus *Beauveria* is reflected in its numerous species, with more than 100 described to date and undoubtedly more awaiting discovery. Each species has evolved to parasitize specific insect

Table 1. Checklist of genera described under the family Cordycipitaceae.

Genus	Report in the Western Ghats
Akanthomyces	Reported
Amphichorda	Not reported
Ascopolyporus	Not reported
Beauveria	Reported
Beejasamuha	Not reported
Blackwellomyces	Not reported
Cordyceps	Reported
Coremiopsis	Not reported
Engyodontium	Not reported
Gibellula	Reported
Granulomanus	Not reported
Hevansia	Not reported
Hyperdermium	Not reported
Isaria	Unpublished data
Lecanicillium	Reported
Leptobacillium	Not reported
Microhilum	Not reported
Parengyodontium	Not reported
Pseudogibellula	Not reported
Samsoniella	Not reported
Simplicillium	Not reported
Torrubiella	Not reported

hosts and this specialization often results in unique adaptations in terms of morphology, behaviour and pathogenicity. Some well-known species within the genus are *Beauveria bassiana* and *Beauveria brongniartii*, each of which exhibits distinct host preferences and environmental tolerances.

One of the key attributes that add to the diversity of *Beauveria* is its ability to produce a wide array of bioactive secondary metabolites. These compounds play a crucial role in the fungi's pathogenicity and can have significant implications for their use in biocontrol strategies against agricultural and forest pests. Additionally, *Beauveria* species have gained attention in biotechnology and pharmaceutical research due to the production of various bioactive molecules with potential applications in medicine and industry. They are widely distributed in nature and are the most common and ubiquitous entomopathogenic fungi, possessing the capability to infect a wide range of insects belonging to diverse orders

Table 2. General morphology of widely studied anamorphic and teleomorphic genera of Cordycipitaceae (Sources: Sung et al., 2007; Spatafora et al., 2011; Vega and Kaya, 2011).

Genus	General morphology	Reference
Anamorphs		
Akanthomyces	White, cream, or flesh-coloured cylindrical, attenuated synnematal growth covered by a hymenium-like layer of phialides producing one-celled catenulate conidia.	Mains, 1950; Samson and Evans, 1974; Hsieh et al., 1997
Beauveria	Mycelium appeared white, dense and powdery. Hyaline, smooth-walled conidia were globose to sub-globose in shape. The size of the conidia varied from 1.80–2.50 μm × 1.70–2.40 μm.	Wargane et al., 2020
Engyodontium	The conidia are hyaline, smooth-walled, subspherical, 2–3 μm × 1.5–2.5 μm.	Simonovicova et al., 2004
Gibellula	Mycelium covering the host, white to cream fluffy, light greyish-brown to violaceous-brown when dried. Synnemata multiple, cylindrical, growing from the abdomen of the host spider, cream to yellowish-white. Conidiophores abruptly narrow to a short distinct neck and form a sub-sphaeroidal vesical.	Mains, 1950; Samson and Evans, 1992; Kuephadungphn et al., 2019
Isaria	White to cream-colored, with less compact mycelial growth. Conidia were ellipsoidal to cylindrical, often slightly curved.	Chhetri et al., 2020
Lecanicillium	Mycelium appeared white and dense	Subramanian et al., 2021
Microhilum	Multiple short conidia from a single denticle.	Wright and Patel, 1992
Simplicilium	Solitary phialides, conidia adhering in globose, slimy heads or imbricate chains.	Zare and Gams 2001
Teleomorphs		
Ascopolyporus	Large sub-globose to polypore-like, bright rusty-red or white-yellow perithecial stromata, which are usually fertile only on the underside of the stroma	Thanakitpipattana et al., 2022
Cordyceps	Orange to brown fleshy stromata arose from the body.	Vega and Kaya, 2011
Hyperdermium	Stromata are flattened or pulvinate and vary in color from white to orange. Perithecia are immersed to sub-immersed with asci.	Sullivan et al., 2000
Torrubiella	White to cream to various shades of brown, production of superficial perithecia and absence of well-developed stipitate stroma.	Johnson et al., 2009

(Guo et al., 2021). It is a 'common resident' of soil (Klingen et al., 2002). Furthermore, the community of fungal entomopathogens is primarily dominated by *B. bassiana*, accounting for 88% of all cadavers (Vega and Kaya, 2011).

It follows a similar life history, characterized by a single reproductive incident. This entomopathogenic fungus is promising in insect biological control (Lacey and Goettel, 1995) and similar scarabs and weevils (Keller, 2000) without harming the non-target fauna (Goettel and Hajek, 2001). These potent fungi first initiate infection and subsequently release all of their infective propagules, known as conidia in a single event following the death of their host insects (Bartlett and Jaronski, 1988; Wraight et al., 2001). The fungus's ability to successfully invade and multiply within a host is contingent upon its capacity to enter the hemocoel (the insect's body cavity containing hemolymph) and either suppress or evade the host's immune response (Gotz, 1991).

The *B. bassiana* has proven highly effective in managing the tobacco cutworm *Spodoptera litura*, a significant pest responsible for substantial crop losses in economically valued crops like cotton, tobacco, soybean, groundnut and various vegetables (Qin et al., 2010; Malarvannan et al., 2010). From the Pulney hills of Tamil Nadu, India, Baskar et al., 2012 documented the growth-inhibiting properties of *B. bassiana* against *Spodoptera litura*. Furthermore, an indigenous strain of the endophytic fungus, *B. bassiana* was isolated from the Ghataprabha region, situated in the Krishna River basin of Karnataka and it exhibited growth-inhibitory effects against the same *Spodoptera litura*. Additionally, Baskar et al., 2012 reported the growth-suppressing impact of *Beauveria bassiana* on coffee berry borer (*Hypothenemus hampei*) infesting coffee plantations in Pulney Hills, Tamil Nadu, India. It is also found to be effective in managing other forest pests in India, viz., *Malacosoma disstria* and *M. americanum* (Smirnoff, 1968), *Hoplocerambyx spinicornis* (Sharma and Joshi, 2004), *Myllocerus viridanus* and *Calopepla leayana* (Sankaran et al., 1989), *Atteva fabriciella* (Agarwal et al., 1988; Ali and Varma, 1994), *Sahyadrassus malabaricus* (Ali and Mathew, 1989), *Hyblaea puera* (Ali et al., 1991; Rajak et al., 1993; Sandhu et al., 1993; Javaregowda and Naik, 2006) and *Indarbela quadrinotata* (Sasidharan, 2004). There is a report on *B. brongniartii* infecting larvae of *Hypsipyla robusta* (Kandaswamy, 1969).

2.3 *Cordyceps*

One of the notable genera within the family is *Cordyceps*, as mentioned at the beginning of this chapter. This particular genus consists of more than 400 species of endoparasites on arthropods in varied climatic situations prevailing worldwide (Stensrud et al., 2005; Sung et al., 2007). *Cordyceps* spp. can parasitize a wide array of insects and their larvae, leading to the development of various morphologically distinct and diverse fruit bodies (Sung et al., 2007; Kepler et al., 2012). The members of the genus *Cordyceps* can be found to infect several insect orders, including Diptera, Hymenoptera, Coleoptera, Lepidoptera, Hemiptera, Isoptera, Orthoptera and in other arthropod classes as well (Patil et al., 2014; Venkatesan et al.,

2009; Varma et al., 2006). Certain species of *Cordyceps* are obligate parasites of ants and serve as significant pathogenic fungi within tropical ecosystems (Evans and Samson, 1984). The prevalence of the disease remains relatively constant or enzootic due in part to the behaviour of infected ant hosts (Evans and Samson, 1982). Infected ants deviate from their usual trails and nest activities, undergoing substantial changes in behaviour as they seek out specific niches. Following infection by *Cordyceps,* ground-dwelling 'Ponerine' ants climb up into vegetation and succumb in exposed positions, clinging to the substrate with their legs and mandibles (Evans, 1988). Interestingly, some *Cordyceps* sp. infecting Lepidoptera larvae and pupae have been employed for medicinal purposes and as a source of food (Hoffman, 1947).

Spores of *Cordyceps* fungal initiate infecting insects by adhering to them, followed by penetration into the host's body. Once inside, the fungus thrives and gradually consumes the host from within, ultimately resulting in the host's demise. Subsequently, the fungus repurposes the remnants of the host's body to encapsulate spores, which are subsequently discharged into the environment. This mechanism enables the fungus to disseminate and infect other potential hosts. In the Indian Subcontinent, the high-altitude regions of the Himalayas, such as Uttarakhand, Bhutan, Nepal and Tibet, serve as potential habitats that sustain a diverse range of *Cordyceps* spp. Such a trend has been documented by various studies (Winkler, 2008; Cannon et al., 2009; Shrestha and Bawa, 2013; Quan et al., 2014; Baral et al., 2015; Negi et al., 2015; Baral, 2017). In contrast, reports of *Cordyceps* occurrences in the Western Ghats and along the west coast of India are relatively rare. The first recorded instance of *Cordyceps unilateralis* parasitizing ants in the Western Ghats dates back to 1961 in Balehonnur, Chikmagalur District, Karnataka (Nag Raj, 1962).

From 2001 to 2008, four different *Cordyceps* species (*Cordyceps forquignoni, C. militaris, C. superficialis* and *Cordyceps blattae*) were reported to infect adult insects such as beetles and their larvae in the Maharashtra State (Nanaware, 2002; Patil et al., 2014). In the year 2011–2012, *Cordyceps natans* was observed parasitising pentatomid bugs, specifically *Nezara viridula,* in Tillari, Maharashtra (Patil et al., 2014). Later, in Sringeri, Chikmagalur, Karnataka, *Cordyceps* sp. reported infected grubs of coconut roots (*Leucopholis conephora*) in plantations (Prathiba, 2015). In a systematic mycological survey carried out in the moist deciduous forests of Megamalai within the Western Ghats region of Tamil Nadu, India, researchers conducted an assessment of *Cordyceps* diversity spanning from June 2007 to February 2008. This assessment was based on the observation of their morphological attributes, the dimensions of their fruiting bodies, the configuration of their pileus (cap) and their specific growth habitats. The Megamalai forest is characterized by an average annual precipitation of approximately 1300 mm, an average temperature of around 27°C and relative humidity levels ranging from 50–75%. These environmental conditions, marked by high humidity and the accumulation of organic matter, provide an optimal environment for the prolific growth of *Cordyceps* spp. (Venkatesan et al., 2009).

2.4 *Gibellula*

Members of this genus are frequently observed growing on spiders. However, there is limited knowledge regarding the full extent of their host range. Describing these interactions poses a significant challenge because the fungus rapidly consumes the parasitized spiders, making it difficult to identify the host. Moreover, crucial taxonomic traits used for identification are often destroyed in the process. Furthermore, the global diversity of *Gibellula* remains uncertain, as do the natural history and phylogenetic relationships of most species within this genus. Interestingly, species of *Gibellula* that infect spiders have been documented in 33 countries. Recently, the presence of *G. clavulifera* and *G. alata* infecting spiders has been reported from the southern Western Ghats (Mendes-Pereira et al., 2023).

2.5 *Isaria*

Members of this genus are some of the most virulent entomopathogens across the globe, but studies from the Western Ghats region are very limited. From the butterfly garden of the KFRI campus for the very first time, a specimen was collected that is morphologically confirmed to be *Isaria* sp. but pending molecular confirmation.

One of the most well-known species is *Isaria fumosorosea* (formerly known as *Paecilomyces fumosoroseus*), which has been extensively studied and applied in biological pest control. It is used as a biological control agent against a wide range of insect pests, including aphids, whiteflies, thrips and various other insects that damage crops. *Isaria fumosorosea*, among other *Isaria* species, is employed as an eco-friendly alternative to chemical pesticides in agriculture. These fungi penetrate their insect hosts using specialized structures and subsequently proliferate within the host's body, leading to the death of the insect. The fungus then grows out of the deceased host and produces spores, which aid in its dispersal to infect other potential hosts.

2.6 *Lecanicillium*

Another cosmopolitan fungus within the family of Cordycipitaceae. The species diversity within the genus *Lecanicillium* is considerable, encompassing various fungal species, each exhibiting specific characteristics, behaviours and host preferences. While the exact number of species is not definitively known and continues to be studied and revised, there are numerous identified species within this genus. A notable species in the genus is *Lecanicillium lecanii* (synonym: *Verticillium lecanii*). This entomopathogen can effectively control a broad spectrum of insects (e.g., aphids, jassids, mealy bugs, mites and whiteflies) (Wang et al., 2007). The typical mode of infection developed by *L. lecanii* is the same as that of other entomopathogenic fungi in the Cordycipitaceae, i.e., the process of infection development involves the initial attachment of fungal conidial spores to the surface of the insect, penetration of the integuments by hyphae and rapid spread in the insect's hemocoel, ultimately leading to the secretion of toxins that results in the death of insects (Gurulingappa et al., 2011). It has been documented in various regions of India for its widespread presence. In the Kodagu District of Karnataka, a study conducted by Bagyaraj et al. (2015) reported the extensive distribution of this fungal species. Furthermore, in Coimbatore District, Tamil Nadu, specifically in the Anamallai region, a separate investigation by Subramaniam et al. (2021) highlighted its efficiency as a biocontrol agent against the pest species *Scirtothrips bispinosus*.

3. Status of Cordycipitaceae in Kerala

3.1 In Different Regions

Kerala State has been divided into northern, central and southern forest circles. Notably, a significant abundance of entomopathogenic fungi has been documented across all of these three forest circles. Furthermore, comprehensive data on the diversity of genera within the Cordycipitaceae family has been collected from various locations, including Wayanad, Silent Valley, Attappady (in the northern region), Vazhani, Peechi, Nelliampathy, Parambikkulam (in the central region), Kottapara and Munnar (in the southern region), covering the entirety of the Kerala portion of the Western Ghats, but by very few research groups. These locations include different ranges of forests, including evergreen, semi-evergreen, moist deciduous and shola.

Fungal infections were observed in insects belonging to a wide array of taxonomic orders, including Coleoptera, Dictyoptera, Diptera, Epimeroptera, Hemiptera, Isoptera, Lepidoptera, Odonata and Orthoptera. Among these, the family Formicidae, within the order Hymenoptera, showed a particularly high prevalence of fungal infections, followed by the family Noctuidae under Lepidoptera. Notably, entomopathogenic fungi were found to be infectious at all stages of an insect's life cycle (Juliya, 2007). Its high time to think about targeted pest management alternatives apart from *Bacillus thuringiensis* (Harish and Krishnan, 2023) and entomopathogens could be a better alternative.

In the northern forest circle, *Beauveria* sp. was found to be associated with the insect orders Coleoptera (Chrysomelidae, Meloidae, Scarabaeidae,

Scolytidae), Hymenoptera (Ichneumonidae), Lepidoptera (Hespiridae, Yponomeutidae), Orthoptera (Acrididae) and in the Southern Forest Circle fungi were observed among the insect orders of Hymenoptera (Membracidae). *B. brongniartii* was a potential entomopathogen on beetles belonging to the Chrysomelidae family that were feeding on teak leaves in Vazhani within the Central Forest Circle (Juliya, 2007).

In the central forest circle of the of the Western Ghats in Kerala, a frequent instance of *Beauveria bassiana* on *Gargara mixta*, feeding on the leaves of *Helicteres isora* and *Terminalia catappa*, was observed throughout various periods at Vazhachal and Peechi. *B. bassiana*, which is known to cause muscardine disease, has been reported to affect numerous insect orders, particularly in the Coleoptera (Bruchidae, Cerambycidae, Chrysomelidae, Curculionidae and Scarabaeidae), Hemiptera (Membracidae), Hymenoptera (Formicidae, Vespidae), Isoptera (Termittidae) and Lepidoptera (Arctiidae) (Agarwal et al., 1985; Sankaran et al., 1989; Ali et al., 1991; Sharma and Joshi, 2004).

Occurrences of *Lecanicillium lecanii* were recorded from central forest circles associated with the insect order Coleoptera (Chrysomelidae), Dictyoptera (Blattidae), Isoptera (Termittidae) and Lepidoptera (Arctiidae) and in the Northern Forest Circle *Lecanicillium lecanii* was isolated from the order Lepidoptera (Pyraustidae).

Both in the northern and central forest circles, frequent occurrences of *Akanthomyces* sp. were noted, particularly affecting insects in the order Hymenoptera (Formicidae). The distribution of *Cordyceps* sp. was also observed to be quite common in the Central, Northern and Southern Forest circles, with a high prevalence among insects belonging to the order Hymenoptera, particularly in Formicidae. *Cordyceps* sp. were notably more abundant in the Silent Valley and Nilambur regions (Northern Forest Circle) of Kerala. Additionally, from Kasaragod District, Kerala, Kumar and Aparna (2014) described *Cordyceps* species infecting coconut root grubs (*Leucopholis conephora*) in coconut plantations. This was followed by a report of another *Cordyceps* sp. associated with the coconut root grubs in Kudlu, Kasaragod, Kerala (Prathiba, 2015). Furthermore, a diverse range of entomopathogens was observed in forest plantations, particularly in the Teak plantation. Among these, *B. brongniartii* was the most frequently recorded entomopathogenic fungus in beetles (Coleoptera: Chrysomelidae) that feed on teak leaves in the Vazhani teak plantation (Varma et al., 2006). *B. bassiana* infecting *Hyblaea puera* was also reported from teak plantations. The occurrence of *Isaria* is also quite common across Kerala because of its cosmopolitan nature, particularly in Lepidoptera genera (personnel observation).

3.2 *Impact of Seasons*

The abundance and diversity of entomopathogenic fungi were notably lower during the dry season compared to the wet season in both forests and plantations (Varma et al., 2006). Typically, a combination of low relative humidity and high temperatures tends to hinder the infection and spread of pathogens, as fungal pathogens require optimal temperatures and higher humidity levels to become pathogenic (Milner and Lutton, 1986; Helyer et al., 1992; Fargues and Luz, 2000; Inglis et al., 2000). However, specific entomopathogenic fungi, such as *B. bassiana* and *M. anisopliae* (Hypocreales: Clavicipitaceae), can still infect their hosts even in conditions of reduced ambient humidity. This ability is presumably due to the presence of sufficient moisture in microenvironments like the boundary layer of leaves and the intersegmental folds of the insect's integuments (Fargues et al., 1997; Ferron, 1997). Interestingly, areas at higher altitudes exhibited a lower incidence of fungal infections. High elevation is directly related to low atmospheric temperature and relative humidity and is hence not very conducive to the growth of fungi (Varma et al., 2006).

3. Dissemination of Cordycipitaceae

Dispersal, or dissemination, is a term used to describe a pathogen's capability to expand and distribute itself within a host population and across the environment. This concept is crucial when examining transmission and is essential for comprehending how infections naturally evolve within the host insect population. The continued existence of any pathogen within a host population over the long term is closely linked to its dispersal capacity. Pathogens with limited dispersal abilities generally have a minimal likelihood of triggering widespread outbreaks, even if they possess high levels of virulence or exhibit efficient environmental survival.

In nature, insect pathogens employ four primary mechanisms for disseminating themselves to susceptible individuals (Fig. 1). Firstly, they use their motility and actions, including the forceful-release of conidial spores from infected hosts. This method is especially crucial for entomopathogenic fungi and can lead to rapid outbreaks when host populations are dense and environmental conditions are favourable. Additionally, these fungi can form resistant resting spores in response to unfavourable conditions, ensuring their survival during periods when no susceptible hosts are available and providing an effective means of dispersal to new host populations in subsequent years. Secondly, dissemination occurs through the behaviour and movement of infected primary hosts. Thirdly, it is facilitated by the behaviour and movement of secondary hosts and non-host carriers.

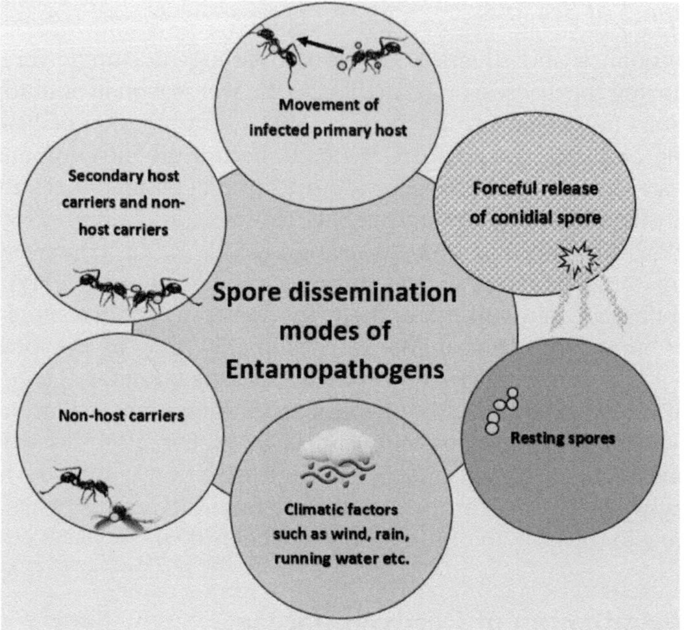

Fig. 1. Mechanism of dissemination of spores of Cordicipitaceae.

Finally, climatic and physical agents such as wind, rain and running water also play a role in the dispersal of insect pathogens. The four mentioned mechanisms collectively contribute to the transmission and distribution of entomopathogens in natural ecosystems (Tanada and Fuxa, 1987).

5. Effect of Environmental Factors

Environmental factors act not discretely but as a complex whole. Therefore, the analysis of a single factor in determining the diversity of fungi is not always successful. This is probably the reason why observations on the importance of environmental factors in judging the pathogenesis and distribution of fungi in natural ecosystems fail. Environmental factors, especially climatic factors, act foremost on the capacity of pathogens in their hosts as well as conidial discharge. These include temperature, relative humidity or moisture, solar radiation, precipitation and wind. These factors, either alone or in combination with some other environmental variables, will influence the disease manifestation in insects by pathogens or their persistence in the environment (Inglis et al., 2000). Average temperature, high relative humidity or moisture and moderate rainfall are some necessary prerequisites for the production and germination of infective spores. Because of this, effective transmission cannot be achieved unless appropriate climatic conditions exist. Entomopathogenic fungi are therefore usually associated with periods of high moisture, often in the form of rain or dew. Solar radiation and temperature extremes are physical factors that inactivate pathogens in all groups. Temperature influences not only the germination of fungal spores and the development of the mycelia but also the speed and quantity of spore production. High temperatures can inactivate fungus before reaching the host insect or may reduce the growth rate within the insect. In contrast, a very low temperature can also reduce or stop germination. So, the influence of these environmental parameters on the survival of pathogens may alter or stop the dissemination of particular diseases within the host population and their distribution within the environment. Wind and rain have similarly been reported to play a major role in the distribution of fungi to susceptible individuals in the local population of terrestrial insects, thereby initiating pathogenesis.

6. Impact of Climate Change

The Western Ghats, one of the largest biodiversity hotspots in India is now facing a serious threat due to the drastic changes in climatic conditions, especially due to the changes in the rainfall pattern and temperature. As per the research conducted by the Energy and Wetland Research Group at the Indian Institute of Science (IISc), Bangalore, which collected rainfall data spanning the last century, the initial patterns revealed are startling (Upadhye, 2016). Recent studies in weather modelling have forecasted an increase in rainfall occurring over a condensed timeframe. The research suggests that the precipitation, which was previously distributed across four months, may now be concentrated within a few days or a week. Ecologists are also expressing concerns regarding temperature fluctuations in the Western Ghats. Even a minor temperature shift in the Ghats has the potential to eradicate a range of both identified and unidentified species. This is similar to the case of entomopathogens. Increasing temperatures might impact the survival and infection of entomopathogens in insects close to the soil surface. With the mean lethal temperature increasing as humidity decreases, the air will become drier. High relative humidity is necessary for the germination of conidia and for the fungal growth and sporulation of insect cadavers (Jaronski, 2010). The impacts of climate change are expected to bring about alterations in precipitation patterns and hydrological processes. These changes would affect the spatial and temporal distribution of runoff, soil moisture and groundwater reserves, leading to an escalation in the occurrence of droughts. The increasing frequency of natural disasters poses a threat to overall biodiversity. Data from the past decade indicates a concerning trend, with simultaneous natural disasters being documented in diverse regions of the Western Ghats. The impact of climate change is equal for all forest types. However, such a catastrophic situation in the Western Ghats leads to serious imbalances in the entire ecosystem.

Conclusion

The Western Ghats, designated as one of the globe's most prominent hotspots of biological diversity, present an intricate tapestry of ecosystems, each offering unique habitats and ecological niches. Within this vibrant mosaic, the Western Ghats harbour an astonishing diversity of entomopathogenic fungi, showcasing the complexity of these microscopic organisms' interactions with their environment. The comprehensive compilation of knowledge about these fungi not only serves as a valuable repository of information but also lays the groundwork for future explorations aimed at unlocking the full potential of these biological wonders. Importantly, the anticipation of a substantial number of undiscovered fungi in tropical regions, including the Western Ghats, is met with a sobering reality of the looming threat of rapid climate change. This environmental upheaval poses a significant risk, potentially eradicating these undiscovered fungi before they can be identified and studied. Thus, the urgency to document and understand the existing fungal diversity becomes not just a scientific endeavour but a race against time to conserve and understand about these species in the face of an increasingly unpredictable climate. The compilation serves not only as a knowledge repository but also as a call to action for sustainable practices and conservation measures to safeguard the fragile ecosystems and their hidden fungal treasures in the Western Ghats.

Acknowledgements

The authors would like to express their gratitude to Prof. K.P. Sudheer, Director of KSCSTE-KFRI, Dr. T.V. Sajeev, Division Programme Coordinator and Registrar. Dr. V. Anitha, Research Coordinator at KSCSTE-KFRI and Mr. Arun V.R., Scientist-B in the Library and Information Division, for the direct and indirect facilitation of the work.

References

Agarwal, G.P., Rajak, R.C., Purnima, K. and Sandhu, S.S. (1985). Studies on entomogenous fungi parasitizing insect pests of teak. J. Trop. For., 1(1): 91–94.

Agarwal, G.P., Rajak, R.C., Sandhu, S.S. and Khan, A. (1988). *Beauveria bassiana*, A potential pathogen of *Atteva fabriciella* Swed. the insect pest of Mahaneem. J. Trop. For., 6: 172–174.

Ali, M.I.M. and Mathew, G. (1989). Occurrence of *Beauveria bassiana* (Balsamo) Vuill. on sapling borer *Sahyadrassus malabaricus* Moore (Lepidoptera: Hepalidae) in Kerala, India. Curr. Sci., 58(16): 931–932.

Ali, M.I.M. and Varma, R.V. (1994). *Beauveria bassiana* new insect pathogen on *Atteva fabriciella* and its comparative efficiency with *Paecilomyces farinosus*. Entomon, 19(3–4): 153–157.

Ali, M.I.M., Varma, R.V. and Sudheendrakumar, V.V. (1991). Evaluation of microbial pathogen for biocontrol against important insect pests of ailanthus and teak. Kerala Forest Research Institute Research Report # 72.

Atlas of Invertebrate Pathogenic Fungi of Thailand. http://www.thai2bio.net/museum/item.php?keyword=*Gibellula*%20pulchra

Aung, O.M., Soytong, K. and Hyde, K.D. (2008). Diversity of entomopathogenic fungi in rainforests of Chiang Mai Province, Thailand. Fungal Divers., 30: 15–22.

Bagyaraj, D.J., Thilagar, G., Ravisha, C., Kushalappa, C.G., Krishnamurthy, K.N. and Vaast, P. (2015). Below ground microbial diversity as influenced by coffee agroforestry systems in the Western Ghats, India. Agric. Ecosyst. Environ., 202: 198–202.

Baral, B. (2017). Entomopathogenicity and biological attributes of Himalayan treasured fungus *Ophiocordyceps sinensis* (Yarsagumba). J. Fungi, 3, 4. https://doi.org/10.3390/jof3010004

Baral, B., Shrestha, B. and Teixeira De Silva, J.A. (2015). A review of Chinese *Cordyceps* with special reference to Nepal, focusing on conservation. Environ. Exper. Biol., 13(2): 61–73.

Bartlett, M.C. and Jaronski, S.T. (1988). Mass production of entomogenous fungi for biological control of insects. pp. 61–85. *In*: Fungi in Biological Control Systems, Burge, M.N. (ed.). Manchester University Press, Manchester, UK.

Baskar, K., Raj, G.A., Mohan, P.M., Lingathurai, S., Ambrose, T. and Muthu, C. (2012). Larvicidal and growth inhibitory activities of entomopathogenic fungus, *Beauveria bassiana* against Asian army worm, *Spodoptera litura* Fab. (Lepidoptera: Noctuidae). J. Entomol., 9(3): 155–162.

Bridge, P.D. and Worland, M.R. (2004). First report of an entomophthoralean fungus on an arthropod host in Antarctica. Polar Biol., 27: 190–192.

Cannon, P.F., Hywel-Jones, N.L., Maczey, N., Norbu, L., Tshitila. et al. (2009). Steps towards sustainable harvest of *Ophiocordyceps sinensis* in Bhutan. Biodivers. Conserv., 18: 2263–2281.

Chen, Y., Yang, B., Li, Z., Yue, Y., tian, Q., Chen, W. et al. (2021). Immune-related genes of *Megalurothrips usitatus* (Bagrall) against *Beauveria brongniartii* and *Akanthomyces attenuatus* identified using RNA Sequencing. Front. Physiol., 2: 671599. https://doi.org/10.3389/fphys.2021.671599

Chhetri, D.R., Chhetri, A., Shahi, N., Tiwari, S., Karna, S.K.L. et al. (2020). *Isaria tenuipes* Peck, an entomopathogenic fungus from Darjeeling Himalaya: Evaluation of *in-vitro* antiproliferative and antioxidant potential of its mycelium extract. BMC Compl. Med. Ther., 20: 1–14.

Dattaraj, H.R., Jagadish, B.R. Sridhar K.R. and Ghate S.D. (2018). Are the scrub jungles of Southwest India potential habitats of *Cordyceps*? KAVAKA, 51: 20–22.

Elienberg, J. (200)2. Biology of fungi from the order of Entomophthorales. PhD Dissertation, Royal Veterinary and Agricultural University, Demark.

Evans, H.C. (1988). Co-evolution of Fungi with Plants and Animals. pp. 149–171. *In*: Pirozynski K.A. and Hawksworth D. L. (eds.). Academic Press, London.

Evans, H.C. and Samson, R.A. (1982). *Cordyceps* species and their anamorphs pathogenic on ants (Formicidae) in tropical forest ecosystems I. The Cephalotes (Myrmicinae) complex. Trans. Br. Mycol. Soc., 79(3): 431–453.

Evans, H.C. and Samson, R.A. (1984). *Cordyceps* species and their anamorphs pathogenic on ants (Formicidae) in tropical forest ecosystems II. The Camponotus (Formicinae) complex. Trans. Br. Mycol. Soc., 82(1): 127–150.

Fargues, J, Goettel, M.S., Smits, N., Ouedraogo, A. and Rougier, M. (1997). Effect of temperature on vegetative growth of *Beauveria bassiana* isolates from different origins. Mycologia, 89(3): 383–392.

Fargues, J. and Luz, C. (2000). Effects of fluctuating moisture and temperature regimes on sporulation of *Beauveria bassiana* on cadavers of *Rhodnius prolixus*. Biocont. Sci. Technol., 8(3): 323–334.

Ferron, P. (1997). Influence of relative humidity on the development of fungal infection caused by *Beauveria bassiana* (Fungi Imperfecti: Moniliales) in imagines of *Acanthoscelides obtectus* (Coleoptera: Bruchidae). Entomophaga, 22: 393–396.

Goettel, M.S. and Hajek, A.E. (2001). Evaluation of non-target effects of pathogens used for management of arthropods. pp. 81–97. *In*: Wajnberg, E., Scott, J.K. and Quimby, P.C. (eds.). Evaluating indirect ecological effects of biological control. Key papers from the symposium ‹Indirect ecological effects in biological control, Montpellier, France, CABI Publishing, Wallingford UK.

Götz, P. (1991). Invertebrate immune response to fungal cell wall components. pp. 317–329. *In*: Fungal Cell Wall and Immune Response. Springer Berlin and Heidelberg.

Guo, C.T., Han, P., Tong, S.M., Ying, S.H. and Feng, M.G. (2021). Distinctive role of *fluG* in the adaptation of *Beauveria bassiana* to insect-pathogenic lifecycle and environmental stresses. Environ. Microbiol., 23(9): 5184–5199.

Gurulingappa, P., Gee, P.M. and Sword, G.A. (2011). In vitro and in planta compatibility of insecticides and the endophytic entomopathogen, *Lecanicillium lecanii*. Mycopathol., 172: 161–168.

Hajek, A.E. (1997). Ecology of terrestrial fungal entomopathogens. pp. 193–249. In: Jones, J.G. (ed.). Advances in Microbial Ecology. Plenum Press, New York.

Harish E.R. and Krishnan, J.U. (2023). Cost-effective media for mass production of *Bacillus thuringiensis* Berliner for the management of taro caterpillar, *Spodoptera litura* Fabricius; J. Entomol. Res., 47(1): 8–15.

Helyer, N., Gill, G., Bywater, A. and Chambers. (1992). Elevated humidities for control of chrysanthemum pests with *Verticillium lecanii*. Pest. Sci., 36(4): 373–378.

Hodge, K.T. (2003). Clavicipitaceous anamorphs. Mycol. Ser., 19: 75–124.

Hoffman, W.E. (1947). Insects as human food. Proceedings of Entomological Society of Washington, 49: 233–237.

Hsieh, L.S., Tzean, S.S. and Wu, W.J. (1997). The genus *Akanthomyces* on spiders from Taiwan. Mycologia, 89(2): 319–324.

Inglis, G.D., Ivie, T.J., Duke, G.M. and Goettel, M.S. (2000). Influence of formulations on rain fastness of *Beauveria bassiana* conidia on potato leaves and Colorado potato beetle larvae. Biol. Cont., 18: 55–64.

Jaronski, S.T. (2010). Ecological factors in the Inundative use of fungal entomopathogen. Biocontrol, 55(1): 159–185. https://doi.org/10.1007/s10526-009-9248-3.

Javaregowda and Naik, K.L. (2006). Influence of white muscardine fungus, *Beauveria bassiana*, (Bals.) Vuill. on teak defoliator, *Hyblaea puera* (Cramer). Insect Environ., 12(1): 3–4.

Johnson, D., Sung, G-H., Hywel-Jones, Luangsa-Srd, J.J., Bischoff, JF. et al. (2009). Systematics and evolution of the genus *Torrubiella* (Hypocreales, Ascomycota). Mycol. Res., 113(3): 279–289.

Juliya, F.R. (2007). Fungal Pathogens Associated with Forest Insects in the Kerala Part of the Western Ghats. PhD Dissertation, Kerala Forest Research Institute, Peechi, Kerala, India.

Kandaswamy, D. (1969). *Hypsiply robusta* Moore, a new host for *Beauveria tenella* (Delacroix) Siemasko. J. Invert. Pathol.,13: 149–150.

Keller, S. (2000). Use of *Beauveria brongniartii* in Switzerland and its acceptance by farmers. IOBC WPRS Bulletin, 23: 67–71.

Kepler, R.M., Sung, G.H., Ban, S., Nakagiri, A., Chen, M. J. et al. (2012). New teleomorph combinations in the entomopathogenic genus *Metacordyceps*. Mycologia, 104(1): 182–197. https://doi.org/10.3852/11-070

Klingen, I., Eilenberg, J. and Meadow, R. (2002). Effects of farming system, field margins and bait insect on the occurrence of insect pathogenic fungi in soils. Agric. Ecosys. Environ., 91(1–3): 191–198. https://doi.org/10.1016/S0167-8809 (01)00227-4.

Kuephadungphan, W., Macabeo, A.P.G., Luangsa-ard, J.J., Tasanathai, K., Thanakitpipattana, D. et al. (2019). Studies on the biologically active secondary metabolites of the new spider parasitic fungus *Gibellula gamsii*. Mycol. Prog., 18(1–2): 135–146. https://doi.org/10.1007/s11557-018-1431-4.

Kumar, T.S. and Aparna, N.S. (2014). *Cordyceps* species as a bio-control agent against coconut root grub, *Leucopholis coneophora* Burm. J. Environ. Res. Develop., 8: 614–618.

Lacey, L.A. and Goettel, M.S. (1995). Current developments in microbial control of insect pests and prospects for the early 21st century. Entomophaga, 40: 3–7.

Mains, E.B. (1950). Entomogenous species of *Akanthomyces, Hymenostilbe* and *Insecticola* in North America. Mycologia, 42(4): 566–589.

Malarvannan, S., Murali, P.D., Shanthakumar, S.P., Prabavathy, V.R. and Nair, S. (2010). Laboratory evaluation of the entomopathogenic fungi, *Beauveria bassiana* against the Tobacco caterpillar, *Spodoptera litura* Fabricius (Noctuidae: Lepidoptera). J. Biopest., 3(1): 126–131.

Mendes-Pereira, T., De Araújo, J.P.M., Kloss, T.G., Costa-Rezende, D.H., De Carvalho, D.S. and Góes-Neto, A. (2023). Disentangling the Taxonomy, Systematics and Life History of the Spider-Parasitic Fungus *Gibellula* (Cordycipitaceae, Hypocreales). J. Fungi, 9(4): 457. https://doi.org/10.3390/jof9040457

Milner, R.J. and Lutton, G.G. (1986). Dependence of *Verticillium lecanii* (Fungi: Hyphomycetes) on high humidities for infection and sporulation using *Myzus persicae* (Homoptera: Aphididae) as host. Environ. Entomol., 15(2): 380–382.

Mongkolsamrit, S., Noisripoom, W., Thanakitpipattana, D., Wutikhun, T., Spatofora, J.W. and Luangsa-Ard, J. (2018). Disentangling cryptic species with *Isaria*-like morphs in Cordycipitaceae. Mycologia, 110: 230–257. https://doi.org/10.1080/00275514.2018.1446651.

Nag Raj, T.R. (1962). Addition to the Indian species of *cordyceps*. Current Science, 7: 301–302.

Nanaware, S. D. 2002. Taxonomical Studies in the Fungi from Western Ghats of Maharashtra. PhD Dissertation., Shivaji Univ, Kolhapur, Maharashtra, India.

Negi, C.S., Joshi, P. and Bohra, S. (2015). Rapid vulnerability assessment of Yartsa Gunbu *Ophiocordyceps sinensis* [Berk.] in Pithoragarh District, Uttarakhand India. Mountain Res. Develop., 35: 382–391.

Patil, A., Dangat, B.T. and Patil, M.S. (2014). *Cordyceps nutans* Pat., A new record to India. Sci. Res. Rep., 4: 64–66.

Prathibha, P.S. (2015). Behavioral studies on palm white grubs *Leucopholis* spp. (Coleoptera: scarabaeidae) and Evolution of new insecticides for their management. PhD Dissertation, University of Agriculture Sciences Bangalore, India.

Qin, Y., Ying, S.H., Chen, Y., Shen, Z.C. and Feng, M.G. (2010). Integration of insecticidal protein Vip3Aa1 into *Beauveria bassiana* enhances fungal virulence to *Spodoptera litura* larvae by cuticle and PerOs infection. Appl. Environ. Microbiol., 76: 4611–4618.

Quan, Q.-M., Chen, L.-L., Wang, X., Li, S., Yang, X.-L. et al. (2014). Genetic diversity and distribution patterns of host insects of caterpillar fungus *Ophiocordyceps sinensis* in the Qinghai-Tibet Plateau. PLoS One, 9(3): e92293.

Rajak, R.C., Agarwal, G.P., Khan, A.R. and Sadhu, S.S. (1993). Susceptibility of teak defoliator (Hyblaea puera Cramer) and teak skeletonizer (*Eutectona machaeralis* Walker) to *Beauveria bassiana* (Bals.) Vuill. Ind. J. Exp. Biol., 31: 80–82.

Roy, H.E., Steinkraus, D.C., Eilenberg, J., Hajek, A.E. and Pell, J. K. (2006). Bizarre interactions and endgames: entomopathogenic fungi and their arthropod hosts. Ann. Rev. Entomol., 51: 331–357.

Samson, R.A. and Evans, H.C. (1974). Notes on entomogenous fungi from Ghana: II. The genus *Akanthomyces*. Acta Bot. Neerl., 23(1): 28–35. https://doi.org/10.1111/j.1438-8677.1974.tb00913.x

Samson, R.A. and Evans, H.C. (1992). New species of *Gibellula* on spiders (Araneida) from South America. Mycologia, 84(3): 300–314.

Samson, R.A., Evans, H.C., Latgé, J.P., Samson, R.A., Evans, H.C. and Latgé, J.P. (1988). Taxonomy of entomopathogenic fungi. *Atlas of entomopathogenic fungi*, 5–16.

Sandhu, S.S., Rajak, R.C. and Agarwal, G.P. (1993). Microbial control agents of forest pest at Jabalpur. Ann. For. Sci., 1(2): 136–140.

Sankaran, K.V., Mohanadas, K. and Ali, M.I.M. (1989). *Beauveria bassiana* (Bals.) Vuill., a possible biocontrol agent against *Myllocerus viridanus* Fabr. and *Calopepla leayana* Latreille in south India. Curr. Sci., 58(8): 467–469.

Sasidharan, K.R. (2004). Studies on the Insect Pest of Casuarina equisetifolia L. in Tamil Nadu and their Management. PhD Dissertation, Forest Research Institute University, Dehra Dun, Uttaranchal, India.

Sharma, N. and Joshi, K.C. (2004). Entomopathogenic fungi attacking Sal heartwood borer, *Hoplocerambyx spinicornis* Newman. Indi. J. For., 27: 133–140.

Shrestha, U.B. and Bawa, K.S. (2013). Trade, harvest and conservation of caterpillar fungus (*Ophiocordyceps sinensis*) in the Himalayas. Biol. Conser., 159: 514–520.

Šimonovičová, A., Gódyová, M. and Kunert, J. (2004). *Engyodontium album*, a new species of microscopic fungi for Slovakia and its keratinolytic activity. Biologia, 59(1): 17–18.

Smirnoff, W.A. 1968. Adaptation of the microsporidian *Thelohania pristiphorae* to the tent caterpillars *Malacosoma disstria* and *Malacosoma americanum*. J. Invert. Pathol., 11(2): 321–325. https://doi.org/10.1016/0022-2011(68)90166-3.

Spatafora, J.W., Sung G-H. and Kepler, R. (2011). *Cordyceps* Working Group. An electronic monograph of *Cordyceps* and related fungi. https://www.yumpu.com/user/cordyceps.us

Stensrud, O., Hywel-Jones, N.L. and Schumacher, T. (2005). Towards a phylogenetic classification of *Cordyceps*: ITS nrDNA sequence data confirm divergent lineages and paraphyly. Mycol. Res., 109(1): 41–56. https://doi.org/10.1017/S095375620400139X

Subramaniam, M.S.R., Babu, A. and Deka, B. (2021). *Lecanicillium lecanii* (Zimmermann) Zare & Gams, as an efficient biocontrol agent of tea thrips, *Scirtothrips bispinosus* Bagnall (Thysanoptera: Thripidae). Egypt. J. Biol. Pest Con., 31(1): https://doi.org/10.1186/s41938-021-00380-y

Sullivan, R.F., Bills, G.F., Hywel-Jones, N.L. and White, J.F. 2000. *Hyperdermium*: a new clavicipitalean genus for some tropical epibionts of dicotyledonous plants. Mycologia, 92(2): 908–918. https://doi.org/10.1080/00275514.2000.1206123

Sung, G-H., Hywel-Jones, N.L., Sung, J.-M., Luangsa-ard, J.J., Shrestha, B. and Spatafora, J.W. 2007. Phylogenetic classification of *Cordyceps* and the clavicipitaceous fungi. Stud. Mycol., 57: 5–59. https://doi.org/doi:10.3114/sim.2007.57.01

Tanada, Y. and Fuxa, J.R. (1987). Epizootiology of Insect Diseases, pp. 166–171. A Wiley-Interscience publication, Canada.

Thanakitpipattana, D., Mongkolsamrit, S., Khonsanit, A., Himaman, W., Luangsa-Ard, J.J. and Pornputtapong, N. (2022). Is Hyperdermium Congeneric with *Ascopolyporus*? Phylogenetic relationships of *Ascopolyporus* spp. (Cordycipitaceae, Hypocreales) and a new genus *Neohyperdermium* on scale insects in Thailand. J. Fungi, 8(8): 516. https://doi.org/10.3390/jof8050516

Tosi, S., Caretta, G. and Humber, R.A. (2004). *Conidiobolus antarticus*, a new species from continental Antarctica. Mycotaxon, 90: 343–347.

Upadhye, A. (2016). Varying rainfall threatens biodiversity in Western Ghats. Indian climate dialogue. https://indiaclimatedialogue.net/2016/03/21/varying-rainfall-threatens-biodiversity-western-ghats/

Van der Putten, W.H., Vet, L.E.M., Harvey, J.A. and Wäckers, F.L. (2001). Linking above- and belowground multitrophic interactions of plants, herbivores, pathogens and their antagonists. Tr. Ecol. Evol., 16(10): 547–554. https://doi.org/10.1016/s0169-5347(01)02265-0

Varma, R.V., Sudhhendrakumar, V.V. and Sankaran, K.V. (2006). Microbial pathogens associated with forest insects in the Kerala part of the Western Ghats with respect to host-parasite relationship and ex-situ conservation. Kerala Forest Research Institute Research Report # 297. ISS0970-8103.

Vega, F.E. and Kaya, H.K. (2011). Insect Pathology. Academic Press, p. 21693.

Venkatesan, S., Ganesan, R. and Muthuchelian, K. (2009). Macrofungal diversity in Megamalai forest, Western ghats, Tamil Nadu, India. J. Pure Appl. Microbiol., 3(1): 319–324.

Wang, L., Huang, J., You, M., Guan, X. and Liu, B. (2007). Toxicity and feeding deterrence of crude toxin extracts of *Lecanicillium* (*Verticillium*) lecanii (Hyphomycetes) against sweet potato whitefly, *Bemisia tabaci* (Homoptera: Aleyrodidae). Pest Manag. Sci., 63(4): 381–387.

Wargane, V.S., Parate, S.R., Bramhankar, S.B., Rakhonde, P.N., Sonune, B.D. et al. (2020). Cultural and morphological characterizations of *Beauveria bassiana*. J. Pharmacog. Phytochem., 9(1): 591–594.

Wijayawardene, N.N., Hyde, K.D., Al-Ani, L.K.T., Tedersoo, L., Haelewaters, D. et al. (2020). Outline of Fungi and fungus-like taxa. Mycosphere. 11(1): 1060–1456.

Winkler, D. (2008). Present and historic relevance of Yartsa Gunbu (*Cordyceps sinensis*). An ancient myco-medicinal in Tibet. Fungi, 1(4): 6–7.

Wraight, S.P., Jackson, M.A. and De Kock, S.L. (2001). Production, Stabilization and Formulation of Fungal Biocontrol Agents. pp. 253–287. *In*: Butt, T.M., Jackson, C. and Magn, N. (eds.). Fungi as Biocontrol Agents, CABI, UK. https://doi.org/10.1079/9780851993560.0253

Wright, P.J. and Patel, V.S. (1992). Morphology and growth in culture of Microhilum oncoperae, the cause of a mycosis of *Oncopera* spp. larvae. Mycol. Res., 96(7): 578–582.

Zare, R. and Gams, W. (2001). The genera *Lecanicillium* and *Simplicillium* gen. nov. Nova Hedw., 73: 1–50.

Bioactive Potential

6

Traditional Knowledge and Pharmacological Potential of *Cordyceps*

*Nadire Özenver** and *Çiğdem Kahraman**

1. Introduction

Cordyceps, a well-known medication administered in China, is an insect parasitic fungus that mainly lives on the head of the larva of *Hepialus armoricanus* Oberthur (Lepidoptera) (Yue et al., 2012). A group of ascomycetous fungi forming a hollow, either as symbionts of the genus *Elaphomyceas* or as endoparasites of arthropods is collectively referred to as 'Cordyceps' (Dong et al., 2015).

The fungus is named 'Dong Chong Xia Cao' in China., This name predominantly makes mentions *Cordyceps sinensis* (Berk.) Sacc. The expression 'cordyceps' (without italics and lowercase) mentions any cordyceps fungus and host complex, while the name 'cordyceps fungus' points to only the fungus that constitutes cordyceps (Zhang et al., 2012). The genus *cordyceps* was introduced to Western civilization in the 17th century. The Italian academic Saccardo termed *Cordyceps* originating from China formally as *Cordyceps sinensis* (Berk.) Sacc. in 1878, which has received broad acceptance to date (Yue et al., 2012). 597 species have been documented as *Cordyceps* until now (http://www.indexfungorum.org/names/Names.asp; retrieved on October 31, 2023).

Hacettepe University, Faculty of Pharmacy, Department of Pharmacognosy, 06100, Ankara, Türkiye.

* Corresponding authors: nadire@hacettepe.edu.tr; cigdemm@hacettepe.edu.tr

The Chinese *Cordyceps* and relevant fungi constitute the largest fungal industry globally and in China. A variety of medicinal and health products, including these fungi, have been improved and broadly commercialized as Chinese *Cordyceps*. The price of Chinese *Cordyceps* has been increasing exponentially due to its medical value, high market demand, limited resources, and inefficient cultivation (Dong et al., 2015). Four genera, *Cordyceps* sensu stricto (Cordycipitaceae), *Metacordyceps* (Clavicipitaceae), *Elaphocordyceps* (Ophiocordycipitaceae), and *Ophiocordyceps* (Ophiocordycipitaceae), establish the *Cordyceps* fungi, each possessing diverse botanical characteristics (Sung et al., 2007; Dong et al., 2015).

An undeniable reality is that the genus *Cordyceps* is gaining growing attention in the areas of both medicine and mycology. Therefore, we aim to provide an overall overview of the genus *Cordyceps*, with a particular focus on its traditional knowledge, phytochemistry, and pharmacological potential based on current scientific knowledge. In this context, we believe that our chapter will make a crucial contribution to the better understanding of the genus Cordyceps, potentially leading to the prevention and/or management of several diseases in the future. In this context, we believe that our chapter will make a crucial contribution to the better understanding of the genus *Cordyceps*, potentially leading to the preclusion and/or management of several diseases in the future.

2. Traditional Knowledge

Cordyceps has been consumed as a stimulant and health supplement in Asian nations (Yue et al., 2012). Its reputation continues to grow and holds more eminence in the public and scientific communities at present. The use of *Cordyceps* for medicinal purposes was formally recorded in the Qing dynasty (Wang, 1955; Zhao, 1765; Dong et al., 2015) to treat kidney and lung disturbances. In 1964, it was officially documented as a drug in the Chinese Pharmacopeia (Dong et al., 2015). The wild or cultivated *Cordyceps* species are grown and harvested for their ascocarp and mycelium in traditional medicine. They are well known and consumed for several purposes, specifically as a diet component or medicinal product. Though a huge number of *Cordyceps* taxa were described, only a small part of them were documented for use in folk medicine. *C. sinensis* and *C. militaris* are among the most commonly used species (Dong et al., 2015).

Cordyceps species were traditionally used as a tonic for longevity, imparting life-force and resilience. They were also employed as therapeutic remedies for conditions such as asthma, chronic bronchitis, tuberculosis, and various skin disorders, including eczema and other dermatological ailments. (Wu et al., 2011). There are many uses of *Cordyceps* in traditional medicine in a number of Asian countries, for example, *C. bassiana* for skin disorders; *C. cicadae* for fever, convulsion, and dizziness; *C. guangdongensis* for fatigue; *C. japonica* for endurance and cancer; *C. jiangxiensis* for snake bite; *C. liangshanensis* for asthma, chronic cough, lumbago and impotence; *C. militaris* for respiratory diseases and as an aphrodisiac and renewing agent; *C. ophioglossoides* as a tonic; *C. pruinosa* for stomach and inflammatory diseases; *C. sinensis* for all diseases as a tonic (Kuo et al., 2002; Kim et al., 2003; Panda and Swain, 2011; Yan et al., 2013a, Olatunji et al., 2018; Wang et al., 2021). The above mentioned Cordyceps spp. have been widely known and used in folk medicine applications in many civilizations. *Cordyceps* and associated products are gaining increased interest in Western countries and other parts of the world, driven by the confirmed biological activities and therapeutic value attributed to them.

3. Diversity of Metabolites

3.1 Nucleosides

Cordycepin (3'-deoxyadenosine), was initially identified in the liquid cultures of *C. militaris* (Cunningham et al., 1950). It stands out as the primary active component and the most abundant nucleoside in various *Cordyceps* sp. Cordycepin has demonstrated diverse biological activities, including antitumoral effects (Yoon et al., 2018), anti-inflammatory properties (Tan et al., 2020), neuroprotective potential (Sun et al., 2020; Zhang et al., 2021a), antioxidant capabilities, insecticidal properties, antimicrobial activity (Qin et al., 2019), and anti-platelet aggregation (Tuli et al., 2013) (Table 1).

Cordyrrole B demonstrated effective inhibition of pancreatic lipase activity and adipocyte differentiation when measured at 100 µM (Kim et al., 2014). Pancreatic lipase activity and cell proliferation tests were conducted on chanhuanoside A-C, new nucleosides isolated from *C. chanhua*. They exhibited inhibitory activity on pancreatic lipase with IC_{50} values of 49.9±6.3, 61.4±7.2, and 62.8±7.1, respectively.

All of them were found to promote the proliferation of RAW264.7 and 293T cell lines in a dose-dependent manner (Dong et al., 2024). Natural nucleosides, N_6-4-methylbutyrat-adenosine and 3'-deoxy-6-*O*-methylinosine, isolated from *C. militaris*, did not exhibit anti-inflammatory activity via LPS-induced NO generation in RAW264.7 cells at a concentration of 100 µM (Xue et al., 2020). 5'-(3''-deoxy-β-D-ribofuranosyl)-3'-deoxyadenosine, along with a known compound, 3'-deoxyadenosine, were isolated from *C. militaris*. 3'-Deoxyadenosine inhibited the NF-κB reporter gene expression induced by TNF-α in HeLa cells in the concentration range of 3-100 µM (Sun et al., 2017).

Table 1. Nucleosides isolated from *Cordyceps* species and mode of action.

Nucleoside	Species	Tested mode of action
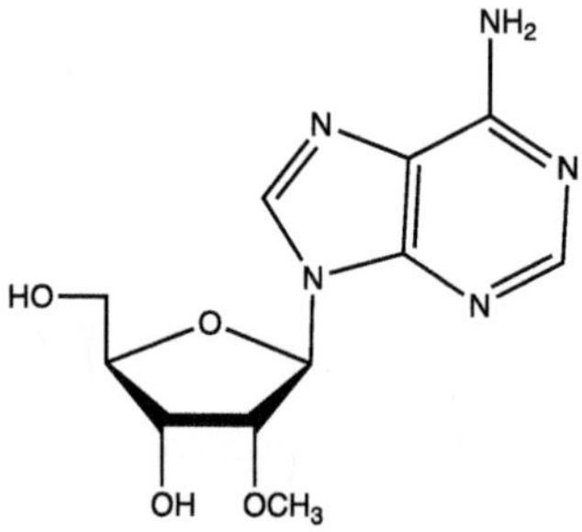 Cordysinin B	*C. sinensis*	-
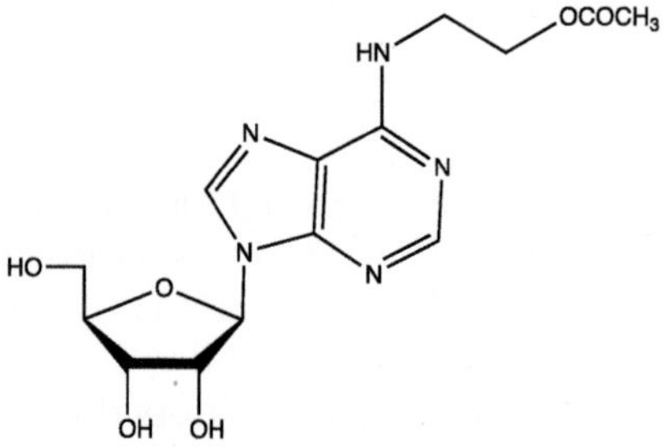 Cordyrrole B	*C. militaris*	Metabolic effects
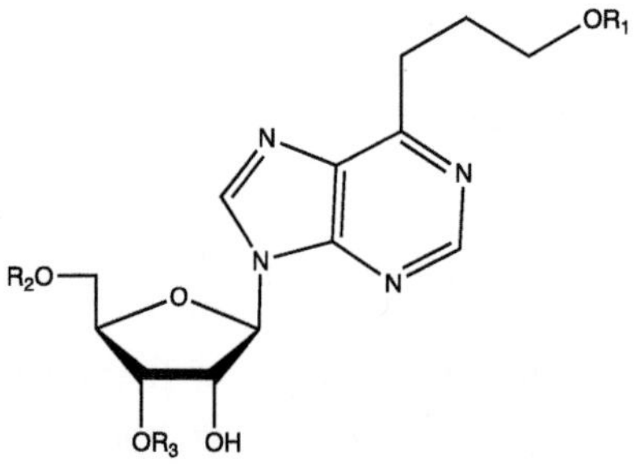 Chanhuanoside A; $R_1 = R_3 = H$, $R_2 = COCH_3$ Chanhuanoside B; $R_1 = R_2 = COCH_3$, $R_3 = H$ Chanhuanoside C; $R_1 = R_3 = COCH_3$, $R_2 = H$	*C. chanhua*	Cell proliferation and metabolic effect

Table 1. contd. ...

Table 1. contd.

Nucleoside	Species	Tested mode of action
N₆-4-methylbutyrat-adenosine	*C. militaris*	Anti-inflammatory activity
3'-deoxy-6-O-methylinosine	*C. militaris*	Anti-inflammatory activity
5'-(3''-deoxy-β-D-ribofuranosyl)-3'-deoxyadenosine	*C. militaris*	Anti-inflammatory activity

3.2 Alkaloids

Cordylactam, a new lactam-fused 4-pyrone, was isolated from *Cordyceps* sp. BCC 12671 and was not evaluated for its biological activity (Isaka et al., 2013) (Table 2). Cordyformamide, isolated from culture broth of *C. brunnearubra,* showed antimalarial activity against *Plasmodium falciparum* K1, a multidrug-resistant strain, with an IC_{50} value of 18 µM (Isaka et al., 2007a). Aurantiamides, cordyceamide A and B, were purified from *C. sinensis* and reported to have cytotoxic activities against L929, A375, and Hela cells (Jia et al., 2009). Cordysinin A, one of the active principles of *C. sinensis*, was demonstrated to have anti-inflammatory activity, inhibiting the generation

Table 2. Alkaloids isolated from *Cordyceps* species and mode of action.

Alkaloids	Species	Tested mode of action
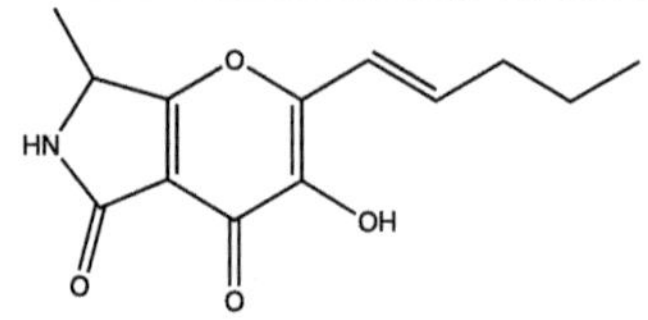 Cordylactam	*Cordyceps* sp.	-
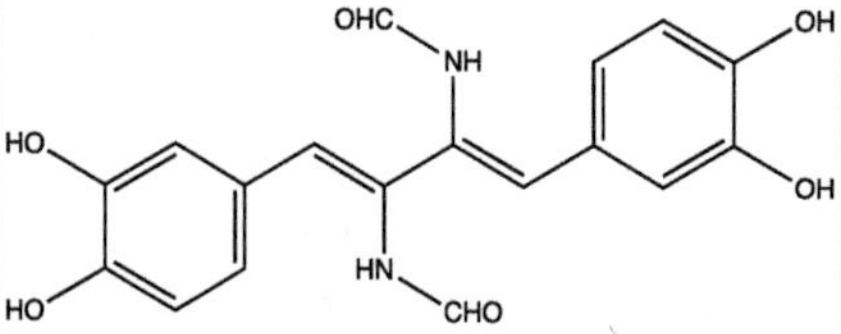 Cordyformamide	*C. brunnearubra*	Antimalarial activity
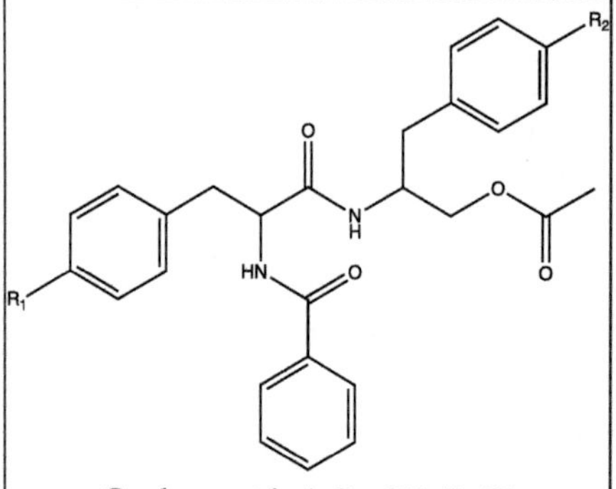 Cordyceamide A; R_1=OH, R_2=H Cordyceamide B; R_1=OH, R_2=OH	*C. sinensis*	Cytotoxicity
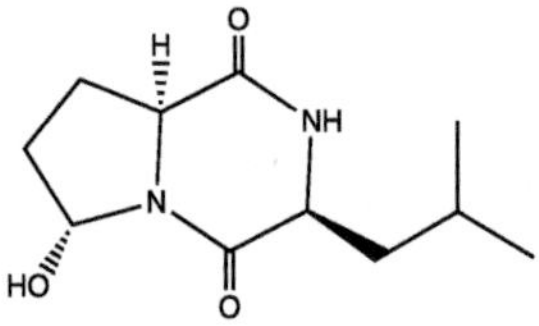 Cordysinin A	*C. sinensis*	Anti-inflammatory activity
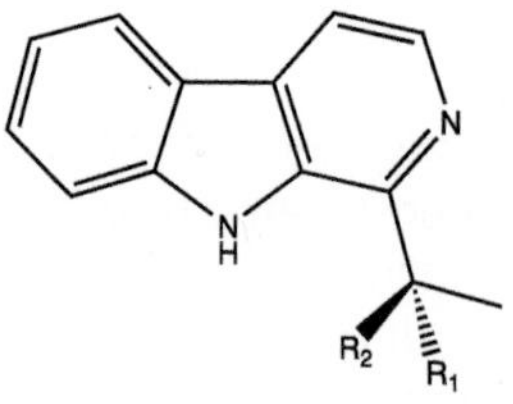 Cordysinin C; R_1=H, R_2=OH Cordysinin D; R_1=OH, R_2=H	*C. sinensis*	Antioxidant activity

Table 2. contd. ...

Table 2. contd.

Alkaloids	Species	Tested mode of action
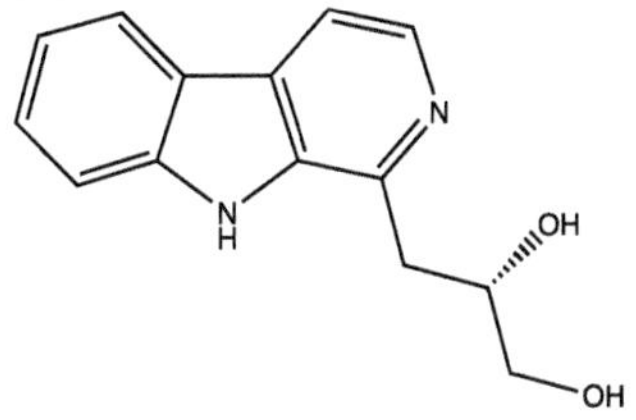 Cordysinin E	*C. sinensis*	Antioxidant activity
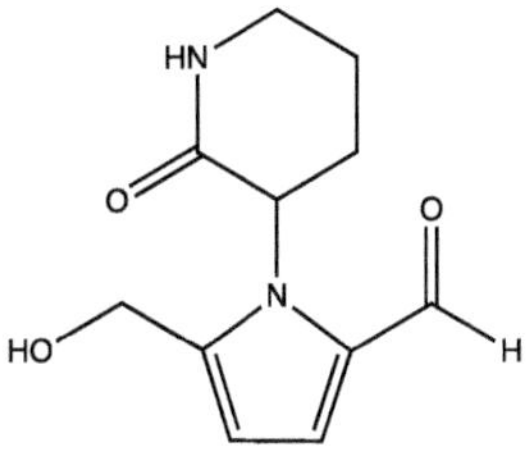 Cordyrrole A	*C. militaris*	Metabolic effects
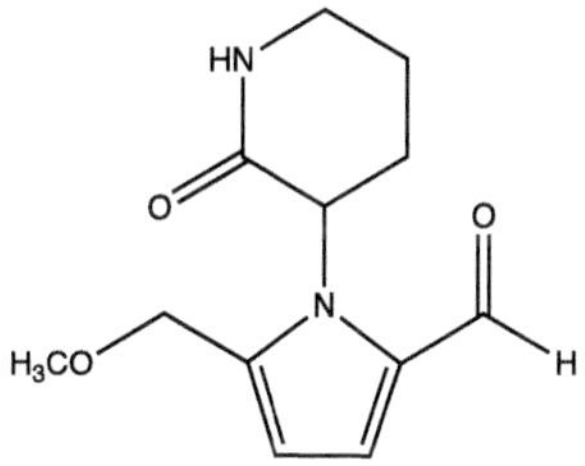 2-carboxaldehyde-5-(methoxymethyl)-1-(2-oxo-3-piperidinyl)-1H pyrrole	*C. militaris*	Anti-inflammatory activity
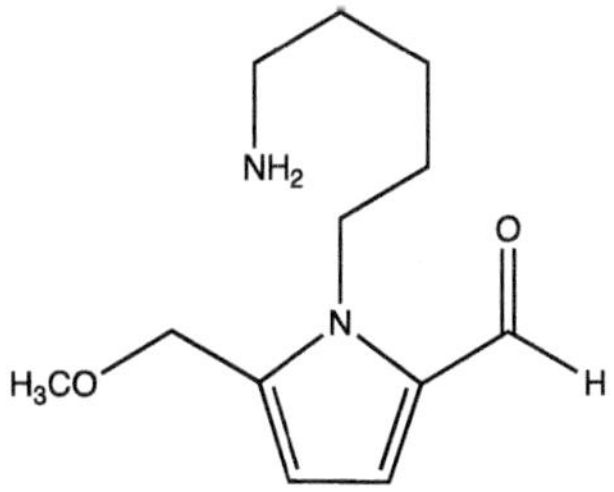 2-carboxaldehyde-1-(4-aminobutyl)-5-(methoxymethyl)-1H-pyrrole	*C. militaris*	Anti-inflammatory activity

…Table 2. contd.

Table 2. contd.

Alkaloids	Species	Tested mode of action
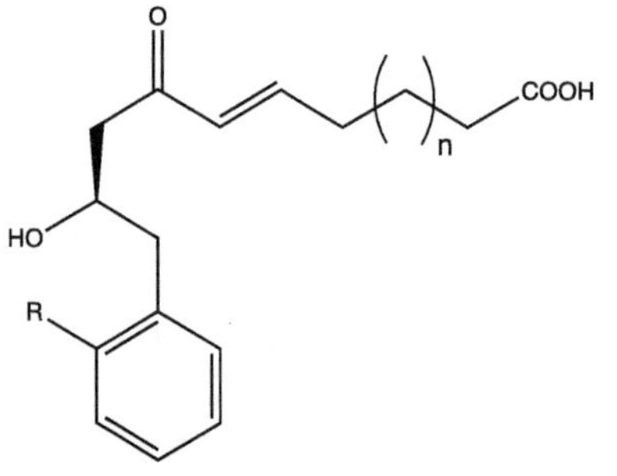 Cordytakaoamide A; R= H, n=2 Cordytakaoamide B; R= OH, n=4	*C. takaomontana*	-
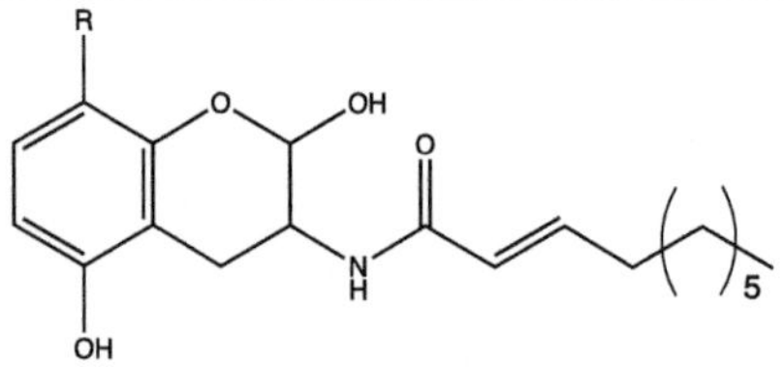 Cordycepamide A; R=OH Cordycepamide B; R=H	*Cordyceps* sp.	Antimicrobial activity, antioxidant activity and cytotoxicity
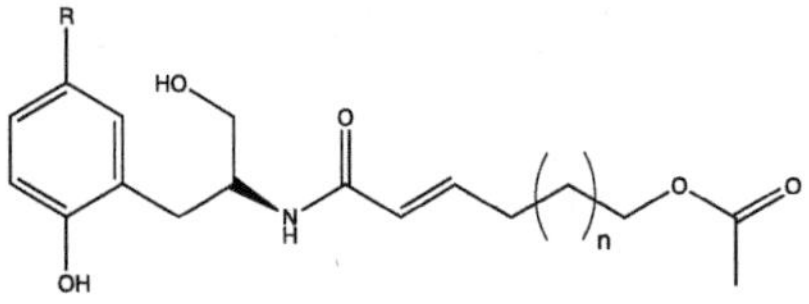 Cordycepamide C; R=OH, n=3 Cordycepamide D; R=H, n=5	*Cordyceps* sp.	Antimicrobial activity, antioxidant activity and cytotoxicity

of superoxide anion and elastase release by human neutrophils in response to N-formyl-methionyl-leucyl-phenylalanine/cytochalasin B with IC_{50} values of 11.34±4.95 and 13.0±23.20 µM, respectively.

Other molecules identified, a nucleoside cordysinin B, and the alkaloids cordysinin C-E, were not tested. Besides, cordysinin C-E did not display antioxidant activity at 500 µM (Table 2) (Yang et al., 2011). Pyrrole alkaloid derivative, Cordyrrole A, showed an effective inhibition on pancreatic lipase activity and adipocyte differentiation measured at 100 µM (Kim et al., 2014). Two pyrrole alkaloid derivatives, 2-carboxaldehyde-5-(methoxymethyl)-1-(2-oxo-3-piperidinyl)-1H-pyrrole and 2-carboxaldehyde-1-(4-aminobutyl)-5-(methoxymethyl)-1H-pyrrole, isolated from *C. militaris*, did not display anti-inflammatory activity via LPS-induced NO production in RAW264.7 cells at a concentration of 100 µM (Xue et al., 2020).

New alkaloids, cordytakaoamide A and B, were isolated from ethyl acetate and *n*-butanol soluble fractions of *C. takaomontana* NBRC 101754, respectively (Hama et al., 2019). Cordycepamides A-D were evaluated for their antimicrobial activities against *Staphylococcus aureus*, *Bacillus subtilis*, and *Escherichia coli*, cytotoxic activities against MCF-7, HepG-2, HT29, and 293T cell lines, and 2,2-diphenyl-1-picrylhydrazyl (DPPH·) scavenging activities. Only cordycepamide D showed scavenging activity against DPPH with an IC_{50} of 51.42 ± 3.08 µM (Fan et al., 2020).

3.3 Pyridones

The *N*-hydroxy-(cordypyridones A and B) and *N*-methoxy-2-pyridone (cordypyridones C and D) derivatives, isolated from *C. nipponica*, were tested for their antimalarial activities (Table 3). Cordypyridone A and B, which were atropisomers of each other, showed higher inhibitory activity against *P. falciparum* (K1, a multidrug-resistant strain) with IC_{50} values of 0.066 and 0.037 µg/ml, respectively, compared to those of Cordypyridone C and D, which were >20 and 7.8 µg/ml (Isaka et al., 2001).

3.4 Cyclopeptides

Cyclopeptides are important metabolites of *Cordyceps* spp. (Table 4). Cycloheptapeptides, cordyheptapeptide A and its minor analogue, cordyheptapeptide B, were isolated from the mycelia extract of the *Cordyceps* sp. BCC 16176 strain and tested for their antimalarial and cytotoxic activities. Cordyheptapeptide A exhibited inhibition against *P. falciparum* K1 with an IC_{50} value of 3.8 µM, unlike cordyheptapeptide B, which was found inactive at a concentration of 10 µM. Cytotoxic studies were conducted with KB, BC, NCI-I118, and Vero cells, and the IC_{50} values of cordyheptapeptide A were 0.78, 0.20, 0.18, and 14 µM against corresponding cells, respectively, while those of cordyheptapeptide B were 2.0, 0.66, 3.1, and 1.6 µM (Isaka et al., 2007b).

3.5 Cyclodepsipeptides

Cyclodepsipeptides, cordycenin A, beauvericin, beauvericin A, beauvericin J, and beauvericin E were subjected to cytotoxic studies, and beauvericin, -A, and -J displayed inhibitory effects against HepG2 and HepG2/ADM with IC_{50} values in the range of 2.82–5.04 µM and 2.40–2.93 µM, respectively (Table 5). A newly isolated cordycenin A from *C. cicadae* was determined to have no cytotoxic effects (Wang et al., 2014). Cardinalisamides A–C, isolated from *C. cardinalis* and tested against *Trypanosoma brucei brucei*, exhibited antitrypanosomal activity with IC_{50} values of 8.56, 8.65, and 8.63 µg/ml. However, they also exhibited cytotoxicity against MRC-5 cells, with IC_{50} values of 18.48, 14.00, and 23.84 µg/ml, respectively

Table 3. Pyridones isolated from *Cordyceps* species and mode of action.

Pyridones	Species	Tested mode of action
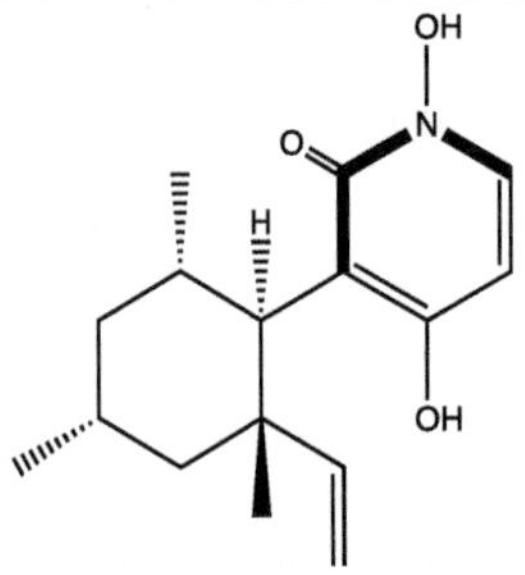 Cordypyridone A	*C. nipponica*	Antimalarial activity
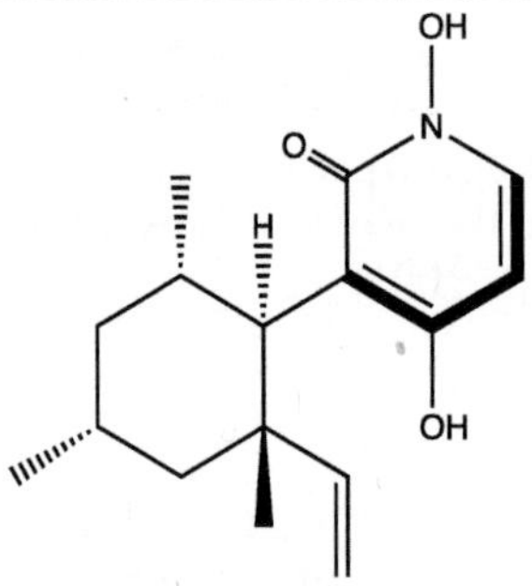 Cordypyridone B	*C. nipponica*	Antimalarial activity
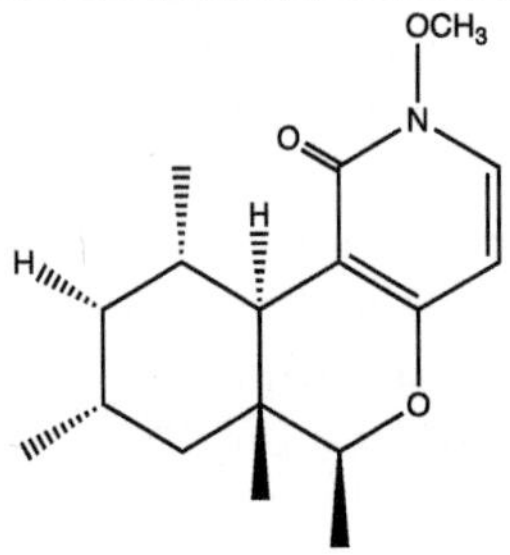 Cordypyridone C	*C. nipponica*	Antimalarial activity
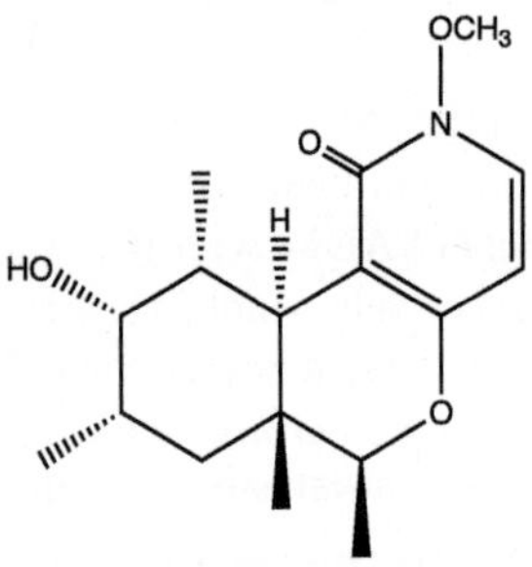 Cordypyridone D	*C. nipponica*	Antimalarial activity

Table 4. Cyclopeptides isolated from *Cordyceps* species and mode of action.

Cyclopeptides	Species	Tested mode of action
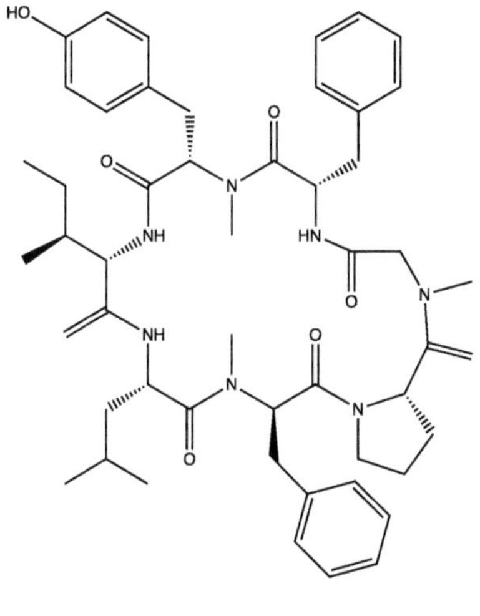 Cordyheptapeptide A	*Cordyceps* sp. BCC 16173 *Cordyceps* sp. BCC 16176	Antimalarial activity and cytotoxicity
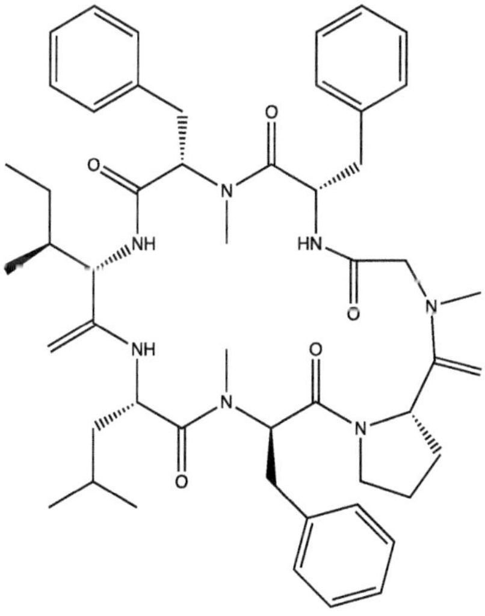 Cordyheptapeptide B	*Cordyceps* sp. BCC 16176	Cytotoxicity

Table 5. Cyclodepsipeptides isolated from *Cordyceps* species and mode of action.

Cyclodepsipeptides	Species	Tested mode of action
Beauvericin E 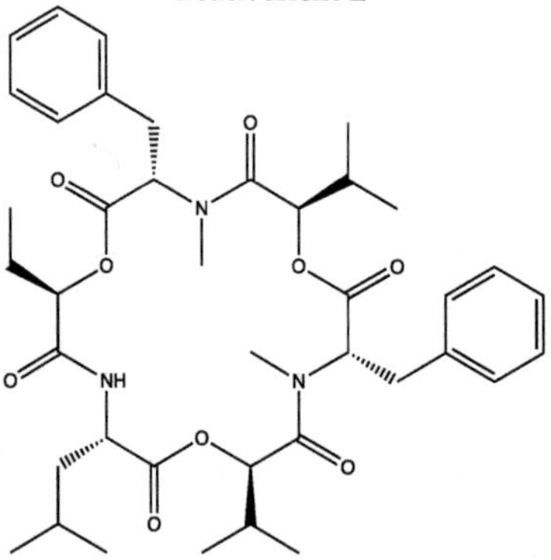	*C. cicadae*	Cytotoxicity (IC_{50}, µg/ml); HepG2, 13.67±2.59; HepG2/ADM, 14.48±1.68
Beauvericin; R_1=H, R_2=CH_3 Beauvericin A; R_1=OH, R_2=CH_2CH_3 Beauvericin J; R_1=OH, R_2=CH_3 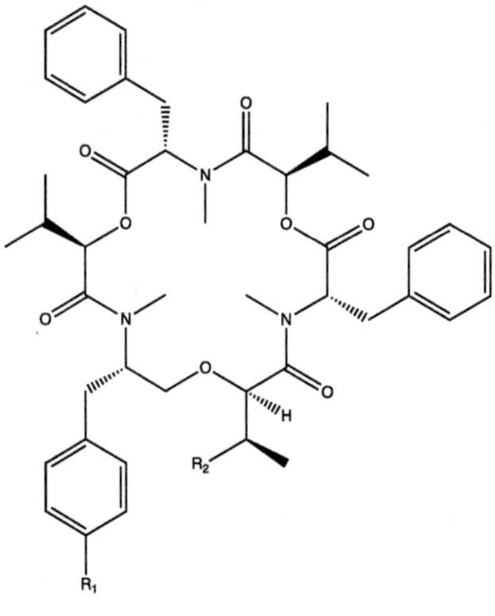	*C. cicadae*	3.12±1.08, 2.40±0.37; 2.81±0.86, 2.93±0.15; 5.04±0.20 and 2.67±0.09, respectively

Table 5. contd. ...

Table 5. contd. ...

Cyclodepsipeptides	Species	Tested mode of action
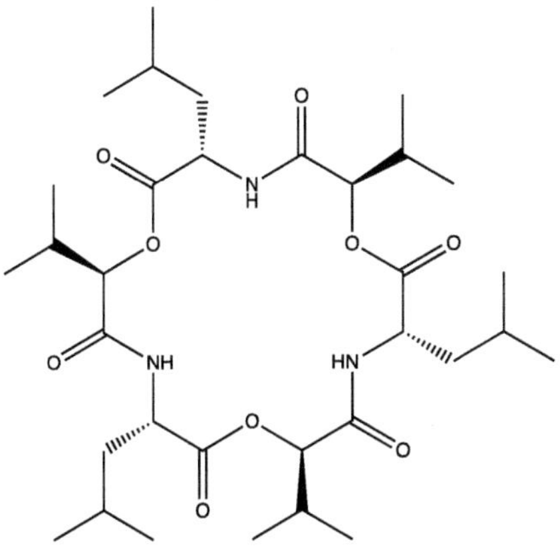 Cordycenin A	*C. cicadae*	Cytotoxicity
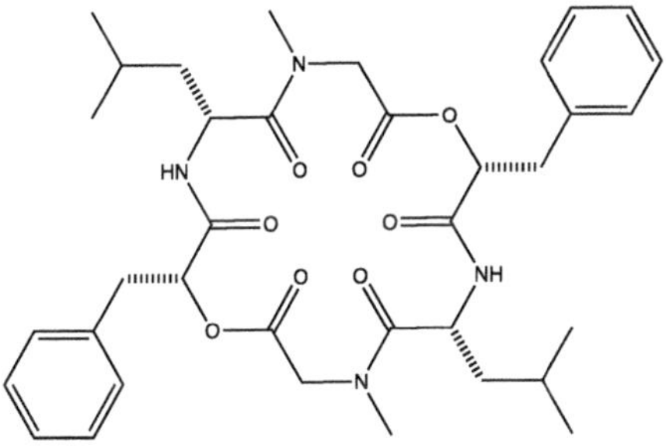 Cardinalisamide A	*C. cardinalis*	Antitrypanosomal activity and cytotoxicity
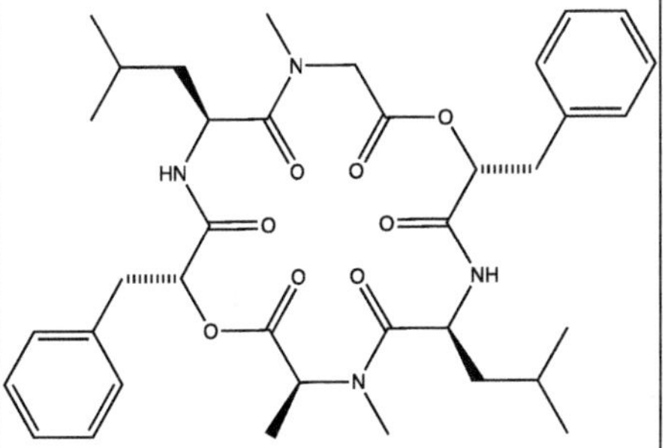 Cardinalisamide B	*C. cardinalis*	Antitrypanosomal activity and cytotoxicity

Table 5. contd. ...

Table 5. contd. ...

Cyclodepsipeptides	Species	Tested mode of action
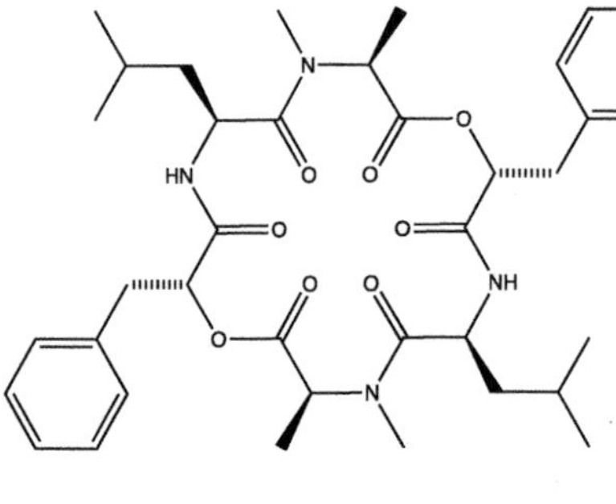 Cardinalisamide C	*C. cardinalis*	Antitrypanosomal activity and cytotoxicity
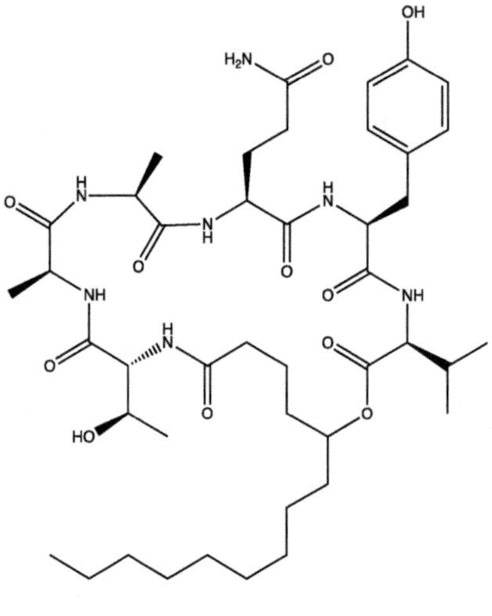 Cordycommunin	*Ophiocordyceps sinensis* (Former: *C. sinensis*)	Antimycobacterial activity

(Umeyama et al., 2014).. Cordycommunin exhibited inhibitory activity against *Mycobacterium tuberculosis* ,with a MIC value of 15 μM. However it did not show activity against *Plasmodium falciparum* K1, *Candida albicans*, and *Magnaporthe grisea*. The IC_{50} value for cytotoxicity against KB cells was found to be 45 μM, but it was inactive against MCF-7, NCI-H187, and Vero cells at a concentration of 88 μM (Haritakun et al., 2010).

3.6 *Macrocyclic Tetralactams*

Novel macrocyclic lactams with unusual skeletons, Gunnilactam A-C, were evaluated for their cytotoxicity against C42B and DU-145 cell lines (Table 6). Gunnilactam A showed selectivity against C42B cells with an IC_{50} value of 5.4 µM (Zheng et al., 2017).

Table 6. Macrocyclic tetralactams isolated from *Cordyceps* species and mode of action.

Macrocyclic tetralactams	Species	Tested mode of action
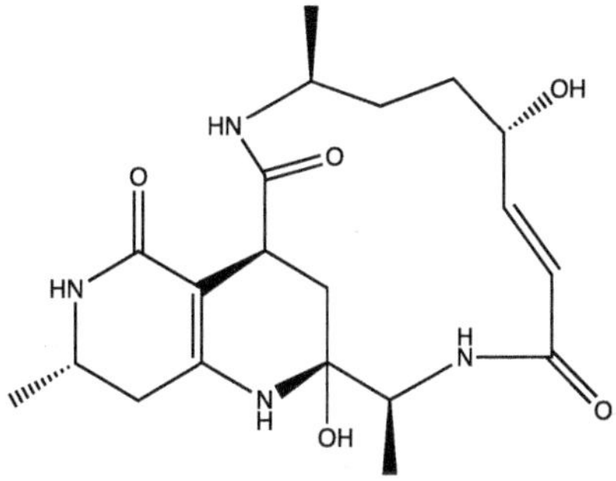 Gunnilactam A	*C. gunnii*	Cytotoxicity
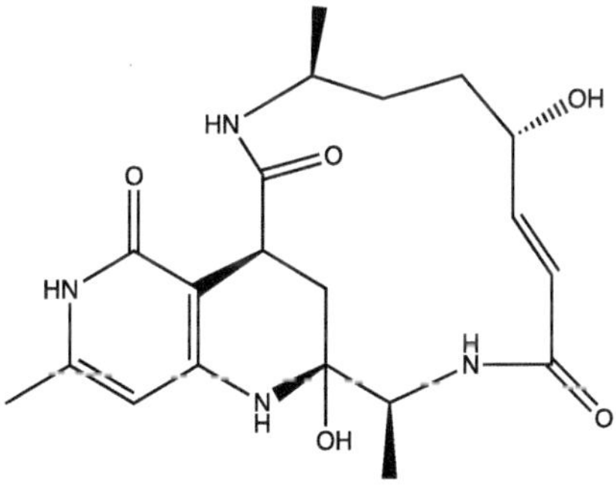 Gunnilactam B	*C. gunnii*	Cytotoxicity
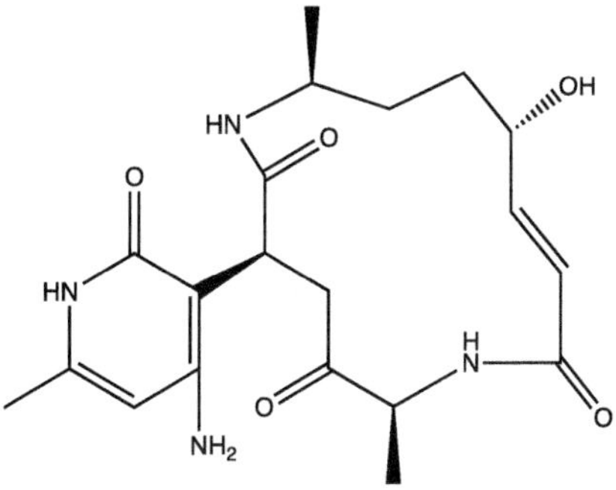 Gunnilactam C	*C. gunnii*	Cytotoxicity

Table 7. Linear peptides isolated from *Cordyceps heteropoda* and mode of action.

Linear peptides	Mode of action
Cicadapeptin I; R_1=CH$_3$, R_2=H Cicadapeptin II; R_1=H, R_2= CH$_3$	Antibacterial activity and antifungal activity

3.7 Linear Peptides

Linear peptides of *Cordyceps* spp. have antibacterial and antifungal activities (Table 7). Cicadapeptins I and II, using 100 µg/disc, produced inhibition zones against *Bacillus cereus*, *B. subtilis*, and *Escherichia coli* in the disc diffusion method with respective diameters of 13 and 12 mm/13 and 11 mm / 16 mm for both. However, both had antifungal activity with a zone diameter of 11 mm against *Botrytis cinerea* only rather than *Colletotrichum fragariae*, *C. gloeosporioides*, and *Fusarium oxysporum* (Krasnoff et al., 2005).

3.8 Bioxanthracene Derivatives

Bioxanthracene derivatives isolated from *Cordyceps* sp. BCC 16173 and *C. pseudomilitaris*, were tested for their antimalarial activity against *P. falciparum* K1, multidrug-resistant strain, and the results are given in Table 8 (Isaka et al., 2007b; Jaturapat et al., 2001).

3.9 Sesquiterpenes and Derivatives

3.9.1 Sesquiterpenes

The cytotoxicities of three unusual and fumagillol analogue sesquiterpenes, cordycepol A-C and cordycol, were evaluated against human tumor cell lines, A549, HepG2, MCF-7, and HeLa, along with a normal cell line, LO$_2$. Cordycepol C was selectively cytotoxic against cancer cell lines with IC$_{50}$ values in the range of 12.0 to 44.8 mg/ml, while cordycepol B was found to be inactive (Sun et al., 2013) (Table 9).

Table 8. Bioxanthracenes isolated from *Cordyceps* species and mode of action.

Bioxanthracenes	Species	Antimalarial activity percentage of inhibition/IC$_{50}$
(1) R$_1$=H, R$_2$=OH, R$_3$=OH **(2)** R$_1$=H, R$_2$=OAc, R$_3$=OAc **(3)** R$_1$=CH$_3$, R$_2$=OH, R$_3$=OH (4) R$_1$=CH$_3$, R$_2$=OH, R$_3$=OAc **(5)** R$_1$=CH$_3$, R$_2$=OAc, R$_3$=OAc **(6)** R$_1$=CH$_3$, R$_2$=H, R$_3$=H **(7)** R$_1$=CH$_3$, R$_2$=H, R$_3$=OH **(8)** R$_1$=CH$_3$, R$_2$=H, R$_3$=OAc	**(1)** *Cordyceps* sp. BCC 16173 **(2)** *Cordyceps* sp. BCC 16173 **(3)** *C. pseudomilitaris* **(4)** *C. pseudomilitaris* **(5)** *C. pseudomilitaris* **(6)** *C. pseudomilitaris* **(7)** *C. pseudomilitaris* **(8)** *C. pseudomilitaris*	<50% at 10 µg/ml; <50% at 10 µg/ml; 4.7 µg/ml; 8.1 µg/ml; 2.2 µg/ml; 8.4 µg/ml; 3.7 µg/ml; and 5.2 µg/ml, respectively
(9) R= OH **(10)** R= H	*C. pseudomilitaris*	1.1 µg/ml and 5.9 µg/ml, respectively

Table 8. contd. ...

Table 8. contd. ...

Bioxanthracenes	Species	Antimalarial activity percentage of inhibition/IC_{50}
(12) R_1= H, R_2=OH **(13)** R_1=CH$_3$, R_2=OH **(14)** R_1=CH$_3$, R_2=H	**(12)** *Cordyceps* sp. BCC 16173 **(13)** *C. pseudomilitaris* **(14)** *C. pseudomilitaris*	3.3 μM, 7.1 μg/ml and 18 μg/ml, respectively

3.9.2 *Amide derivatives*

The anti-inflammatory activities of cordycepiamides were tested in terms of the inhibition of nitric oxide (NO) production stimulated by lipopolysaccharide in RAW 264.7 murine macrophages. Cordycepiamide A, B and D inhibited the NO release by 6.5%, 11.2%, and 17.4% at a concentration of 10 μM, respectively. Cell viability was > 90% (Chang et al., 2017) (Table 10).

3.9.3 *Cerebrosides*

Three cerebrosides, cordycerebroside A, soyacerebroside I, and glucocerebroside were analyzed for their inhibitory effect on inducible nitric oxide synthase (iNOS) and cyclooxygenase-2 (COX-2) protein expression in lipopolysaccharide-induced RAW 264.7 murine macrophages (Table 11). Soyacerebroside I was found to have a stronger inhibitory effect on NO production with an IC_{50} value of 8.4 μg/ml than the other two compounds with IC_{50} values >10 μg/ml. The mechanism underlying the inhibition of pro-inflammatory proteins was, in part, reduced phosphorylation of p65, a subunit of NF-κB (Chiu et al., 2016).

In another study, cordycerebroside B was isolated from *C. militaris* along with glucocerebroside, cordycerebroside A, and soyacerebroside I, and tested for their anti-PTP1B enzyme activities. Cordycerebroside B inhibited the enzyme activity with IC_{50} values of 4.68±0.18 μM which were

Table 9. Sesquiterpenes isolated from *Cordyceps* species and mode of action.

Sesquiterpenes	*Species*	Mode of action
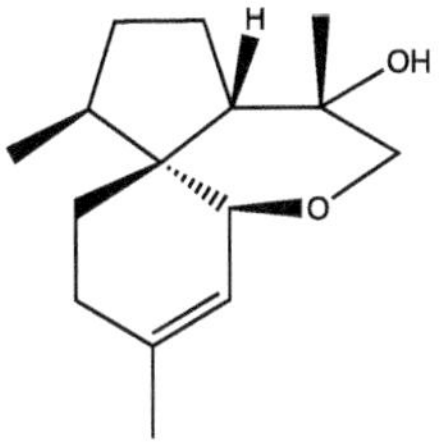 Cordycepol A	*C. ophioglossoides*	Cytotoxicity (IC_{50}, µg/ml); A549: > 80; MCF-7: > 80; HepG2: > 80; HeLa: > 80; LO2: > 80
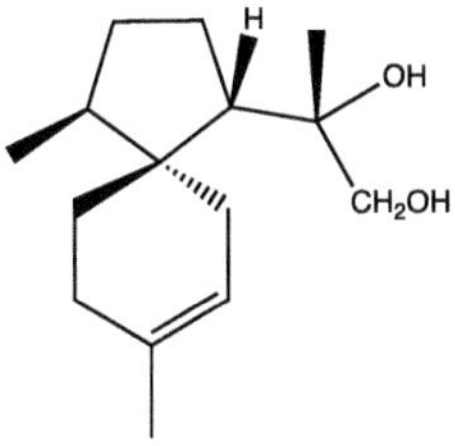 Cordycepol B	*C. ophioglossoides*	Cytotoxicity
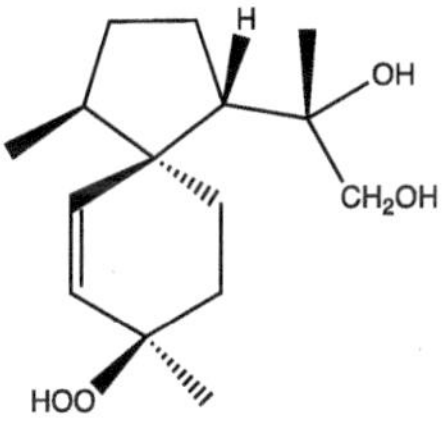 Cordycepol C	*C. ophioglossoides*	Cytotoxicity (IC_{50}); A549: 44.8 ± 2.4; MCF-7: 30.8 ± 3.0; HepG2: 33.0 ± 3.9; HeLa: 12.0 ± 1.8; LO2: > 80
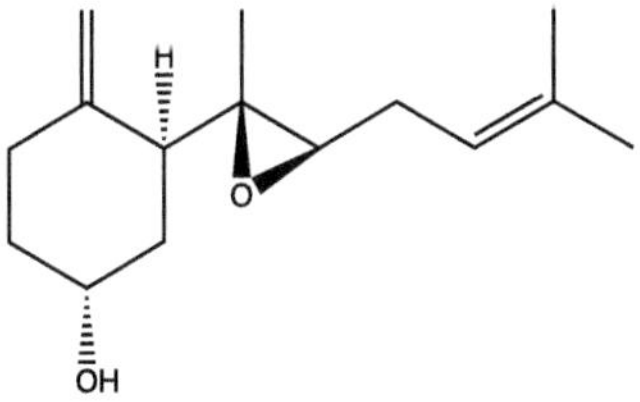 Cordycol	*C. ophioglossoides*	Cytotoxicity (IC_{50}); MCF-7: > 80 A549: 40.2, 2.44; HepG2: 36.2 4.0; HeLa: 21.1, 7.1; LO2: > 80

lower than those of glucocerebroside (16.93±1.08 µM), cordycerebroside A (10.43±0.64 µM), and soyacerebroside I (18.92±1.65 µM). None of the compounds displayed cytotoxicity against PC12 cells (Sun et al., 2019).

Table 10. Sesquiterpenes (amide derivatives) isolated from *Cordyceps* species and mode of action.

Amide derivatives	Species	Mode of action
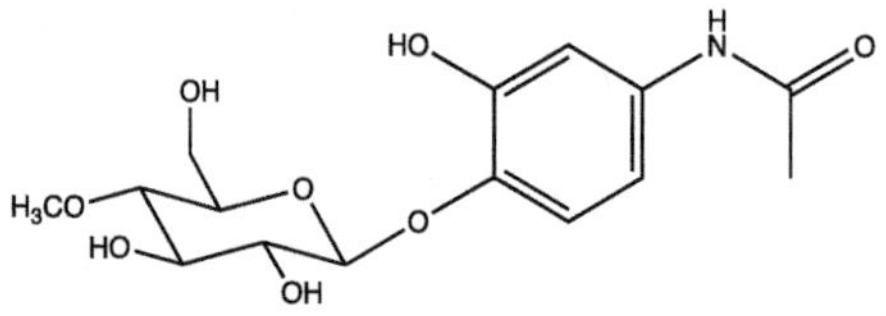 Cordycepiamide A	*C. ninchukispora*	Anti-inflammatory activity
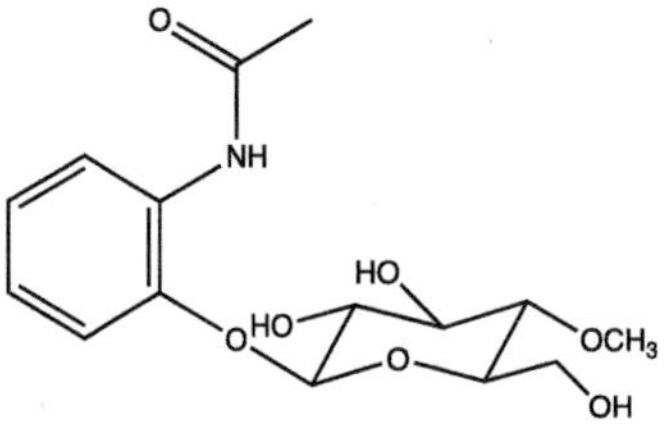 Cordycepiamide B	*C. ninchukispora*	Anti-inflammatory activity
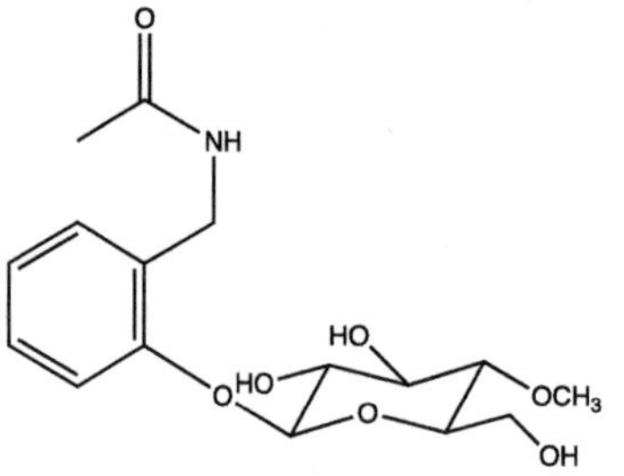 Cordycepiamide C	*C. ninchukispora*	Anti-inflammatory activity
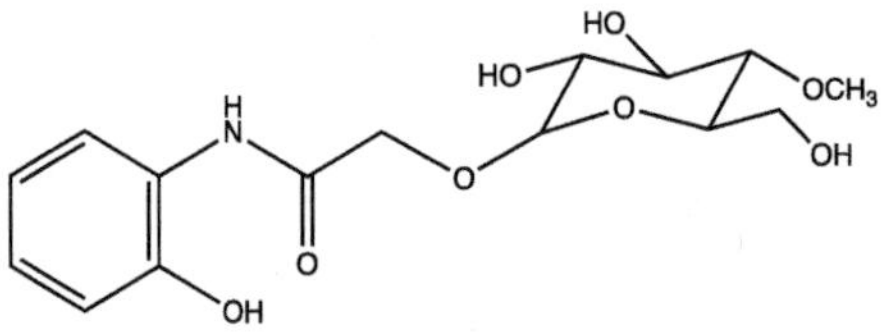 Cordycepiamide D	*C. ninchukispora*	Anti-inflammatory activity

3.10 Phenolics and Polyketides

Annullatin A-E, isolated from *C. annulata,* was tested for its agonist activity on the cannabinoid receptors CB1 and CB2. While annulantin A and B were shown to be CB1 agonists, annulantin C was shown to be a CB2 inverse agonist (Asai et al., 2012) (Table 12). Cordycepsidone A exhibited greater inhibitory activity against *Giberella fujikuroi,* and *Pythium ultimum,* with

Table 11. Sesquiterpenes (cerebrocides) isolated from *Cordyceps* species and mode of action.

Cerebrosides	Species	Mode of action
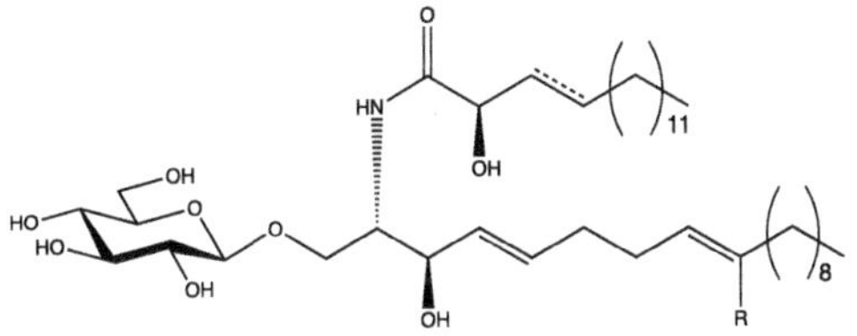 Cordycerebroside A; R= CH$_3$, $\Delta^{3',4}$ Soyacerebroside I; R= H Glucocerebroside; R= CH$_3$	*C. militaris*	Anti-inflammatory activity
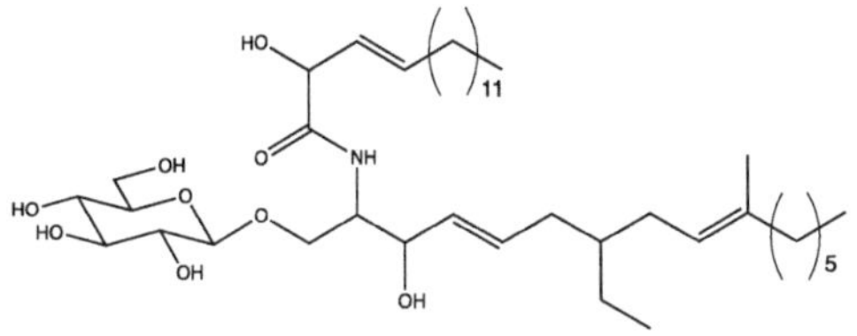 Cordycerebroside B	*C. militaris*	Anti-PTP1B activity

MIC values of 8.3 µg/ml and 1.2 µg/ml, respectively, than the corresponding values of >50 and 25.0 ± 0.1 for Cordycepsidone B. Both compounds, at 50 µg/ml, were found to be inactive against *Staphylococcus aureus*, *Escherichia coli*, and *Pseudomonas aeruginosa* (Varughese et al., 2012).

Daidzein 7-*O*-β-D-glucoside 4''-*O*-methylate **(1)**, glycitein 7-*O*-β-D-glucoside 4''-*O*-methylate **(2)**, genistein 7-*O*-β-d-glucoside 4''-*O*-methylate **(3)**, and genistein 4'-*O*-β-D-glucoside 4''-*O*-methylate **(4)** were determined to be novel isoflavones methylglycosides of *C. militaris* (Choi et al., 2010). Anthraquinones, morakotins A-E, were isolated from *C. morakotii*, and only morakotins C and D were assayed in antimicrobial and cytotoxicity studies. Morakotin C exhibited antifungal activity against *C. albicans* with IC$_{50}$ values of 25.87 µg/ml and antibacterial activity with MIC values of 12.5 µg/ml, while morakotin D demonstrated antibacterial activity against *B. cereus* and *S. aureus* with MIC values of 3.13 µg/ml and 6.25 µg/ml, respectively (Wang et al., 2019a).

Fumosoroseanosides A and B were found to have antibacterial and antifungal activity against *E. coli* and *C. albicans* at 10 to 30 µg per paper disk. Fumosoroseain A increased the lifespan of *Caenorhabditis elegans* by 11.3% at a concentration of 200 µM (Wei et al., 2022a). New diphenyl ether glycosides (cordyol A-B) and diphenyl ether (cordyol C) were isolated from *Cordyceps* sp. BCC 1861 for conducting biological activity studies. Cordyol C displayed antiviral activity against herpes simplex virus type 1 (HSV-

Table 12. Phenolics and polyketides isolated from *Cordyceps* species and mode of action.

Phenolics and Polyketides	Species	Tested mode of action
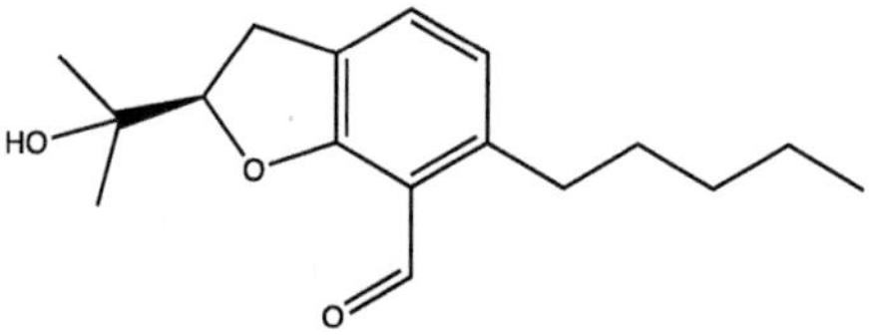 Annullatin A	*C. annulata*	Effect on cannabinoid receptors CB1 and CB2
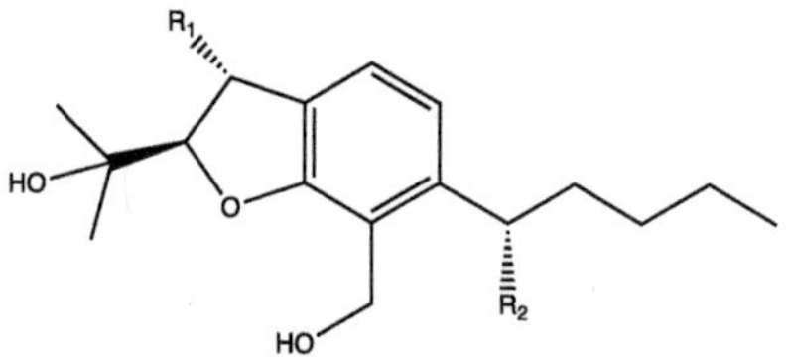 Annullatin B; R_1=R_2=H Annullatin C; R_1=OH, R_2=H	*C. annulata*	Effect on cannabinoid receptors CB1 and CB2
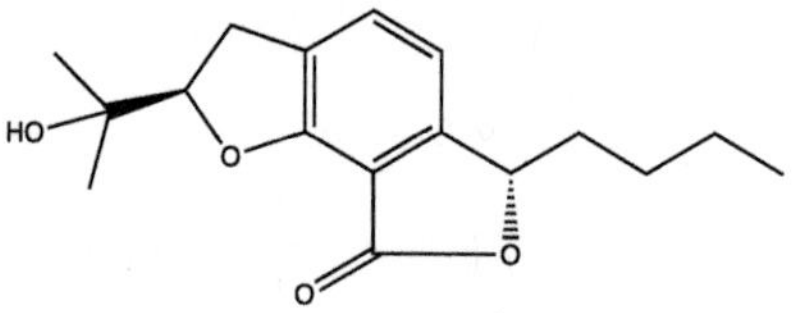 Annullatin D	*C. annulata*	Effect on cannabinoid receptors CB1 and CB2
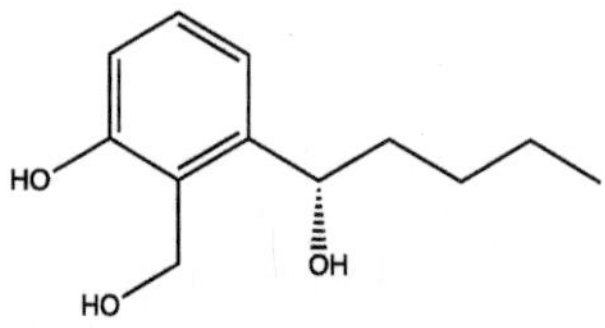 Annullatin E	*C. annulata*	Effect on cannabinoid receptors CB1 and CB2
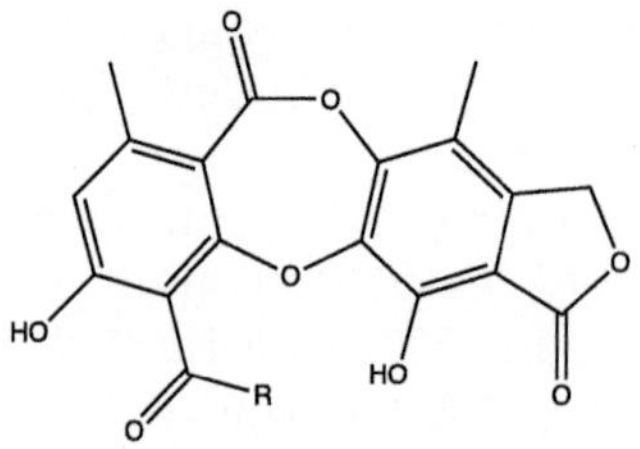 Cordycepsidone A; R=H Cordycepsidone B; R=OH	*C. dipterigena*	Antifungal activity

Table 12. contd. ...

Table 12. contd. …

Phenolics and Polyketides	*Species*	Tested mode of action
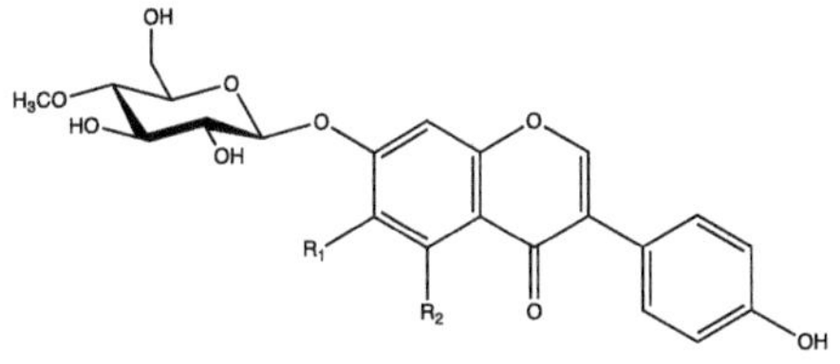 (1) R_1=H, R_2=H (2) R_1=OCH$_3$, R_2=H (3) R_1=H, R_2=OH	*C. militaris*	-
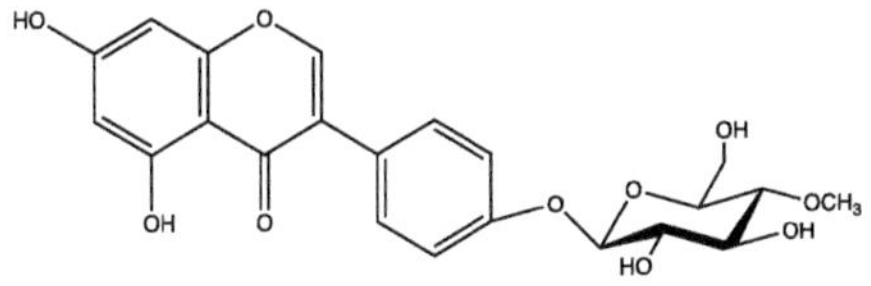 (4) Genistein 4′-O-β-d-glucoside 4″-O-methylate	*C. militaris*	-
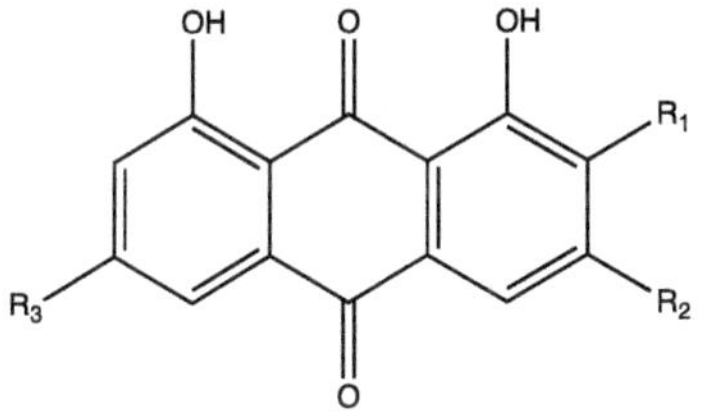 Morakotin A; R_1, R_2=OCH$_3$, R_3=OH Morakotin B; R_1, R_2, R_3=OCH$_3$	*C. morakotii*	-
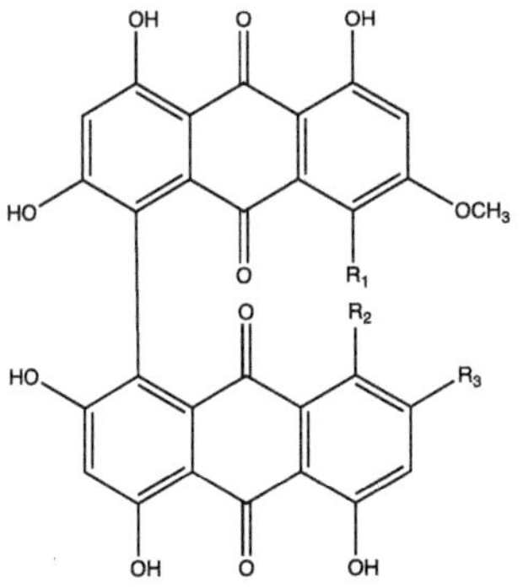 Morakotin C; R_1, R_2= OH, R_3= OCH$_3$ Morakotin D; R_1, R_2= H, R_3= OH	*C. morakotii*	-

Table 12. contd. …

Table 12. contd. ...

Phenolics and Polyketides	Species	Tested mode of action
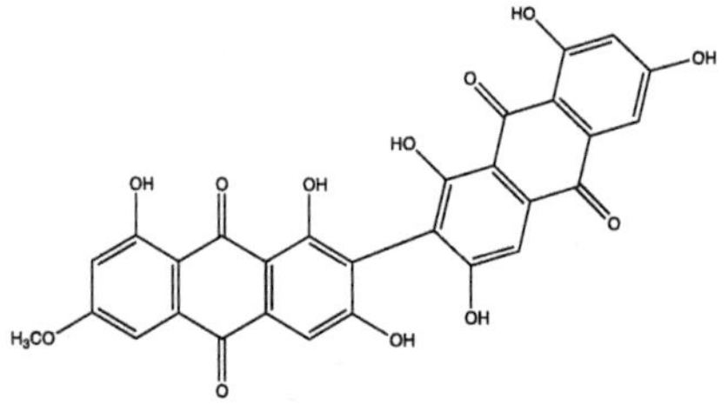 Morakotin E	*C. morakotii*	-
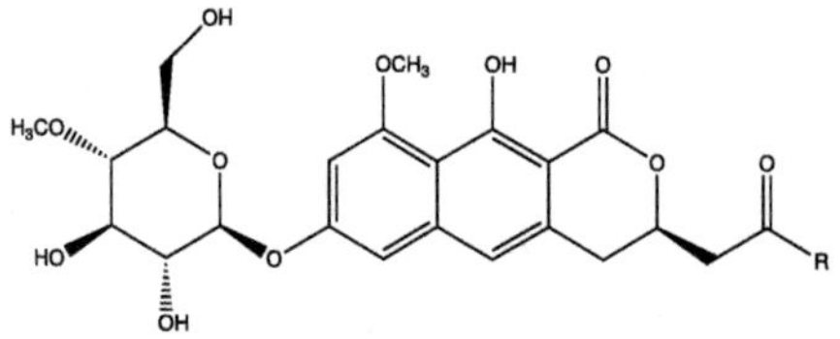 Fumosoroseanoside A; R=CH$_3$ Fumosoroseanoside B; R=OCH$_3$	*C. fumosorosea*	Antimicrobial activity and antiaging activity
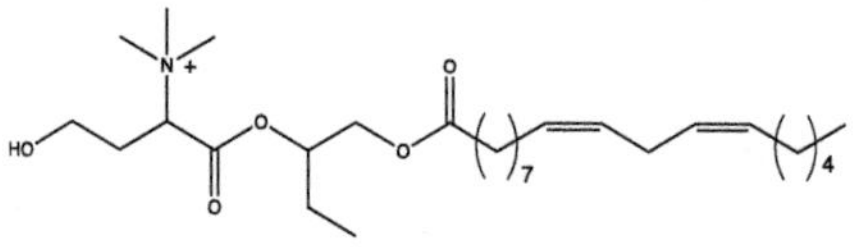 Fumosoroseain A	*C. fumosorosea*	Antimicrobial activity and antiaging activity
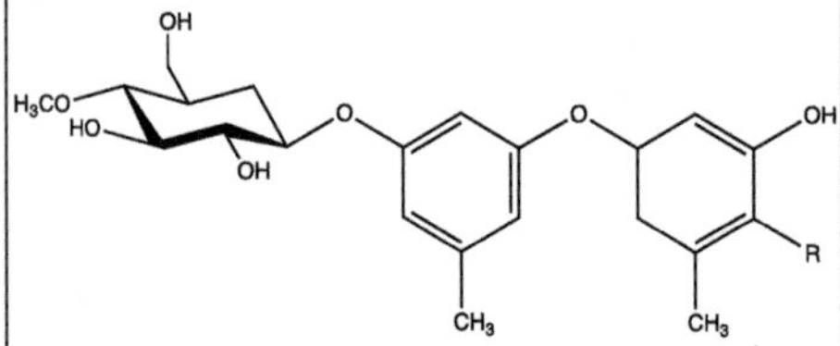 Cordyol A; R=H Cordyol B; R=CO$_2$CH$_3$	*Cordyceps* sp.	Antiviral activity, antimicobacterial actikvity, antimalarial activity and cytotoxicity
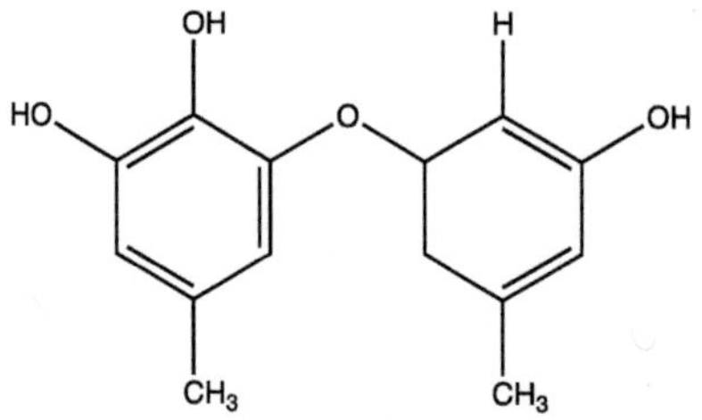 Cordyol C	*Cordyceps* sp.	Antiviral activity, antimycobacterial activity, antimalarial activity and cytotoxicity

Table 12. contd. ...

Table 12. contd. ...

Phenolics and Polyketides	*Species*	Tested mode of action
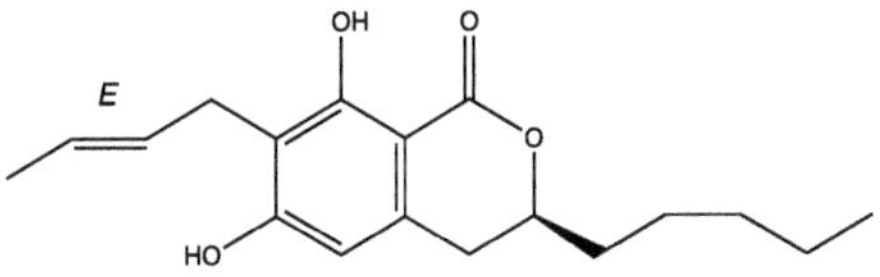 Bifusicoumarin A	*C. bifusispora*	Cytotoxicity
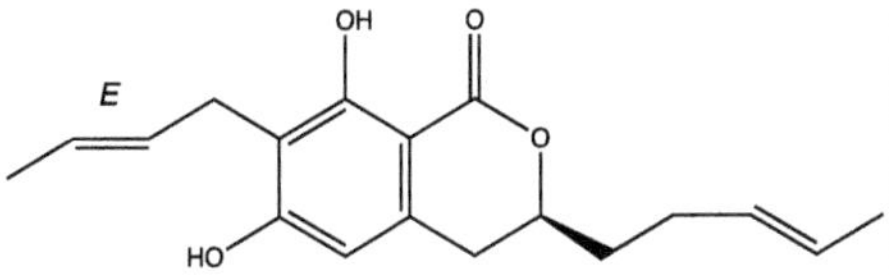 Bifusicoumarin B	*C. bifusispora*	Cytotoxicity
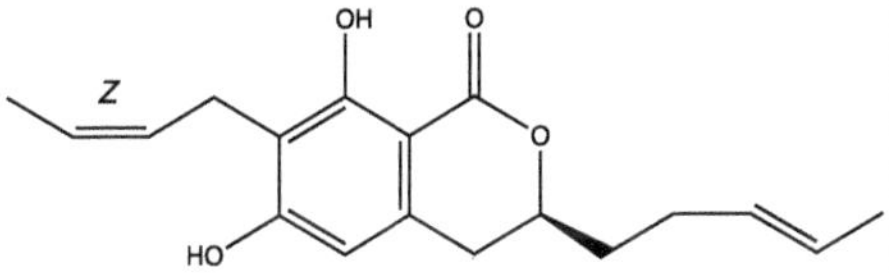 Bifusicoumarin C	*C. bifusispora*	Cytotoxicity
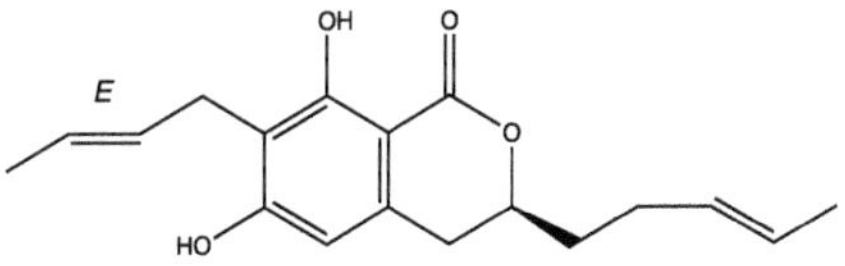 Bifusicoumarin D	*C. bifusispora*	Cytotoxicity

1) with an IC_{50} value of 1.3 µg/ml, along with cytotoxicity against BC and NCI-H187 cell lines with IC_{50} values of 8.65 and 3.72 mg/ml, respectively. Cordyol A showed antimycobacterial activity against *Mycobacterium tuberculosis* with a MIC value of 100 µg/ml. The inhibitory activity of all compounds on *Plasmodium falciparum* K1 was determined to be higher than the MIC value of 10 µg/ml (Bunyapaiboonsri et al., 2007).

Newly described coumarins, bifusicoumarin A-D, were assessed for their toxicity against CCRF-CEM, CEM-ADR5000, MDA-MB231-pcDNA3, HCT116 (p53[+/+]), and HEK293 cell lines. They exhibited inhibition against CCRF-CEM and CEM-ADR5000IC50 cell lines in the IC_{50} ranges of 23.89–67.74 and 22.53–67.74 µM, respectively (Elshamy et al., 2023).

Table 13. Other compounds from *Cordyceps* species.

2-((2-hydroxyethyl)amino)benzoic acid	E-dec-2-enamide
Cordyglycoside A	Cordycepene

3.11 Other Compounds

2-((2-hydroxyethyl)amino)benzoic acid and E-dec-2-enamide were isolated from *n*-butanol soluble fractions of *C. takaomontana* NBRC 101754. This marked the first isolation of these compounds from a natural source (Hama et al., 2019) (Table 13). Cordyglycoside A, a newly isolated glycoside metabolite from *Cordyceps* sp., exhibited no antimicrobial activities against *S. aureus, B. subtilis,* and *E. coli,* cytotoxic activity against MCF-7, HepG-2, HT29, and 293T cell lines, and DPPH scavenging activity (Fan et al., 2020).

A novel cordycepic pigment, cordycepene, was isolated from *C. militaris* and displayed scavenging activity against DPPH radicals with an IC_{50} value of 0.81 mg/ml. The inhibitory effect of *C. takaomontana* NBRC 101754 on the senescence of human skin fibroblast cells was demonstrated by promoting the antioxidant defense system (Tang et al., 2019).

4. Pharmacological Potential

Existing evidence has pointed out the pharmacological potential of *Cordyceps* species and *Cordyceps*-based agents or products. In the following sections, we organized and compiled the biological activity investigations of *Cordyceps*-based products.

4.1 Anti-Inflammatory Activity

Inflammation is a component of the body's defense mechanism in which the immune system perceives and eradicates destructive and unfamiliar stimuli and initiates the curative procedure. Inflammation can be acute or chronic based on the conditions and functions in the etiology of many conditions, such as cancer, diabetes, and cardiovascular diseases (Pahwa et al., 2023). Based on the traditional use of *C. sinensis* in the intervention of airway ailments such as lung inflammation and asthma, the antiallergic rhinitis and the anti-asthmatic influences of a water extract obtained from the mycelium culture of *C. sinensis* (Cs-4) were investigated in a mouse model of ovalbumin (OVA)-induced allergic rhinitis and rat model of the anti-asthmatic effect, respectively. OVA-sensitized and challenged mice under Cs-4 treatment were repressed. The inhibition was linked to IgE/OVA-IgE and interleukin (IL)-4/IL-13 reductions in the nasal fluid. Cs-4 treatment also led to decreased airway responsiveness and improved scratching behavior in capsaicin-challenged rats. It also alleviated plasma IgE levels in addition to IgE and eosinophil peroxidase levels in the bronchoalveolar fluid. Further, interleukin levels were reduced in rat lung tissue under CS-4 treatment, all confirming the potential of CS-4 in the attenuation of inflammatory conditions such as asthma and allergic rhinitis (Chen et al., 2020).

Likewise, investigating the chloroform and *n*-butanol fractions of methanol extract from *C. sinensis* (CS) for their anti-inflammatory properties showed that the extracts blocked the increased release of NO and cytokines tumor necrosis factor (TNF)-α and interleukin (IL)-12 in LPS/IFN-γ-triggered murine peritoneal macrophages, proposing a CS-associated anti-inflammatory response (Rao et al., 2007). The anti-inflammatory effect of *C. guangdongensis* on chronic bronchitis was examined. The hot-water extract from *C. guangdongensis* was shown to alleviate inflammatory cell infiltration and possess remarkable anti-inflammatory activity, which was attributed to the polysaccharide content of the extract (Yan et al., 2014).

Investigating the hot water extract of *C. militaris* fruiting bodies (CMWE) in terms of its anti-inflammatory effect revealed that NO production together with tumor necrosis factor-α (TNF-α) and interleukin-6 (IL-6) secretions were reduced in LPS-stimulated RAW 264.7 cells, implying the anti-inflammatory function of *C. militaris* (Jo et al., 2010). Similarly, another study by Chiu et al. (2016) indicated that the hydroalcoholic fraction of *C. militaris* fruiting bodies possessed a potent anti-inflammatory activity. The bioactivity-guided fraction of the fraction led to the isolation of eight compounds, among which cordycerebroside, soyacerebroside, and glucocerebroside hindered the increase of pro-inflammatory iNOS protein and attenuated COX-2 protein levels in LPS-induced RAW264.7 macrophages (Chiu et al., 2016).

The ethyl acetate and an enriched ergosterol sub-fraction of the ethyl acetate fraction of *C. militaris* (CE3) were examined in terms of their nitric oxide (NO) generation capacity on LPS-stimulated BV2 microglia cells, and these extracts were shown to reduce nitric oxide production, pointing out the anti-inflammatory potential of *C. militaris* (Nallathamby et al., 2015). Evaluating the anti-inflammatory properties of *C. militaris*, a linear β-(1→3)-D-glucan-purified alkaline extract was observed to exhibit better anti-inflammatory activity through inhibiting IL-1β, TNF-α, and COX-2 expression. The β-(1→3)-D-glucan was found to display an identical effect, suggesting that this polymer is the strongest anti-inflammatory compound available in the polysaccharide extracts of *C. militaris*. Furthermore, the isolated β-(1→3)-D-glucan was disclosed to possess anti-inflammatory activity against lipopolysaccharide (LPS)-induced peritonitis in mice (Smiderle et al., 2014).

Similar to the anti-inflammatory property of the polysaccharide content, the anti-inflammatory capacity of exopolysaccharide (EPS) generated by Cs-HK1 mycelial fermentation culture obtained from the fruiting body of *C. sinensis* was investigated. EPS notably repressed NF-κB and various pro-inflammatory factors' release, including NO, TNF-α, and IL-1β, *in vitro*. Furthermore, the oral administration of EPS remarkably obstructed the release of substantial inflammatory cytokines *in vivo* murine model (Li et al., 2020a). Crude polysaccharides (CP) and water-soluble nondigestible polysaccharides (NDPs) gained from the fruiting bodies of cultivated *C. cicadae* were surveyed for their anti-inflammatory properties. Both CP and NDP exhibited anti-inflammatory activity. Specifically, NDP endowing with stronger inhibition on NO, IL-1β, and TNF-α generation exhibited a greater anti-inflammatory response in LPS-induced RAW264.7 macrophages (Yang et al., 2019).

Jiao et al. (2022) carried out research on *C. militaris* extract (CME) to assess its impact on gouty arthritis (GA). CME was demonstrated to attenuate the inflammatory progress of GA and amend the initiation of GA. The background mechanism was considered to be associated with decreased levels of pro-inflammatory factors IL-1β, IL-6, TNF-α, and caspase-1 and enhanced levels of IL-10. Furthermore, transcriptome analysis described four major genes contributing to the regulation of inflammation: CCL7, CSF2RB, IL-1β, and LIF (Jiao et al., 2022).

4.2 Antimicrobial Activity

Many investigations have revealed the *Cordyceps*-linked antibacterial, antifungal, and antimalarial activities with examples that follow.

Examining fruiting bodies and fermented mycelia of *C. militaris* in terms of their antimicrobial action revealed that fruiting body extract exert extensive antibacterial and antifungal effects on all tested microorganisms, whereas fermented mycelia extract displayed selective antimicrobial activity, indicating the property of *C. militaris* as an antimicrobial agent (Dong et al., 2014). Investigating the potential of *C. militaris* against skin infective bacteria disorders pointed out that the ethanolic and aqueous extracts of *C. militaris* possessed antibacterial impacts against *Staphylococcus aureus*, *Cutibacterium acnes*, *Pseudomonas aeruginosa*, and methicillin-resistant *S. aureus*, which provided a basis for the accomplishment of bacteria-induced skin disorders (Eiamthaworn et al., 2022).

Varughese et al. (2012) found that *C. dipterigena* diligently hindered the mycelial development of the pathogenic fungus *Gibberella fujikuroi*. Cordycepsidone A and cordycepsidone B, two new depsidone metabolites obtained from the potato dextrose agar (PDA) culture extract of *C. dipterigena*, were unraveled to account for the antifungal property (Varughese et al., 2012). Joshi et al. (2015) studied the ethanol, chloroform, methanol, acetone, and aqueous extracts of *C. militaris* in terms of their antibacterial properties. The ethanolic extract suppressed *E. coli* at most, while the acetone extracts chiefly inhibited *S. aureus*, providing *C. militaris* as an alternate source.

A water-soluble polysaccharide of *C. cicadae* was demonstrated to illustrate a potent antibacterial effect against *Escherichia coli*, *Bacillus subtilis*, *Salmonella paratyphi*, *Staphyloccocus aureus*, and *Pseudomonas aeruginosa* through bacterial cell wall injury and enhanced cell permeability, causing the loss of cell components (Zhang et al., 2017). Cordycepin, one of the principal contents of *Cordyceps* species, was found to exhibit an efficient inhibition towards a number of gram-positive and gram-negative bacteria, which was found to occur by the disruption of bacterial cell membranes and the interaction of cordycepin with bacterial genomic DNA to interrupt cellular actions (Jiang et al., 2019). Similarly, beauvericin, another compound isolated from *C. tenuipes*, displayed antimicrobial impacts against *Staphylococcus aureus* and *Bacillus subtilis* (Yoneyama et al., 2022).

4.3 Antioxidant Activity

Oxidative stress contributes to the etiology of many disturbances. *Cordyceps* species holding a prominent role in the management of assorted situations exhibited remarkable antioxidant capacities, contributing to their other pharmacological activities. Ji et al. (2009) demonstrated that CSE promoted the function of antioxidant enzymes including catalase, superoxide dismutase, and glutathione peroxidase and diminished lipid peroxidation levels and monoamine oxidase activity in mice aged by D-galactose, which was estimated to be the substantial mechanism behind the CSE-induced antiaging activity (Ji et al., 2009).

Compared to the antioxidant effects of fruiting bodies and solid-state fermented rice (FRE) from two wild-type strains of *C. militaris*, the FRE of the Zhangzhou strain (FRE-Z) was shown to display a broad antioxidant capacity against the oxidation of DPPH, hydroxyl, and superoxide radicals (Wu et al., 2019). Evaluating the antioxidant properties of four polysaccharide fractions isolated from *C. kyushuensis* in terms of DPPH radical assay, they all displayed remarkable DPPH scavenging activity dose-dependently (Su et al., 2020).

Dong et al. (2014) disclosed the high antioxidant capacity of the fruiting body and fermented mycelia extracts of *C. militaris in vitro* based on the assays for the ascertainment of total antioxidant capacity, DPPH· radical scavenging activity, chelating ability on ferrous ions, and reducing power ability. The fruiting bodies possessed greater DPPH· radical scavenging property, while the fermented mycelia displayed potent total antioxidant capacity, reducing power, and chelating capacity, suggesting that they acted in separate ways (Dong et al., 2014).

Further scientific evidence providing more knowledge about the antioxidant potential of *Cordyceps* species was indicated in many investigations before (Tuli et al., 2014; Yue et al., 2012; Zhu et al., 2020; Das et al., 2021).

4.4 *Antitumour, Anti-Proliferative and Anti-Metastatic Activity*

Many studies focusing on the discovery of the potential of *Cordyceps* species against cancer confirmed the use of *Cordyceps* species for the management of cancer as a traditional remedy. Several studies were provided below.

Investigating the combination of aqueous extract of *C. sinensis* (AECS) and *cis* diamminedichloroplatinum (II) (DPP) on non-small cell lung cancer (NSCLC) cells, Huo et al. (2017) observed that the combination approach enhanced the efficacy of DDP-dependent chemotherapy and alleviated treatment-associated toxicity, probably by suppressing IκBα/NFκB and AKT/MMP2/MMP9 pathways, suggesting the potential of joint use of AECS together with DDP in the treatment of NSCLC (Huo et al., 2017). In another study, Jo et al. (2020a) examined the impact of *C. militaris* extract (CME) on the proliferation of carboplatin-resistant SKOV-3 cells and uncovered that *C. militaris* repressed carboplatin-resistant SKOV-3 cell proliferation and promoted apoptosis, possibly via the upregulation of ATF3/TP53 signaling and the activation of CHOP/PUMA, pointing out the capacity of CME for overcoming carboplatin resistance in human ovarian cancer (Jo et al., 2020b).

Validating the ethnopharmacological use of *C. taii* as an anticancer agent demonstrated that the chloroform extract of *C. taii* (CFCT) dose-time-dependently inhibited the cell viability of A549 and SGC-7901 cells. CFCT alone, or in the case being jointly used with cyclophosphamide (CTX) caused tumor tissue necrosis and further suppressed the lung metastasis of melanoma B16F10 in tumor-bearing C57BL/6 mice, confirming that *C. taii* is a natural product from which anticancer chemo preventive representatives may be developed (Liu et al., 2015). The confirmation study of the use of *C. militaris* for anticancer purposes in folk medicine was executed on an ethanolic extract of *C. militaris* (Cm-EE) on a xenograft mouse model bearing murine T cell lymphoma (RMA) cell-derived cancer (Park et al., 2017). The *C. militaris*-provided group attenuated cancer size and mass in comparison to the control group. In addition, Cm-EE hindered the viability of cultured RMA cells and increased the number of proapoptotic cells. The regulation of p85/AKT-dependent or GSK3β-related caspase-3-dependent apoptosis was estimated to be the mechanism behind the anticancer potential of *C. militaris* (Park et al., 2017). Compared to the antiproliferative effects of fruiting bodies and solid-state fermented rice (FRE) from two wild-type strains of *C. militaris*, the FRE of the Zhangzhou strain (FRE-Z) dose-dependently hindered the cell proliferation of MCF-7 and MDA-MB-231 cells. The influence of FRE-Z on MCF-7 was displayed to be due to an early stage of apoptosis, which occurred through p53 activation (Wu et al., 2019).

In addition to the fungi themselves, their bioactive metabolites were revealed to exert antitumor, anti-proliferative, and anti-metastatic activity. The substantial composition of *Cordyceps* species includes polysaccharides, sterols, and adenosine (Jędrejko et al., 2021; Gu et al., 2020; Borde and Singh 2022; Yang et al., 2009). Many studies conducted on the constituents of *Cordyceps* species pointed out the possible anticancer properties of these agents. To exemplify, one of the major bioactive constituents of *Cordyceps* species, cordycepin (3′-deoxyadenosine) was found to act against drug-resistant tumor cells based on gene expression and drug sensitivity profiling (Özenver et al., 2021). Cordycepin inhibited the proliferation, migration, and metastatic actions of many cancers, including bladder, cervical, gastric, testicular, human malignant melanoma, human liver, non-small cell lung, colon, breast, esophageal, tongue, and ovarian cancers (Lee et al., 2010; Kim et al., 2019; Yang et al., 2020; Chang et al., 2020; Zhang et al., 2023a; Luo et al., 2019; Guo et al., 2020; Wei et al., 2022b; Lee et al., 2019; Xu et al., 2019; Zheng et al., 2020; Liu et al., 2020; Jang et al., 2019; Wang et al., 2019b; Li et al., 2019; Tania et al., 2019; Zhang et al., 2022a).

Polysaccharides obtained from *Cordyceps* species were further proven to act against cancerous cells, as indicated in a variety of investigations in the following: The polysaccharides obtained by pre-soaking ultrasonic water extraction of artificially cultured *C. cicadae* (CCP) suppressed the cell viability of Hela cells by S phase arrest and apoptosis (Xu et al., 2021). Qi et al. (2020) demonstrated the action of *C. sinensis* polysaccharide (CSP) on HCT116 human colon cancer cells and observed that CSP obstructed the spread of HCT116 cells by leading apoptosis and blocking autophagy flux, which might be accomplished by PI3K-AKT-mTOR and AMPK-mTOR-ULK1 signaling (Qi et al., 2020). Another study to assess the impact of wild *Cordyceps* polysaccharides (WCP) on H22 liver cancer and the associated mechanism uncovered that WCP restrained the proliferation of H22 tumors by inducing the apoptosis of tumor cells, probably affecting the IL-10/STAT3/Bcl2 and Cyto-c/Caspase8/3 signaling pathways.

Many studies are in existence, unraveling the probable anticancer action of *Cordyceps* species and *Cordyceps*-associated content. The comprehensive evaluation of these investigations leads us to the conclusion that *Cordyceps* species may hold anticancer potential either themselves or via their constituents. The mechanism behind their anticancer potential may be that: they promote immunological activity (Das et al., 2021); (1), blocking the germination of blood vessels (Li et al., 2020b; Li et al., 2022); (2), leading tumor cell apoptosis and affecting the regulation of signaling pathways (Jo et al., 2020a and 2020b; Xie et al., 2019); (3), restricting RNA and protein synthesis (Wu 2005); (4), exhibiting antioxidant activity (Lan et al., 2022); (5), influencing the nucleic acids of tumors adversely (Liao et al., 2015); (6), impairing tumor-inspiring viruses (Du et al., 2016); (7) and antimutagen effects (Cho et al., 2003) (9).

4.5 *Effect on the Cardiovascular System*

Cordyceps species have been used as standardized traditional Chinese medicine interventions for the management of cardiovascular disorders such as arrhythmia, atrioventricular blockade, and refractory bradycardia (Hao et al., 2017; Wang et al., 2022). To provide evidence, Wang et al. (2022) analyzed the randomized controlled trials of patients with bradycardia treated with *Cordyceps*. The outcomes are worthy in that *Cordyceps* application markedly induced a positive effect on arrhythmia treatment, which was substantially linked to the adrenergic signaling in cardiomyocytes and the PI3K-Akt signaling pathway (Wang et al., 2022). *C. sinensis* (CS) is also a traditional Chinese medicine remedy and has been used for cardiovascular diseases as well as arrhythmia and hypertension (Yan et al., 2013b). At this point, research by Yan et al. (2013b) was conducted to observe the probable impact of *C. sinensis'* extracts on ischemia-reperfusion. CS extracts were demonstrated to have a repressing function in ischemic contracture, and adenosine receptor stimulation was predicted to be involved in the attenuation of contracture in hearts pretreated with CS (Yan et al., 2013b).

Another study by Wu et al. (2018) assessed the preventive function of fermented CS against doxorubicin (DOX)caused cardiotoxicity and the background mechanisms. Fermented CS was elucidated to reduce DOXinduced cardiotoxicity by blocking myocardial hypertrophy and myocardial injury, improving systolic activity and the antioxidant enzyme system, and promoting cardiac energy metabolism, which confirmed the potency of fermented CS as an agent contributing to the management of DOXinduced cardiac problems (Wu et al., 2018). Likewise, Wu et al. (2015) tested the fermented CS in DOX-administered rats and blocked DOX-associated oxidative stress (OS) responses (e.g., it markedly raised the activities of catalase, glutathione peroxidase, and the total superoxide dismutase activity in cardiac tissue, the scavenging activity of O_2^- in serum, and lowered the malondialdehyde capacity in liver and cardiac tissues). Therefore, the capacity of fermented CS for the protection of heart diseases induced by OS (Wu et al., 2015).

4.6 Effect on the Liver

Emerging evidence has pointed out that *Cordyceps species* and *Cordyceps*-based agents possess hepatoprotective and ameliorating effects on the liver. The fermented *C. militaris* extract by *Pediococcus pentosaceus*ON188 (ONE) was studied against nonalcoholic fatty liver diseases (NAFLD) in mice fed a high-fat diet (HFD). ONE was shown to lower lactate dehydrogenase, aspartate transaminase, and alanine transaminase levels in addition to its reducing effect on lipid droplets and triglyceride levels. Besides, the sizes of adipocytes were markedly lowered dose-dependently in epididymal fat tissue. ONE was shown to arouse fatty acid oxidation via SPHK2 in the liver and to block lipogenesis in adipose tissue, confirming its function against liver dysfunction and obesity (Tran et al., 2019).

C. sinensis mycelia significantly reduced inflammatory cell infiltration and lessened collagen accumulation in the liver tissue. Thus, CCl_4-induced liver fibrosis was ameliorated in mice, likely by the regulation of the Toll-like receptor 4 (TLR4)/nuclear transcription factor-κ (NF-κB) signaling pathway and inhibiting angiopoietin-like protein 4 (ANGPTL4) expression (Zhang et al., 2022b). Peng et al. (2016) announced that cultured mycelium *C. sinensis* (CMCS), when administered to rats markedly ameliorated liver function, alleviated liver inflammation and fibrosis, and improved hepatic natural killer cells in CCl_4-treated mice with liver inflammation (Peng et al., 2016). Ascertaining the efficacy and safety of *C. militaris* in Korean adults with mild liver disturbance by Heo et al. (2015) demonstrated that *C. militaris* may be consumed as a functional food in patients with mild liver dysfunction safely and is believed to prevent the advancement of fatty liver or cirrhosis through precluding lipid collection in hepatocytes (Heo et al., 2015). When evaluating the cytoprotective capacity of *C. sinensis*, the levels of AST, TNF-α, and NO decreased, contrary to the enhanced levels of IL-10 and SOD in mice with D(+)-galactosamine (GalN)/lipopolysaccharide (LPS)-induced fulminant hepatic failure receiving *C. sinensis* pre-treatment. *C. sinensis* further reduced the apoptosis of hepatocytes, suggesting *C. sinensis* as a potent anti-inflammatory, antioxidant, and anti-apoptotic agent to attenuate liver damage (Cheng et al., 2014).

Another study conducted on the extracellular polysaccharide from *C. militaris* (CEP-1) ameliorated the function of malondialdehyde, glutathione peroxidase, and superoxide dismutase in serum and organs, repaired the physiological indicators of total albumin, creatinine, protein, and blood urea nitrogen in serum, and lowered the action of aspartate aminotransferase and lactate dehydrogenase in the liver and kidney of mice poisoned by Pb^{2+}. CEP-I lessened the Pb^{2+} content in mice poisoned by Pb^{2+}, demonstrating the protective impact of the constituents of *C. militaris* in the liver (Song and Zhu 2020). Zhang et al. (2021) investigated two polysaccharides (CPA-1 and CPB-2) from *C. cicadae* in a high fructose/high fat diet (HF/HFD) that triggered obesity and metabolic disorders in rats. CPA-1 and CPB-2 markedly lowered serum and hepatic lipid contents, liver performance, and pro-inflammatory cytokines (IL-1β, IL-6, and TNF-α) in addition to the attenuation of hepatic oxidative stress in the liver, implying the defensive activity of *C. cicadae* polysaccharides to repair metabolic disturbances (Zhang et al., 2021a).

A polysaccharide from *C. militaris* (EPCMa) was shown to display a hepatoprotective effect by triggering the Nrf2 pathway for the attenuation of ROS and lipid peroxidation and repressing NF-κB p65 and IκBα phosphorylation, finally acquiring the hepatoprotective activity in CCl_4-induced liver injury, proposing EPCMa as a natural agent against liver damage (Zhao et al., 2022).

4.7 *Effect on the Renal System*

Scientific knowledge has pointed out the renoprotective effects of fermented *C. sinensis* in the diabetic mouse (Liu et al., 2022; Zhang et al., 2023b; Kan et al., 2012). Examining the modes of action of fermented *C. sinensis*-associated nephroprotective effects, the enhancement of proliferation, and the inhibition of apoptosis of renal proximal tubular cells, likely through caspase-3, Bax, VEGFA, and PTEN, were observed (Zhang et al., 2023b). Examining the probable benefits of *C. cicadae* polysaccharides (CCP) in rats induced by diabetic nephropathy (DN), CCP administration was shown to markedly retard the progression of renal interstitial fibrosis, possibly by blocking the inflammatory response and regulating gut microbiota dysbiosis (Yang et al., 2020).

Similarly, *C. cicadae*-associated renoprotection on hypertensive nephropathy was linked possibly to the modulation of SIRT1-induced autophagic stress through the pathway of FOXO3a and oxidative stress (Cai et al., 2022). A nucleoside/nucleobase-abundant extract from *C. sinensis* (CS-N) remarkably alleviated the abnormal renal functional parameters, repaired histopathological alterations, and suppressed the epithelial-mesenchymal transition (EMT) and extracellular matrix (ECM) collection by modulating p38/ERK signaling pathways, suggesting the potential of CS-N on experimental diabetic renal fibrosis (Dong et al., 2019).

4.8 Immunomodulatory Activity

The immunomodulatory actions of various extracts of *Cordyceps* species, their-containing polysaccharides, and their other constituents, such as cordycepin, were announced based on the emerging literature data (Lee et al., 2020; Das et al., 2021; Fan et al., 2021). For instance, Su et al. (2020) evaluated the influence of polysaccharides purified from *C. kyushuensis*-cultured stroma in terms of their likely immunomodulatory action and demonstrated that they encouraged the increase of mouse splenocytes, boosted peritoneal macrophages, and enhanced the levels of IL-2 and TNF-α in serum, proposing immunomodulatory activities of these fractions (Su et al., 2020).

4.9 Neuroprotective Activity

Inquired about the neuroprotective action of *C. cicadae* mycelium extract (CCME) in rats with optic nerve crush. The CCME was shown to induce neuroprotective effects via exhibiting antiapoptotic and anti-inflammatory properties (Wen et al., 2022). Cordycepin obtained from *C. militaris* has been researched for the determination of its role in Parkinson's disease (PD) due to the fact that the onset and progression of PD depends on neuroinflammation and cordycepin achieves anti-inflammatory and neuroprotective properties, which were proven by diverse scientific data (Tan et al., 2020; Govindula et al., 2021; Zhang et al., 2021a). Investigating the neuroprotective effects of cordycepin demonstrated that cordycepin induced neuroprotection through the regulation of TLR4/NF-κB/NLRP3-linked pyroptosis, provides the potential of cordycepin against PD in the future (Sun et al., 2020).

The administration of the ethanolic extract of *C. militaris* in amyloid-beta $(A\beta)_{1-42}$-induced AD mouse models improved new direction awareness and novel object realization. Besides, the CM extract suppressed nitric oxide generation and lipid peroxidation in the brain, liver, and kidney, emphasizing the defensive function of CM towards the development of AD (He et al., 2019). Chang (2023) demonstrated the influence of deep ocean water (DOW)-cultured *C. cicadae* (DCC) in rats with brain injury and memory impairment and unraveled that DCC and its metabolite N6-(2-hydroxyethyl)-adenosine (HEA) promoted retention ability and displayed persuasive antioxidant and free radical scavenging activity in an *in vivo* rat model of aging. DOW-cultured *C. cicadae* further eased the release of inflammatory factors. Thus, DOW-cultured *C. cicadae* possessing antioxidant, anti-inflammatory, and neuroprotective influences suggested an alternate agent for age-associated brain injury and cognitive deterioration (Chang et al., 2023).

4.10 Other Effects

Cordyceps species have gained rising popularity and importance due to their traditional and experimentally validated functions and commercial uses in many cultures. Emerging evidence has confirmed a wide range of biological functions of *Cordyceps* species to date. Besides, *Cordyceps*-linked investigations to unravel their biological activities and modes of action are still ongoing at present. Together with the above-mentioned specialities of the *Cordyceps* genus, the hypoglycemic (Zhang et al., 2006, Yu et al., 2015, Sun et al., 2021, Zhao et al., 2018, Shang et al., 2020, Wang et al., 2023, Borde and Singh 2022), hypolipidemic (Borde and Singh 2022, Kalita et al., 2022, Liu et al., 2019, Yang et al., 2021, Gangzheng et al., 2023), antiaging (Ji et al., 2009, Zhu et al., 2020, Shao et al., 2023), anti-viral (Verma 2022, Panya et al., 2021, Lee et al., 2014, Zhu et al., 2016), and other biological activities were ascertained (Yue et al., 2008, Sohn et al., 2012, Lin et al., 2007, Borde and Singh 2022, Tuli et al., 2014).

5. Clinical Trials

Emerging data and promising biological outcomes of *Cordyceps* species have brought them to clinical trials. At present, a number of clinical investigations have been carried out on *Cordyceps* species. The influence of *C. cicadae* extract on visual acuity and visual fatigue, as well as against high intraocular pressure and dry eye, the effect of *C. sinensis* mycelium culture on immunity promotion, and the impact of *C. sinensis* on cardiovascular fitness and cognitive function, have been examined in clinical trials (NIH, 2023). In clinical research, *Cordyceps* species hold undeniable potential with diverse biological activities.

Conclusion

Natural medications have come to the forefront for the last few decades due to their broad therapeutic profiles in the fields of the pharmaceutics industry, their applications in the area of complementary and alternative medicine, and their persuasive safety profile in comparison to synthetic ones. *Cordyceps* species have gained enhanced value recently due to having been used for centuries in traditional medicine applications and the pharmacological features of *Cordyceps*-based products confirmed by experimental studies. The engrossing potency of these therapeutic mushrooms on human health has to be further examined, and cultivation techniques should be developed to gain sufficient amounts of these fungi to validate their ethnopharmacological and commercial uses. The undeniable reality is that *Cordyceps* taxa may still be endowed with undefined pharmacological effects, which may enable the development of new therapeutic applications of *Cordyceps*-associated products for prospective therapeutic applications.

References

Asai, T., Luo, D., Obara, Y., Taniguchi, T., Monde et al. (2012). Dihydrobenzofurans as cannabinoid receptor ligands from *Cordyceps annullata*, an entomopathogenic fungus cultivated in the presence of an HDAC inhibitor. Tetrahedron Lett., 53(17): 2239–2243.

Borde, M. and Singh, S.K. (2022) 'Prospects of Cordycepin and Polysaccharides Produced by *Cordyceps*'. pp. 93–107. *In*: Rajpal, V.R., Singh, I. and Navi, S.S. (eds.). Fungal diversity, ecology and control management, Singapore: Springer Nature Singapore.

Bunyapaiboonsri, T., Yoiprommarat, S., Intereya, K. and Kocharin, K. (2007). New diphenyl ethers from the insect pathogenic fungus *Cordyceps* sp. BCC 1861. Chem Pharm Bull, 55(2): 304–7.

Cai, Y., Feng, Z., Jia, Q., Guo, J., Zhang, P. et al. (2022). *Cordyceps cicadae* ameliorates renal hypertensive injury and fibrosis through the regulation of SIRT1-mediated autophagy. Front. Pharmacol., 12:801094. https://doi.org/10.3389/fphar.2021.801094

Chang, C.Y., Yang, P.X., Yu, T.L. and Lee, C.L. (2023). *Cordyceps cicadae* NTTU 868 mycelia fermented with deep ocean water minerals prevents D-galactose-induced memory deficits by inhibiting oxidative inflammatory factors and aging-related risk factors. Nutrients, 15(8): 1968.

Chang, H.S., Cheng, M.J., Wu, M.D., Chan, H.Y., Hsieh, S.Y et al. (2017). Secondary metabolites produced by an endophytic fungus *Cordyceps ninchukispora* from the seeds of *Beilschmiedia erythrophloia* Hayata. Phytochem. Lett., 22: 179–184.

Chang, M.M., Hong, S.Y., Yang, S.H., Wu, C.C., Wang, C.Y. et al (2020). Anti-cancer effect of cordycepin on FGF9-induced testicular tumorigenesis. Int. J. Mol. Sci., 21(21): 8336.

Chen, J., Chan, W.M., Leung, H.Y., Leong, P.K., Yan, C.T.M. et al. (2020). Anti-inflammatory effects of a *Cordyceps sinensis* mycelium culture extract (Cs-4) on rodent models of allergic rhinitis and asthma. Molecules, 25(18): 4051.

Cheng, Y.J., Cheng, S.M., Teng, Y.H., Shyu, W.C., Chen, H.L et al. 2014. *Cordyceps sinensis* prevents apoptosis in mouse liver with D-galactosamine/lipopolysaccharide-induced fulminant hepatic failure. Am. J Chinese Med., 42(02): 427–441.

Chiu, C.P., Liu, S.C., Tang, C.H., Chan, Y., El-Shazly et al. (2016). Anti-inflammatory cerebrosides from cultivated *Cordyceps militaris*. J. Agric. Food Chem., 64(7): 1540–1548.

Cho, M.A., Lee, D.S., Kim, M.J., Sung, J.M. and Ham, S.S. et al. (2003). Antimutagenicity and cytotoxicity of cordycepin isolated from *Cordyceps militaris*. Food Sci. Biotechnol., 12: 472–475.

Choi, J.N., Kim, J., Lee, M.Y., Park, D.K., Hong, Y.S. et al. (2010). Metabolomics revealed novel isoflavones and optimal cultivation time of *Cordyceps militaris* fermentation. J. Agric. Food Chem., 58(7): 4258–4267.

Cunningham, K.G., Manson, W.I.L.L.I.A.M., Spring, F.S. and Hutchinson, S.A et al. (1950). Cordycepin, a metabolic product isolated from cultures of *Cordyceps militaris* (Linn.) Link. Nature, 166(4231): 949–949.

Das, G., Shin, H.S., Leyva-Gómez, G., Prado-Audelo, M.L.D., Cortes, H., Singh et al. (2021). *Cordyceps* spp.: a review on its immune-stimulatory and other biological potentials. Front. Pharmacol., 11:602364. doi: 10.3389/fphar.2020.602364

Dong, C., Guo, S., Wang, W. and Liu, X. (2015). *Cordyceps* industry in China. Mycology, 6(2): 121–129.

Dong, C.H., Yang, T. and Lian, T. (2014). A comparative study of the antimicrobial, antioxidant, and cytotoxic activities of methanol extracts from fruit bodies and fermented mycelia of caterpillar medicinal mushroom *Cordyceps militaris* (Ascomycetes). Int. J. Med., Mushrooms, 16(5): 485–95.

Dong, M., Zhao, C., Huang, Y., Zheng, K., Bao, G., Hu, F., Peng, F., Chen, M., Li, Z. and Lu, R. (2024). Metabolites analysis and new bioactive compounds from the medicine food homology product of *Cordyceps chanhua* on artificial media. J. Pharm. Biomed. Anal., 237: 115749.

Dong, Z., Sun, Y., Wei, G., Li, S. and Zhao, Z. et al. (2019). A Nucleoside/nucleobase-rich extract from *Cordyceps sinensis* inhibits the epithelial–mesenchymal transition and protects against renal fibrosis in diabetic nephropathy. Molecules, 24(22): 4119.

Du, Y., Yu, J., Du, L., Tang, J. and Feng, W.H. et al. (2016). Cordycepin enhances Epstein–Barr virus lytic infection and Epstein–Barr virus-positive tumor treatment efficacy by doxorubicin. Cancer Lett., 376(2): 240–248.

Eiamthaworn, K., Kaewkod, T., Bovonsombut S. and Tragoolpua, Y. (2022). Efficacy of *Cordyceps militaris* extracts against some skin pathogenic bacteria and antioxidant activity. J. Fungi, 8(4).

Elshamy, A.I., Mohamed, T.A., Yoneyama, T., Noji, M., Ban, S. et al. (2023). Bifusicoumarins A-D: cytotoxic 3S-dihydroisocoumarins from the entomopathogenic fungus *Cordyceps bifusispora* (NBRC 108997). Phytochem, 212: 113743.

Fan, H.B., Zou, Y., Han, Q., Zheng, Q.W., Liu, Y.L et al. (2021). *Cordyceps militaris* immunomodulatory protein promotes the phagocytic ability of macrophages through the TLR4-NF-κB Pathway. Int. J. Mol. Sci., 22(22): 12188.

Fan, W., Li, E., Ren, J., Wang, W., Liu, X. et al. (2020). Cordycepamides A–E and cordyglycoside A, new alkaloidal and glycoside metabolites from the entomopathogenic fungus *Cordyceps* sp. Fitoterapia, 142: 104525.

Gangzheng, W., Chengyuan, S., Qiuju, H., Chenghua, Z., Min, L et al. (2023). Effect and correlation of *Cordyceps guangdongensis* ethanolic extract on obesity, dyslipidemia and gut microbiota dysbiosis in high-fat diet mice. J. Funct. Foods, 107: 105663.

Govindula, A., Pai, A., Baghel, S. and Mudgal, J. (2021). Molecular mechanisms of cordycepin emphasizing its potential against neuroinflammation: an update. Eur. J. Pharmacol., 908: 174364.

Gu, L., Yu, T., Liu, J. and Lu, Y. (2020). Evaluation of the mechanism of *Cordyceps* polysaccharide action on rat acute liver failure. Arch. Med. Sci., 16(5): 1218–1225.

Guo, Z., Chen, W., Dai, G. and Huang, Y. (2020). Cordycepin suppresses the migration and invasion of human liver cancer cells by downregulating the expression of CXCR4. Int. J. Mol. Med., 45(1): 141–150.

Hama, M., Elshamy, A.I., Yoneyama, T., Kasai, Y., Yamamoto, H., Tanigawa, K. et al. (2019). New alkaloidal metabolites from cultures of entomopathogenic fungus *Cordyceps takaomontana* NBRC 101754. Fitoterapia, 139: 104364.

Hao, P., Jiang, F., Cheng, J., Ma, L., Zhang, Y. et al. (2017). Traditional Chinese Medicine for cardiovascular disease: evidence and potential mechanisms. J. Am. Coll. Cardiol., 69(24): 2952–2966.

Haritakun, R., Sappan, M., Suvannakad, R., Tasanathai, K. and Isaka, M. et al. (2010). An antimycobacterial cyclodepsipeptide from the entomopathogenic fungus *Ophiocordyceps communis* BCC 16475. J. Nat. Prod., 73(1): 75–78.

He, M.T., Lee, A.Y., Kim, J.H., Park, C.H., Shin et al. (2019). Protective role of *Cordyceps militaris* in Aβ1–42-induced Alzheimer's disease in vivo. Food Sci. Biotechnol., 28(3): 865–872.

Heo, J.Y., Baik, H.W., Kim, H.J., Lee, J.M., Kim, H.W et al. (2015). The efficacy and safety of *Cordyceps militaris* in Korean adults who have mild liver dysfunction. J. Clin. Nutr., 7(3): 81–86.

Huo, X., Liu, C., Bai, X., Li, W., Li, J. et al. (2017). Aqueous extract of *Cordyceps sinensis* potentiates the antitumor effect of DDP and attenuates therapy-associated toxicity in non-small cell lung cancer via IκBα/NFκB and AKT/MMP2/MMP9 pathways. RSC Adv., 7(60): 37743–37754.

Isaka, M., Boonkhao, B., Rachtawee, P. and Auncharoen, P. (2007a). A xanthocillin-like alkaloid from the insect pathogenic fungus *Cordyceps brunnearubra* BCC 1395. J. Nat. Prod., 70(4): 656–658.

Isaka, M., Chinthanom, P., Rachtawee, P., Somyong, W., Luangsa-ard, J.J et al. (2013). Cordylactam, a new alkaloid from the spider pathogenic fungus *Cordyceps* sp. BCC 12671. Phytochem. Lett., 6(2): 162–164.

Isaka, M., Srisanoh, U., Lartpornmatulee, N. and Boonruangprapa, T. (2007b). ES-242 Derivatives and cycloheptapeptides from *Cordyceps* sp. Strains BCC 16173 and BCC 16176. J. Nat. Prod., 70(10): 1601–1604.

Isaka, M., Tanticharoen, M., Kongsaeree, P. and Y. Thebtaranonth, Y. (2001). Structures of cordypyridones A–D, antimalarial N-hydroxy- and N-methoxy-2-pyridones from the insect pathogenic fungus *Cordyceps nipponica*. J. Org. Chem., 66(14): 4803–4808.

Jang, H.J., Yang, K.E., Hwang, I.H., Huh, Y.H., Kim, D.J. et al. (2019). Cordycepin inhibits human ovarian cancer by inducing autophagy and apoptosis through Dickkopf-related protein 1/β-catenin signaling. Am. J. Transl. Res., 11(11): 6890.

Jaturapat, A., Isaka, M., Hywel-Jones, N.L., Lertwerawat, Y., Kamchonwongpaisan et al. (2001). Bioxanthracenes from the insect pathogenic fungus. *Cordyceps pseudomilitaris* BCC 1620. I. Taxonomy, fermentation, isolation and antimalarial activity. J. Antibiot., 54(1): 29–35.

Jędrejko, K. J., Lazur, J. and Muszyńska, B. (2021). *Cordyceps militaris*: an overview of its chemical constituents in relation to biological activity. Foods, 10(11): 2634.

Ji, D.B., Ye, J., Li, C.L., Wang, Y.H., Zhao, J. et al. (2009). Antiaging effect of *Cordyceps sinensis* extract. Phytother. Res., 23(1): 116–122.

Jia, J.M., Tao H.H. and Feng, B.M. (2009). Cordyceamides A and B from the culture liquid of *Cordyceps sinensis* (BERK.) SACC. Chem. Pharm. Bull., 57(1): 99–101.

Jiang, Q., Lou, Z.,Wang H. and Chen, C. (2019). Antimicrobial effect and proposed action mechanism of cordycepin against *Escherichia coli* and *Bacillus subtilis*. J Microbiol, 57(4): 288–297.

Jiao, C., Liang, H., Liu, L., Li, S., Chen, J. et al. (2022). Transcriptomic analysis of the anti-inflammatory effect of *Cordyceps militaris* extract on acute gouty arthritis. Front. Pharmacol., 13:1035101. doi: 10.3389/fphar.2022.1035101

Jo, W.S., Choi, Y.J., Mm, H.J., Lee, J.Y., Nam, B.H. et al. (2010). The anti-inflammatory effects of water extract from *Cordyceps militaris* in murine macrophage. Mycobiology, 38(1): 46–51.

Joshi, M., Sagar A. and Pathania, P. (2015). FTIR studies and antimicrobial potential of *Cordyceps militaris* (a highly medicinal fungus) against Hospital Pathogens. Microbiology, 4(5).

Kalita, P., Ahmed, A B, Sen S. and Chakraborty, R. (2022). A comprehensive review on polysaccharides with hypolipidemic activity: occurrence, chemistry and molecular mechanism. Int. J. Biol. Macromol., 206: 681–698.

Kan, W.C., Wang, H.Y., Chien, C.C., Li, S.L., Chen, Y.C. et al. (2012). Effects of extract from solid-state fermented *Cordyceps sinensis* on type 2 diabetes mellitus. Evid. Based Complement. Alternat. Med., 2012: 743107.

Kim, K.M., Kwon, Y.G., Chung, H.T., Yun, Y.G., Pae, H.O. et al. (2003). Methanol extract of *Cordyceps pruinosa* inhibits *in vitro* and *in vivo* inflammatory mediators by suppressing NF-κB activation. Toxicol. Appl Pharmacol, 190(1): 1–8.

Kim, S.B., Ahn, B., Kim, M., Ji, H.J., Shin, S.K. et al. (2014). Effect of *Cordyceps militaris* extract and active constituents on metabolic parameters of obesity induced by high-fat diet in C58BL/6J mice. J. Ethnopharmacol., 151(1): 478–484.

Kim, S.O., Cha, H.J., Park, C., Lee, H., Hong, S.H. et al. (2019). Cordycepin induces apoptosis in human bladder cancer T24 cells through ROS-dependent inhibition of the PI3K/Akt signaling pathway. Biosci. Trends, 13(4): 324–333.

Krasnoff, S.B., Reátegui, R.F., Wagenaar, M.M., Gloer, J.B. and Gibson, D.M. et al. (2005). Cicadapeptins I and II: New aib-containing peptides from the entomopathogenic fungus *Cordyceps heteropoda*. J. Nat. Prod., 68(1): 50–55.

Kuo, Y.C., Lin, L.C., Don, M.J., Liao, H.F., Tsai, Y.P. et al. (2002). Cyclodesipeptide and dioxomorpholine derivatives isolated from the insect-body portion of the fungus *Cordyceps cicadae*. J. Chin. Med., 13(4): 209–219.

Lan, Y.H., Lu, Y.S., Wu, J.Y., Lee, H.T., Srinophakun, P. et al. (2022). *Cordyceps militaris* reduces oxidative stress and regulates immune T cells to inhibit metastatic melanoma invasion. Antioxidants, 11(8).

Lee, C.T., Huang, K.S., Shaw, J.F., Chen, J.R., Kuo, W.S. et al. (2020). Trends in the immunomodulatory effects of *Cordyceps militaris*: total extracts, polysaccharides and cordycepin. Front. Pharmacol., 11: 575704. doi: 10.3389/fphar.2020.575704

Lee, D., Lee, W.Y., Jung, K., Kwon, Y.S., Kim, D. et al (2019). The inhibitory effect of cordycepin on the proliferation of MCF-7 breast cancer cells, and its mechanism: an investigation using network pharmacology-based analysis. Biomolecules, 9(9): 414.

Lee, E.J., Kim W.J. and Moon, S.K. (2010). Cordycepin suppresses TNF-alpha-induced invasion, migration and matrix metalloproteinase-9 expression in human bladder cancer cells. Phytother. Res., 24(12): 1755–1761.

Lee, H.H., Park, H., Sung, G.H., Lee, K., Lee, T. et al. (2014). Anti-influenza effect of *Cordyceps militaris* through immunomodulation in a DBA/2 mouse model. J. Microbiol., 52(8): 696–701.

Li, L.Q., Song, A.X., Yin, J.Y., Siu, K.C., Wong, W.T et al. (2020a) Anti-inflammation activity of exopolysaccharides produced by a medicinal fungus *Cordyceps sinensis* Cs-HK1 in cell and animal models. Int. J. Biol. Macromol., 149: 1042–1050.

Li, S.Z., Ren, J.W., Fei, J., Zhang, X.D. and Du, R.L. et al. (2019). *Cordycepin induces* Bax-dependent apoptosis in colorectal cancer cells. Mol. Med. Rep., 19(2): 901–908.

Li, W., Li, J., Hu, X., Xu, L., Liu, X. et al. (2022). Tumour suppression by Chinese *Cordyceps* extract via antiangiogenic activity. Mycol. Prog., 21(9): 76.

Li, Z., Guo, Z., Zhu, J., Bi, S., Luo, Y. et al. (2020b). *Cordyceps militaris* fraction inhibits angiogenesis of hepatocellular carcinoma *in vitro* and *in vivo*. Pharmacogn. Mag., 16(67): 169–176.

Liao, Y., Ling, J., Zhang, G., Liu, F., Tao et al. (2015). Cordycepin induces cell cycle arrest and apoptosis by inducing DNA damage and up-regulation of p53 in Leukemia cells. Cell Cycle, 14(5): 761–771.

Lin, W.H., Tsai, M.T., Chen, Y.S., Hou, R.C.W., Hung, H.F., Li et al. (2007). Improvement of sperm production in subfertile boars by *Cordyceps militaris* supplement. Am. J. Chinese Med., 35(04): 631–641.

Liu, C., Qi, M., Li, L., Yuan, Y., Wu, X. et al. 2020. Natural cordycepin induces apoptosis and suppresses metastasis in breast cancer cells by inhibiting the Hedgehog pathway. Food Funct., 11(3): 2107–2116.

Liu, R.-M., Dai, R., Luo, Y. and Xiao, J.H. (2019). Glucose-lowering and hypolipidemic activities of polysaccharides from *Cordyceps taii* in streptozotocin-induced diabetic mice. BMC Complement Altern. Med., 19(1): 230.

Liu, R.M., Zhang, X.J., Liang, G.Y., Yang, Y.F., Zhong, J.J. et al. (2015). Antitumor and antimetastatic activities of chloroform extract of medicinal mushroom *Cordyceps taii* in mouse models. BMC Complement Altern. Med., 15(1): 216.

Liu, W., Gao, Y., Zhou, Y., Yu, F., Li, X et al (2022). Mechanism of *Cordyceps sinensis* and its extracts in the treatment of diabetic kidney disease: a review. Front. Pharmacol., 13:881835. doi: 10.3389/fphar.2022.881835

Luo, L., Ran, R., Yao, J., Zhang, F., Xing, M. et al. (2019). Se-enriched *Cordyceps militaris* inhibits cell proliferation, induces cell apoptosis, and causes G2/M phase arrest in human non-small cell lung cancer cells. Onco Targets Ther., 12: 8751–8763.

Nallathamby, N., Guan-Serm, L., Vidyadaran, S., Malek, S.N.A., Raman et al. (2015). Ergosterol of *Cordyceps militaris* attenuates LPS induced inflammation in BV2 microglia cells. Nat. Prod. Commun., 10(6): 885–886.

NIH. (2023). ClinicalTrials.gov [Online]. Available: https://clinicaltrials.gov/search?term=Cordyceps&page=3 [Accessed on 16.11.2023].

Olatunji, O.J., Tang, J., Tola, A., Auberon, F., Oluwaniyi, O. et al. (2018). The genus Cordyceps: an extensive review of its traditional uses, phytochemistry and pharmacology. Fitoterapia, 129: 293–316.

Özenver N., Boulos, J. C. and Efferth, T. (2021). Activity of cordycepin from *Cordyceps sinensis* against drug-resistant tumor cells as determined by gene expression and drug sensitivity profiling. Nat. Prod. Commun., 16(2): 1934578X21993350.

Pahwa, R., Goyal A. and Jialal, I. (2023). Chronic Inflammation, Treasure Island (FL): StatPearls Publishing.

Panda, A. K. and Swain, K.C. (2011). Traditional uses and medicinal potential of *Cordyceps sinensis* of Sikkim. J. Ayurveda Integr. Med., 2(1): 9–13.

Panya, A., Songprakhon, P., Panwong, S., Jantakee, K., Kaewkod, T. et al. (2021). Cordycepin inhibits virus replication in dengue virus-infected Vero cells. Molecules, 26(11): 3118.

Park, J.G., Son, Y.J., Lee, T.H., Baek, N.J., Yoon, D.H. et al. (2017). Anticancer efficacy of *Cordyceps militaris* ethanol extract in a xenografted leukemia model. Evid. Based Complement. Alternat. Med., 2017: 8474703.

Peng, Y., Huang, K., Shen, L., Tao, Y.Y. and Liu, C.H. et al. (2016). Cultured mycelium *Cordyceps sinensis* alleviates CCl4-induced liver inflammation and fibrosis in mice by activating hepatic natural killer cells. Acta Pharmacol. Sin., 37(2): 204–216.

Qi, W., Zhou, X., Wang, J., Zhang, K., Zhou, Y et al. (2020). *Cordyceps sinensis* polysaccharide inhibits colon cancer cells growth by inducing apoptosis and autophagy flux blockage via mTOR signaling. Carbohydr. Polym., 237: 116113.

Qin, P., Li, X., Yang, H., Wang, Z.Y. and D. Lu, D. et al. (2019). Therapeutic potential and biological applications of cordycepin and metabolic mechanisms in cordycepin-producing fungi. Molecules, 24(12).

Rao, Y. K., Fang, S.H. and Tzeng, Y.M. (2007). Evaluation of the anti-inflammatory and anti-proliferation tumoral cells activities of *Antrodia camphorata*, *Cordyceps sinensis*, and *Cinnamomum osmophloeum* bark extracts. J. Ethnopharmacol., 114(1): 78–85.

Shang, X.L., Pan, L.C., Tang, Y., Luo, Y., Zhu, Z.Y. et al. (2020). 1H NMR-based metabonomics of the hypoglycemic effect of polysaccharides from *Cordyceps militaris* on streptozotocin-induced diabetes in mice. Nat. Prod. Res., 34(10): 1366–1372.

Shao, L., Jiang, S., Li, Y., Yu, L., Liu, H. et al. (2023). Aqueous extract of *Cordyceps cicadae* (Miq.) promotes hyaluronan synthesis in human skin fibroblasts: a potential moisturizing and anti-aging ingredient. Plos One, 18(7): e0274479.

Smiderle, F.R., Baggio, C.H., Borato, D.G., Santana-Filho, A.P., Sassaki, G.L. et al. (2014). Anti-inflammatory properties of the medicinal mushroom *Cordyceps militaris* might be related to its linear $(1{\rightarrow}3)$-β-D-glucan. Plos one, 9(10): e110266.

Sohn, S.H., Lee, S.C., Hwang, S.Y., Kim, S.W., Kim, I.W. et al. (2012). Effect of long-term administration of cordycepin from *Cordyceps militaris* on testicular function in middle-aged rats. Planta Med., 78(15): 1620–1625.

Song, Q. and Z. Zhu, Z. (2020). Using *Cordyceps militaris* extracellular polysaccharides to prevent Pb 2+-induced liver and kidney toxicity by activating Nrf2 signals and modulating gut microbiota. Food Funct., 11(10): 9226–9239.

Su, J., Sun, J., Jian, T., Zhang, G. and Ling, J. (2020). Immunomodulatory and antioxidant effects of polysaccharides from the parasitic fungus *Cordyceps kyushuensis*. Biomed. Res. Int., 2020: 8257847.

Sun, H., Yu, X., Li, T. and Zhu, Z. (2021). Structure and hypoglycemic activity of a novel exopolysaccharide of *Cordyceps militaris*. Int. J. Biol. Macromol., 166: 496–508.

Sun, J., Jin, M., Zhou, W., Diao, S., Zhou, Y. et al. (2017). A new ribonucleotide from *Cordyceps militaris*. Nat. Prod. Res., 31(21): 2537–2543.

Sun, J., Xu, J., Wang, S., Hou, Z., Lu, X. et al. (2019). A new cerebroside from *Cordyceps militaris* with anti-PTP1B activity. Fitoterapia, 138: 104342.

Sun, Y., Huang, W.M., Tang, P.C., Zhang, X., Zhang, X.Y. et al. (2020). Neuroprotective effects of natural cordycepin on LPS-induced Parkinson's disease through suppressing TLR4/NF-κB/NLRP3-mediated pyroptosis. J. Funct. Foods, 75: 104274.

Sun, Y., Zhao, Z., Feng, Q., Xu, Q., Lü, L. et al. (2013). Unusual apirodecane sesquiterpenes and a fumagillol analogue from *Cordyceps ophioglossoides*. Helv. Chim. Acta, 96(1): 76–84.

Sung, G.H., Hywel-Jones, N.L., Sung, J.M., Luangsa-Ard, J.J., Shrestha, B. et al. (2007). Phylogenetic classification of *Cordyceps* and the clavicipitaceous fungi. Stud. Mycol., 57: 5–59.

Tan, L., Song, X., Ren, Y., Wang, M., Guo, C. et al. (2020). Anti-inflammatory effects of cordycepin: a review. Phytother Res., 35(3): 1284–1297.

Tang, H., Chen, C., Zou, Y., Lou, H., Zheng, Q. et al. (2019). Purification and structural characterization of a novel natural pigment: cordycepene from edible and medicinal mushroom *Cordyceps militaris*. Appl. Microbiol. Biotechnol., 103(19): 7943–7952.

Tania, M., Shawon, J., Saif, K., Kiefer, R., Khorram, M.S. et al. (2019). Cordycepin downregulates Cdk-2 to interfere with cell cycle and increases apoptosis by generating ROS in cervical cancer cells: in vitro and in silico study. Curr. Cancer Drug Targets, 19(2): 152–159.

Tran, N.K.S., Kim, G.T., Park, S.H., Lee, D., Shim, S.M. et al. (2019). Fermented *Cordyceps militaris* extract prevents hepatosteatosis and adipocyte hypertrophy in high fat diet-fed mice. Nutrients, 11(5): 1015.

Tuli, H. S., Sandhu S. S. and Sharma, A. K. (2014). Pharmacological and therapeutic potential of *Cordyceps* with special reference to Cordycepin. 3 Biotech, 4(1): 1-12.

Tuli, H.S., Sharma, A.K., Sandhu S.S. and Kashyap, D. (2013). Cordycepin: a bioactive metabolite with therapeutic potential. Life Sci., 93(23): 863–869.

Umeyama, A., Takahashi, K., Grudniewska, A., Shimizu, M., Hayashi, S. et al. (2014). In vitro antitrypanosomal activity of the cyclodepsipeptides, cardinalisamides A-C, from the insect pathogenic fungus *Cordyceps cardinalis* NBRC 103832. J. Antibiot., 67(2): 163–166.

Varughese, T., Rios, N., Higginbotham, S., Arnold, A.E., Coley, P.D. et al. (2012). Antifungal depsidone metabolites from *Cordyceps dipterigena*, an fndophytic fungus antagonistic to the phytopathogen *Gibberella fujikuroi*. Tetrahedron Lett., 53(13): 1624–1626.

Verma, A.K. (2022). Cordycepin: a bioactive metabolite of *Cordyceps militaris* and polyadenylation inhibitor with therapeutic potential against COVID-19. J. Biomol. Struct. Dyn., 40(8): 3745–3752.

Wang, A. (1955). Synopsis of Materia media.

Wang, J., Zhang, D.M., Jia, J.F., Peng, Q.L., Tian, H.Y. et al. (2014). Cyclodepsipeptides from the ascocarps and insect-body portions of fungus *Cordyceps cicadae*. Fitoterapia, 97: 23–27.

Wang, L., Sun, H., Yang, M., Xu, Y., Hou, L. et al. (2022). Bidirectional regulatory effects of *Cordyceps* on arrhythmia: clinical evaluations and network pharmacology. Fron. Pharmacol., 13: 948173. doi: 10.3389/fphar.2022.948173

Wang, M., Kornsakulkarn, J., Srichomthong, K., Feng, T., Liu, J.K. et al. (2019a). Antimicrobial anthraquinones from cultures of the ant pathogenic fungus *Cordyceps morakotii* BCC 56811. J. Antibiot., 72(3): 141–147.

Wang, Y., Dai, Y.D., Yang, Z.L., Guo, R., Wang, Y.B. et al. (2021). Morphological and molecular phylogenetic data of the Chinese medicinal fungus *Cordyceps liangshanensis* reveal its new systematic position in the family Ophiocordycipitaceae. Mycobiology, 49(4): 297–307.

Wang, Y., Lv, Y., Liu, T.S., Di Yan, W., Chen, L.Y., Li, Z.H., Piao, Y.S., An, R.B., Lin, Z.H. and Ren, X.S. (2019b). Cordycepin suppresses cell proliferation and migration by targeting CLEC2 in human gastric cancer cells via Akt signaling pathway. Life Sci., 223: 110-119.

Wang, Y., Zeng, T., Li, H., Wang, Y. , Wang J. et al. (2023). Structural characterization and hypoglycemic function of polysaccharides from *Cordyceps cicadae*. Molecules, 28(2): 526.

Wei, C., Khan, M.A., Du, J., Cheng, J., Tania, M. et al. (2022b). Cordycepin inhibits triple-negative breast cancer cell migration and invasion by regulating EMT-TFs SLUG, TWIST1, SNAIL1, and ZEB1. Front. Oncol., 12: 898583.

Wei, J., Zhou, X., Dong, M., Yang, L., Zhao, C. et al. (2022a). Metabolites and novel compounds with anti-microbial or antiaging activities from *Cordyceps fumosorosea*. AMB Express, 12(1): 40.

Wen, Y.T., Jhou, B.Y., Hsu, J.H., Fu, H.I., Chen, Y.L. et al. (2022). Neuroprotective Effects of *Cordyceps cicadae* (Ascomycetes) mycelium extract in the rat model of optic nerve crush. Int. J. Med. Mushrooms, 24(2): 41–48.

Wu, G. (2005). Study progress in research tumour with aweto. Guiding J TCM, 11: 80-82.

Wu, G., Li, L., Sung, G.H., Kim, T.W., Byeon, S.E. et al. (2011). Inhibition of 2,4-dinitrofluorobenzene-induced atopic dermatitis by topical application of the butanol extract of *Cordyceps bassiana* in NC/Nga mice. J. Ethnopharmacol., 134(2): 504–509.

Wu, H.C., Chen, S.T., Chang, J.C., Hsu, T.Y., Cheng, C.C. et al. (2019). Radical scavenging and antiproliferative effects of cordycepin-rich Ethanol extract from brown rice-cultivated *Cordyceps militaris* (Ascomycetes) mycelium on breast cancer cell lines. Int. J. Med. Mushrooms, 21(7): 657–669.

Wu, R., Gao, J.P., Wang, H.L., Gao, Y., Wu, Q. et al. (2015). Effects of fermented *Cordyceps sinensis* on oxidative stress in doxorubicin treated rats. Pharmacogn. Mag., 11(44): 724–731.

Wu, R., Yao, P.A., Wang, H.L., Gao, Y., Yu, H.L. et al. (2018). Effect of fermented *Cordyceps sinensis* on doxorubicin-induced cardiotoxicity in rats. Mol. Med. Rep., 18(3): 3229–3241.

Xie, H., Li, X., Chen, Y., Lang, M., Shen, Z. et al. (2019). Ethanolic extract of *Cordyceps cicadae* exerts antitumor effect on human gastric cancer SGC-7901 cells by inducing apoptosis, cell cycle arrest and endoplasmic reticulum stress. J. Ethnopharmacol., 231: 230–240.

Xu, J., Tan, Z.C., Shen, Z.Y., Shen X.J. and Tang S.M. et al. (2021). *Cordyceps cicadae* polysaccharides inhibit human cervical cancer hela cells proliferation via apoptosis and cell cycle arrest. Food Chem. Toxicol., 148: 111971.

Xu, J.C., Zhou, X.P., Wang, X.A., Xu, M.D., Chen, T. et al. (2019). Cordycepin induces apoptosis and G2/M phase arrest through the ERK pathways in esophageal cancer cells. J. Cancer, 10(11): 2415.

Xue, Y., Wu, L., Ding, Y., Cui, X., Han, Z. et al. (2020). A new nucleoside and two new pyrrole alkaloid derivatives from *Cordyceps militaris*. Nat. Prod. Res., 34(3): 341–350.

Yan, W., Li, T., Lao, J., Song, B. and Shen, Y. et al. (2013a). Anti-fatigue property of *Cordyceps guangdongensis* and the underlying mechanisms. Pharm. Biol., 51(5): 614–620.

Yan, W., Li, T. and Zhong, Z. (2014). Anti-inflammatory effect of a novel food *Cordyceps guangdongensis* on experimental rats with chronic bronchitis induced by tobacco smoking. Food Funct., 5(10): 2552–2557.

Yan, X.F., Zhang, Z.M., Yao, H.Y., Guan, Y., Zhu, J.P. et al. (2013b). Cardiovascular protection and antioxidant activity of the extracts from the mycelia of *Cordyceps sinensis* act partially via adenosine receptors. Phytother. Res., 27(11): 1597–1604.

Yang, C.H., Su, C.H., Liu S.C. and Ng T.T. (2019). Isolation, anti-inflammatory activity and physico-chemical properties of bioactive polysaccharides from fruiting bodies of cultivated *Cordyceps cicadae* (Ascomycetes). Int. J. Med. Mushrooms, 21(10): 995–1006.

Yang, F. Q., Feng, K., Zhao J. and Li, S.P. (2009). Analysis of sterols and fatty acids in natural and cultured *Cordyceps* by one-step derivatization followed with gas chromatography-mass spectrometry. J. Pharm. Biomed. Ana., 49(5): 1172–8.

Yang, J., Dong, H., Wang, Y., Jiang, Y., Zhang, W. et al. (2020b). Cordyceps cicadae polysaccharides ameliorated renal interstitial fibrosis in diabetic nephropathy rats by repressing inflammation and modulating gut microbiota dysbiosis. Int. J. Biol. Macromol., 163: 442–456.

Yang, M.L., Kuo, P.C., Hwang, T.L. and Wu, T.S. (2011). Anti-inflammatory principles from *Cordyceps sinensis*. J. Nat. Prod., 74(9): 1996–2000.

Yang, S.X., Bai, X.L., Hu X.F. and Cao, L. (2020a). Water extracts of *Cordyceps sinensis* inhibits proliferation and metastasis via regulating cell cycle and matrix metalloproteinases in melanoma. Life, 3(2): 51.

Yang, X., Lin, P., Wang, J., Liu, N., Yin, F. et al. (2021). Purification, characterization and anti-atherosclerotic effects of the polysaccharides from the fruiting body of *Cordyceps militaris*. Int. J. Biol. Macromol., 181: 890–904.

Yoneyama, T., Elshamy, A.I., Yamada, J., El-Kashak, W.A., Kasai, Y. et al. (2022). Antimicrobial metabolite of *Cordyceps tenuipes* targeting MurE ligase and histidine kinase via *in silico* study. Appl. Microbiol. Biotechnol., 106(19): 6483–6491.

Yoon, S.Y., Park, S.J. and Park, Y.J. (2018). The anticancer properties of cordycepin and their underlying mechanisms. Int. J. Mol. Sci., 19(10).

Yu, S.H., Chen, S.Y.T., Li, W.S., Dubey, N.K., Chen, W.H., Chuu, J.J. et al. (2015). Hypoglycemic activity through a novel combination of fruiting body and mycelia of *Cordyceps militaris* in high-fat diet-induced type 2 diabetes mellitus mice. J. Diabetes Res., 2015: 723190.

Yue, G.G.L., Bik-San Lau, C., Fung, K.P., Leung, P.C. and Ko, W.H. et al. (2008). Effects of *Cordyceps sinensis*, *Cordyceps militaris* and their isolated compounds on ion transport in Calu-3 human airway epithelial cells. J. Ethnopharmacol., 117(1): 92–101.

Yue, K., Ye, M., Zhou, Z., Sun, W. and Lin, X. et al. (2012). The genus *Cordyceps*: a chemical and pharmacological review. J. Pharm. Pharmacol., 65(4): 474–493.

Zhang, L., Zhang, D., Xu, Y., Yang, H., Qi, S. et al. (2022b). Cultured mycelia of *Cordyceps sinensis* exerts a protective effect on a mouse model of liver fibrosis by inhibiting the Toll-like receptor 4/nuclear transcription factor-κB signaling pathway and angiopoietin-like protein 4. J. Clin. Hepatol., 8(7): 1540.

Zhang, G., Huang, Y.,Bian, Y., Wong, J.H., Ng, T.B. et al. (2006). Hypoglycemic activity of the fungi *Cordyceps militaris*, *Cordyceps sinensis*, *Tricholoma mongolicum*, and *Omphalia lapidescens* in streptozotocin-induced diabetic rats. Appl. Microbiol. Biotechnol., 72(6): 1152–1156.

Zhang, S.R., Pan, M., Gao, Y.B., Fan, R.Y., Bin, X.N. et al. (2023a). Efficacy and mechanism study of cordycepin against brain metastases of small cell lung cancer based on zebrafish. Phytomedicine, 109: 154613.

Zhang, X., Li, J., Yang, B., Leng, Q., Li, J. et al. (2021b). Alleviation of liver dysfunction, oxidative stress, and inflammation underlines the protective effects of polysaccharides from *Cordyceps cicadae* on high sugar/high fat diet-induced metabolic syndrome in rats. Chem. Biodivers., 18(5): e2100065.

Zhang, X., Zhou, X., Gao, M., Lyu, Y., Wang, Y. et al. (2022a). Cordycepin inhibits the proliferation and migration of human gastric cancer cells by suppressing lipid metabolism via AMPK and MAPK activation. Xi bao yu fen zi mian yi xue za zhi=Chinese Journal of cellular and Molecular Immunology, 38(6): 513–521.

Zhang, X.L., Huang, W.M., Tang, P.C., Sun, Y., Zhang, X. et al. (2021a). Anti-inflammatory and neuroprotective effects of natural cordycepin in rotenone-induced PD models through inhibiting Drp1-mediated mitochondrial fission. Neurotoxicology, 84: 1–13.

Zhang, Y., Li, E., Wang, C., Li, Y. and Liu, X. et al. (2012). *Ophiocordyceps sinensis*, the flagship fungus of China: terminology, life strategy and ecology. Mycology, 3(1): 2–10.

Zhang, Y., Wu, Y.T., Zheng, W., Han, X.X., Jiang, Y.H. et al. (2017). The antibacterial activity and antibacterial mechanism of a polysaccharide from *Cordyceps cicadae*. J. Funct. Foods, 38: 273–279.

Zhang, Y., Xu, L., Lu, Y., Zhang, J., Yang, M et al. (2023b). Protective effect of *Cordyceps sinensis* against diabetic kidney disease through promoting proliferation and inhibiting apoptosis of renal proximal tubular cells. BMC Complement. Med. Ther., 23.

Zhao, H., Deng, B., Li, D., Jia, L. and Yang, F. et al. (2022). Enzymatic-extractable polysaccharides from *Cordyceps militaris* alleviate carbon tetrachloride-induced liver injury via Nrf2/ROS/ NF-κB signaling pathway. J. Funct. Foods, 95: 105152.

Zhao, H., Lai, Q., Zhang, J., Huang, C. and Jia, L. et al. (2018). Antioxidant and hypoglycemic effects of acidic-extractable polysaccharides from *Cordyceps militaris* on type 2 diabetes mice. Oxid. Med. Cell. Longev., 2018: 9150807.

Zhao, X.M. (1765). Supplement to compendium of Materia Medica. Reprinted from the: 123-124.

Zheng, Q., Sun, J., Li, W., Li, S. and Zhang, K. et al. (2020). Cordycepin induces apoptosis in human tongue cancer cells *in vitro* and has antitumor effects *in vivo*. Arch. Oral Biol., 118: 104846.

Zheng, Y., Zhang, J., Wei, L., Shi, M., Wang, J. et al. (2017). Gunnilactams A–C, macrocyclic tetralactams from the mycelial culture of the entomogenous fungus *Paecilomyces gunnii*. J. Nat. Prod., 80(6): 1935–1938.

Zhu, Y., Ma, L., Hu, Q., Li, J., Chen, Y. et al. (2016). [*In Vitro* Anti-HIV-1 activity of *Cordyceps sinensis* Extracts]. Bing Du Xue Bao, 32(4): 417–22.

Zhu, Y., Yu, X., Ge, Q., Li, J., Wang, D. et al. (2020). Antioxidant and anti-aging activities of polysaccharides from *Cordyceps cicadae*. Int. J. Biol. Macromol., 157: 394–400.

7

Bioactive Secondary Metabolites and Pharmaceutical Prospects of *Cordyceps*

Samay Tirkey, Srishti Verma and *Kamlesh Kumar Shukla**

1. Introduction

Cordyceps, a genus of parasitic fungi with a rich historical legacy, has captured the attention of cultures across the world for centuries. Its usage dates back to ancient civilizations such as traditional Chinese medicine, where it was revered for its purported health benefits. The earliest records of the utilization of *Cordyceps* can be traced to texts like the "Ben Cao Gang Mu" (Compendium of Materia Medica) (Chiu et al., 2016), a comprehensive pharmacopoeia compiled during the Ming dynasty. Descriptions in this text emphasized the potential of *Cordyceps* to enhance vitality, boost immunity, and address various ailments. Revered as a "cure-all" in certain folklore, *Cordyceps* garnered a mystical aura due to its unique growth habit-emerging from the bodies of insect hosts in a remarkable symbiotic relationship. This blend of historical reverence and intriguing growth patterns has elevated *Cordyceps* to a symbol of both traditional wisdom and contemporary scientific curiosity (Steinhardt et al., 2018).

One of the key factors that underlines the pharmaceutical importance of *Cordyceps* lies in the intricate chemistry it possesses, specifically the bioactive secondary metabolites it produces. *Cordyceps* is a prolific source

School of Studies in Biotechnology, Pandit Ravi Shankar Shukla University, Raipur (Chhattisgarh), India.
* Corresponding author: kshukla26@gmail.com

of these compounds, which are small molecules not directly involved in the primary growth and development of the organism but play critical roles in ecological interactions and defence mechanisms. These bioactive metabolites hold immense potential for pharmaceutical applications owing to their diverse range of biological activities. *Cordyceps* is known to contain compounds such as polysaccharides, alkaloids, phenols, polyphenols, terpenoids, nucleosides, sterols, and cyclic peptides. These compounds have exhibited various pharmacological activities, including immunomodulatory, antioxidant, anti-inflammatory, antimicrobial, and anticancer effects (Zheng et al., 2014; Chirivi et al., 2017; Olatunji et al., 2018; Liu et al., 2019). Researchers have been particularly intrigued by cordycepin and adenosine, nucleoside derivatives with promising antimicrobial and antitumor properties. As scientific understanding of these compounds deepens, their potential to be developed into novel therapeutic agents becomes increasingly apparent, offering a bridge between traditional herbal remedies and modern pharmacology.

Cordyceps are characterized by the formation of distinctive stromata, elongated club-shaped structures that emerge from parasitized insect hosts, showcasing a wide range of colors such as white, yellow, orange, brown, or black (Sung et al., 2007; Wen et al., 2017). The fungus exhibits specificity to its host insect, and the mycelium, a thread-like structure, grows within the host before stroma development. Ascospores, produced within sac-like structures, contribute to the reproductive cycle. *Cordyceps* fungi play a vital ecological role by controlling insect populations and are found in various habitats, including forests, grasslands, and alpine meadows. Beyond their ecological significance, some *Cordyceps* species, such as *Cordyceps sinensis*, hold cultural and medicinal importance, particularly in traditional Chinese medicine, where they are valued for their purported health benefits (Liu et al., 2011). The diverse morphology and ecological roles of *Cordyceps* species make them a subject of interest in both scientific research and traditional practices. This chapter discusses mainly bioactive metabolites and pharmaceutical potential of *Cordyceps* along with morphological features and methods of characterization.

2. Classical and Molecular Methods

The identification of *Cordyceps* species can be achieved through a combination of classical and molecular methods. Classical methods involve the observation of morphological and ecological characteristics, while molecular methods involve the analysis of genetic material (Wen et al., 2017; Mongkolsamrit et al., 2020). Here's an overview of both approaches.

2.1 Classical Methods

2.1.1 Morphological Characteristics

Stromata. Observation of stromatal features, including size, shape, color, and the presence of hairs or other structures.

Ascospores. Examination of ascospores under a microscope for size, shape, and other morphological characteristics.

Host Insect. Identification of the host insect species, which may provide clues about the specific *Cordyceps* species.

2.1.2 Cultural Characteristics

Growth Characteristics. Examination of cultural characteristics, such as the growth rate, texture, and color of the mycelium in laboratory cultures.

Ecological Considerations

Habitat. Study of the ecological niche, including the type of vegetation, altitude, and climate in which the *Cordyceps* species is found.

2.2 Molecular Methods

2.2.1 DNA Sequencing

ITS Region. The Internal Transcribed Spacer (ITS) section of the fungal rDNA is often sequenced for molecular identification. It provides species-specific information and is widely used in fungal taxonomy.

SSU and LSU Regions. Sequencing of the Small Subunit (SSU) and Large Subunit (LSU) regions of the rDNA can also contribute to species identification.

2.2.2 PCR-Based Techniques

PCR Amplification. Polymerase Chain Reaction (PCR) can be used to amplify specific DNA regions for further analysis. For *Cordyceps*, primers targeting specific regions of the rDNA can be employed.

Restriction Fragment Length Polymorphism (RFLP). Analyzing the pattern of DNA fragments produced by digesting amplified DNA with restriction enzymes can aid in species differentiation.

2.3 Phylogenetic Analysis

Constructing Phylogenetic Trees. Comparative analysis of DNA sequences allows the construction of phylogenetic trees, helping to determine the evolutionary relationships among different *Cordyceps* species.

2.3.1 DNA Barcoding

COI Gene. The Cytochrome c Oxidase subunit I (COI) gene is often used as a DNA barcode for fungal identification. It provides a standardized region for comparison across species (Zhang et al., 2020a).

2.3.2 Genome Analysis

Whole-Genome Sequencing: Advances in genomics enable the sequencing of the entire genome of *Cordyceps* species, providing comprehensive genetic information.

2.3.3 Real-time PCR (qPCR)

Quantitative PCR. Quantitative PCR can be used to quantify the amount of specific DNA sequences, aiding in the quantification of different *Cordyceps* species in mixed samples.

3. Types of Bioactive Secondary Metabolites

Cordyceps, a genus of parasitic fungi with a long history of traditional use in various cultures, has gained modern scientific attention for its diverse array of bioactive secondary metabolites. These compounds, often synthesized by the fungi in response to environmental cues or as a means of interaction with their hosts, have shown remarkable potential for various pharmaceutical applications. The classification and diversity of these bioactive secondary metabolites in *Cordyceps* highlight their significance as a source of novel therapeutic agents. Diversity of secondary metabolites and their bioactive potential have been presented in Table 1.

3.1 Polysaccharides

Polysaccharides are complex carbohydrate molecules that play a crucial role in the immunomodulatory properties attributed to *Cordyceps*. These compounds have been extensively studied for their potential to enhance immune responses, regulate inflammation, and promote overall health. *Cordyceps* polysaccharides are mainly composed of monosaccharides (e.g., rhamnose, ribose, arabinose, xylose, mannose, glucose, galactose, mannitol, fructose, and sorbose), and they are divided into α-glucans and β-glucans, with varying degrees of branching and molecular weights. These polysaccharides have shown promise in improving immune function, antioxidant activity, and antitumor effects, such as CPS-3, CBP-1, CMP-S1, CMP-W1, LCMPs-II, P70-1, CPMN Fr III, CPSN Fr II, CMPS II, CBPS II, CMN1, CP2-c2-s2, and acid polysaccharide (APS), from *Cordyceps militaris* (Zhang et al., 2019). APS1, CPS-1, CPS-2, neutral mannoglucan, CME-1, Cordyglucans, CS-F10, Cordysinocan, from *Cordyceps sinensis* (Liu et al., 2015a).

Table 1. Diversity of secondary metabolites in different species of *Cordyceps* and their bioactive potential (NA, Not Available).

Cordyceps Species	Metabolites	Bioactive potentials	Reference
Cordyceps annulata Kobayasi & Shimizu	Annullatin E, tetrahydrocannabinol, 2,3-dihydrobenzofurans	Cannabinoid receptor agoinst	Zhang et al., 2020b
C. bassiana Z.Z. Li, C.R. Li, B. Huang & M.Z. Fan	Nicotinamide, 3-methyluracil, 1,7-dimethylxanthine, nudifloric acid, mannitol	Antiproliferative, anti-obesity, anti-angiogenic, anti-nociceptive activities, Apoptosis-inducing, Free radical scavenging, Anti-inflammatory	Olatunji et al., 2018
C. cardinalis G.H. Sung & Spatafora	Cyclodepsipeptides, cardinalisamides A–C	Antitrypanosomal activity	Umeyama et al., 2014
C. cicadae (Miq.) Massee	Ergosterol, ergosterol peroxide, beauvericin, cordycecin, bassiatin, 4-hydroxy-3-methoxyphenyl-4-O-methyl-β-glucopyranoside, 4-hydroxy-2-methoxyphenyl-4-O-methyl-β-glucopyranoside, 3-methoxy-4-[(O-methyl-β-gluco-pyranosyl)oxy]-benzoic acid, 5-methoxycinnamic acid −3-O-(4′-O-methyl-β-glucopyranoside, ergothioneine, aminobutyric acid, myriocin, 5-(2-hydroxyethyl)-2-furanacetic acid, mevalonolactone, diomorpholine,	Cytotoxicity, renoprotective, immunomodulatory, antitumor, hypoglycemic, anti-inflammatory, Insecticidal, neuroprotective.	Olatunji et al., 2018; Liu et al., 2019
C. formosana Kobayasi & Shimizu	Rugulosin	Cytotoxicity, antidiabetic, antidepressant	Lu et al., 2014a; Wang et al., 2015
C. guangdongensis T.H. Li, Q.Y. Lin & B. Sung	Polysaccharides	Anti-inflammatory, anti-fatigue	Yan et al. 2013
C. gunnii (Berk.) Berk.	Cordycepic acid	Antioxidant, antitumor, immunomodulatory	Xiao et al., 2009
C. heteropoda Kobayasi	Cicadapeptin	Antifungal, antibacterial	Krasnoff et al., 2005; Doan et al., 2017

C. japonica Lloyd	Polysaccharide	Antitumor, immuno-stimulating, antioxidant, cytotoxicity	Suh et al., 2017
C. militaris (L.) Fr.	Ergosta-7,22-diene-3,5,6-triol, cordycerebroside A, soyacerebroside I, glucccerebroside	Anticancer, antitumor antioxidant, anti-inflammatory, antihyperlipidemic, hepatoprotective, acetylcholinesterase inhibition, anti-HCV, immunomodulatory, antihyperglycemic, anti-obesity, antimicrobial, anti-allergic, hypouricemic activity, HIV-1 protease inhibiting.	Chirivi et al., 2017; Olatunji et al., 2018; Liu et al., 2019
C. ninchukispora (C.H. Su & H.H. Wang) G.H. Sung, J.M. Sung, Hywel-Jones & Spatafora	Cordycepiamide A-D, N-(2-hydroxybenzyl) acetamide, (–)-Syringaresinol, Phenolic groups	Anti-inflammatory, antioxidant	Chang et al., 2017; Olatunji et al., 2018
C. nipponica Kobayasi	Cordypyridone A-D, 1-dehydroxycordypyridone A	Antimalarial	Isaka et al., 2001; Wongsa et al., 2005; Jessen and Gademann, 2010
C. ophioglossoides (J.F. Gmel.) Fr.	Cordycepol A-C, cordycol, decane sesquiterpenes, cordycepol, fumagillol analogue, cordycol	Antioxidant, antitumour, neuroprotective, antifungal, cytotoxic activity	Yamada et al., 1984; Sun et al., 2013; Olatunji et al., 2018
C. pruinosa Petch	Cordycepin, polysachharides	Anti-inflammatory, antioxidant, anticancer	Meng et al., 2015
C. scarabaeidicola Kobayasi	NA	Anticoagulant activity, anticomplementary activity, immunomodulatory activity, antioxidant activity	Yu et al., 2003; Olatunji et al., 2018;

Table 1. contd. …

Table 1. contd. …

Cordyceps Species	Metabolites	Bioactive potentials	Reference
C. *sinensis* (Berk.) Sacc.	Ergosterol, ergosteryl-3-O-β- D-glucopyranoside, 5α,8α-epidioxy-22E-ergosta-6,9-(11)-22-trien-33β-ol, 5α,6α-epoxy-5α-ergosta-7,22-dien-3β-ol, 5α,8α-epidioxy-24(R)-methylcholesta-6,22-dien-3β-D-glucopyranoside, ergosterol peroxide, ergosta-4,6,8(14),22-tetraen-3-one, H1-A, 22-dihydro-ergosteryl-3-O-β-D-glucopyranoside, cordysinin A-E, Flazin, Perlolyrine, 3′,4′,7-trihydroxyisoflavone, diadzein, 6,7,2′,4′,5′-pentamethoxyflavone, glycitein-7-O-β-D-glucoside-4′-O-methylate, Iso-sinensetin, salicylic acid, p-methoxyphenol, p-methoxybenzoic acid, 4-hydroxyacetophenone, vanillic acid, methyl-p-hydroxyphenylacetate, Syringic acid, p-hydroxyphenylacetic acid, p-hydroxybenzoic acid, protocatechuic acid Ophicordin 2-furancarboxylic acid, nicotinic acid, succinic acid	Renoprotective activity, antitumor activity, anti-inflammatory activity, anticancer activity, immunomodulatory activity, anti-fibrotic activity, anti-diabetic activity, anti-arteriosclerosis activity, anti-hypertensive activity, radio-protective activity, anti-thrombotic activity, hepatoprotective activity, anti-fatigue activity.	Zheng et al., 2014; Olatunji et al., 2018; Liu et al., 2019
C. *sobolifera* (Hill ex Watson) Berk. & Broome	NA	Renoprotective, HIV-1 reverse transcriptase-Inhibitory activity	Liu et al., 2001
C. *sphecocephala* (Klotzsch ex Berk.) Berk. & M.A. Curtis	Polysachharide-peptide complex	Anti-asthmatic activity, anticancer activity	Oh et al., 2008; Heo et al., 2010
C. *taii* Z.Q. Liang & A.Y. Liu	Deacetylcytochalasin C, zygosporin D, Helvolic acid, cordycepic acid, Polysaccharide	Antitumor activity, antimetastatic activity, Anticancer activity, antioxidant activity, antimutagenic activity	Xiao et al., 2009; Liu et al., 2019;
C. *unilateralis* (Tul. & C. Tul.) Sacc.	Erythrostominone, deoxyerythrostominone, 4-O-methyl erythrostominone, epierythrostominol, deoxyerythrostominol and 3,5,8-trihydroxy-6-methoxy-2-(5-oxohexa-1,3-dienyl)-1,4-naphthoquinone	Antimalarial activity, cytotoxic activity	Kittakoop et al., 1999

Polysaccharides are one of the major bioactive compounds in *Cordyceps*. They consist of complex carbohydrate chains with varying degrees of branching. The structural features of *Cordyceps* polysaccharides contribute to their immunomodulatory, antioxidant, and antitumor activities. These compounds often contain unique sugar moieties and glycosidic linkages that differentiate them from other fungal polysaccharides. The biosynthesis of polysaccharides in *Cordyceps* involves the activity of enzymes that catalyze the assembly of sugar monomers into complex carbohydrate chains. Enzymes responsible for glycosyl transfer and branching determine the structural diversity of these polysaccharides, influencing their bioactivity.

3.2 *Phenolic Compounds*

Cordyceps are known to contain various beneficial phenolic compounds, including quercetin, which reduces oxidative stress by scavenging free radicals; resveratrol, which contributes to their potential anti-aging and cardiovascular benefits; catechins, which have strong antioxidant properties that offer cardiovascular protection and potential anticancer effects; and rutin, a flavonoid glycoside known for its potent antioxidant properties and potential anti-inflammatory and vasoprotective effects (Liu et al., 2019). Moreover, compounds like caffeic acid, syringic acid, ferulic acid, and ellagic acid, all possessing antioxidant and anti-inflammatory properties, contribute to the overall health-promoting effects of *Cordyceps*. Kaempferol and myricetin, two additional flavonoids, offer antioxidant benefits and potential cardiovascular and anticancer protective effects. Furthermore, the presence of benzoic acid, protocatechualdehyde, chlorogenic acid, syringaldehyde, and vanillin adds to the antimicrobial, anti-inflammatory, and cardiovascular benefits associated with *Cordyceps* such as *C. militaris*, *C. sinensis*, *C. ophjoglossoides*, *C. unilateralis*, *C. cicadae*, *C. caloceroides*, *C. takaomontana*, making them a rich source of various health-promoting compounds (Zheng et al., 2014; Chirivi et al., 2017; Olatunji et al., 2018; Liu et al., 2019).

3.3 *Sterols and Terpenoids*

This is a structurally diverse and large group of metabolites with over 35,000 entries (Jansen and Shenvi, 2014). Sterols and terpenoids are another class of bioactive secondary metabolites found in *Cordyceps*. These compounds are well known for their anti-inflammatory, antioxidant, and cholesterol-lowering properties. *Cordyceps* species have been shown to contain ergosterol, a common fungal sterol, as well as various terpenoids with potential health benefits. Ergosteryl-3-O-β-D-glucopyranoside, 5α,8α-epidioxy-22E-ergosta-6,9-(11)-22-trien-33β-ol,   5α,6α-epoxy-5α-ergosta-7,22-dien-3β-ol,   Ergosta-4,6,8(14),22-tetraen-3-one,   Fungisterol, β-sitosterol-3-O-acetate, Stigmasterol3-O-acetate, H1-A, β-sitosterol, daucosterol, Cholesterol, Cholesteryl palmitate, 22-dihydro-ergosteryl-3-O-β-D-glucopyranoside, Dihydrobrassicasterol, from *Cordyceps sinensis*, Ergosta-7,22-diene-3,5,6-triol, from *Cordyceps militaris*, Ergosterol, Ergosterol peroxide, from *Cordyceps cicadae*, Jiangxienone, Docosanoic acid campesterol ester, Hexadecanoic acid ergosterol ester from *Cordyceps jiangxiensis* (Olatunji et al., 2018). Terpenes like Cordycepol A, Cordycepol B, Cordycepol C, Cordycol from *Cordyceps ophioglossoides* (Sun et al., 2013; Zheng et al., 2014; Olatunji et al., 2018; Liu et al., 2019). Figure 1 illustrates the chemical structures of key secondary metabolites found in *Cordyceps* species.

Sterols, including ergosterol, and terpenoids found in *Cordyceps* contribute to its antioxidative and anti-inflammatory properties. These compounds often contain characteristic functional groups and ring structures that are associated with their bioactivity. Terpenoids and sterols are synthesized through the mevalonate pathway, which involves the sequential addition of isoprene units. Enzymes in this pathway catalyze reactions that lead to the formation of diverse terpenoid and sterol structures.

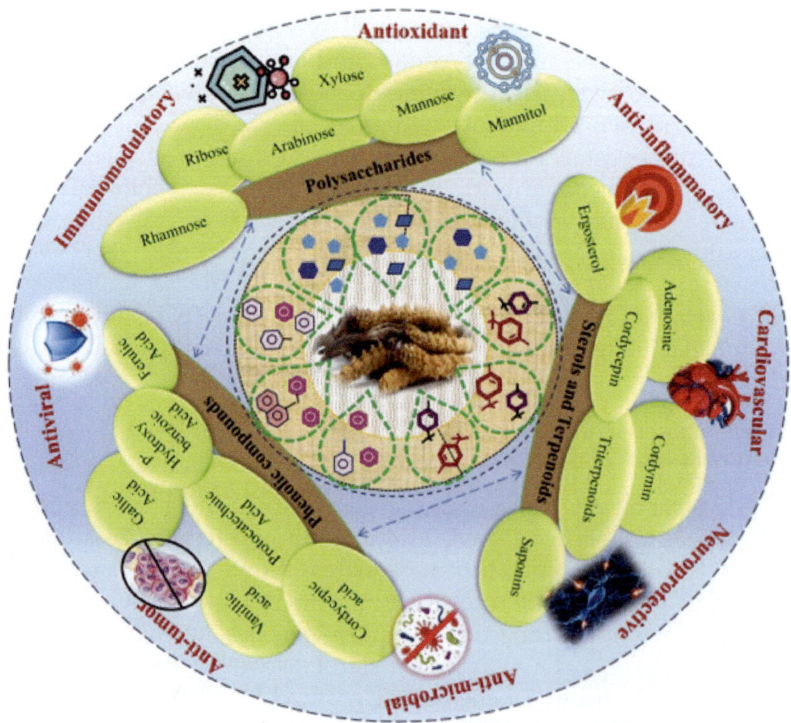

Fig. 1. Secondary metabolites across *Cordyceps* species.

3.4 Peptides and Proteins

Cordyceps also produces various peptides and proteins that contribute to its bioactivity. Some of these peptides exhibit antioxidant, antitumor, and immunomodulatory properties. These compounds are often explored for their potential to regulate cell growth and modulate immune responses. Those explored are, Cordycommunin from *C. communis* (Haritakun et al., 2010), Beauvericin A, B, E, J, Cordycecin A, from *Cordyceps cicadae* (Wang et al., 2014), Cycloaspeptide A, C, F, G, from *C. sinensis* (Zhang et al., 2009), Cardinalisamide A, B, C, from *C. cardinalis* (Umeyama et al., 2014), Cicadapeptin I, II, from *C. heteropoda* (Krasnoff et al., 2005), *Cordyceps* idone A, B, from *C. dipterigena* (Varughese et al., 2012), Destruxin A, A2, B, B2, from *C. indigotica* (Asai et al., 2012), Cordycepeptide A, *C. militaris* (Jiang et al., 2000) and Cordycepoid A from *C. bifusispora* (Lu et al., 2013).

Cordyceps produces various peptides and proteins, such as cordycepin and cordymin, which are involved in its biological effects. These molecules possess unique amino acid sequences and three-dimensional structures that contribute to their diverse range of pharmacological activities, including antitumor and immunomodulatory effects. Peptides and proteins in *Cordyceps* are synthesized through the translation of specific genes into amino acid sequences, followed by post-translational modifications. These modifications, such as phosphorylation and glycosylation, contribute to the unique structural features and biological activities of these molecules.

3.5 Nucleosides and Nucleotides

Cordyceps species are known to contain various nucleosides and nucleotides, including adenosine, cordycepin, uridine, and guanosine. Cordycepin, also known as 3'-deoxyadenosine, is particularly noteworthy for its antimicrobial, antiviral, and antitumor activities. It has been studied extensively for its potential to inhibit DNA and RNA synthesis, making it a target for antiviral drug development. About 16 nucleotides, nucleosides, and nucleobases have been found in *Cordyceps sinensis* and *Cordyceps militaris* such as guanosine monophosphate (GMP), adenosine monophosphate (AMP), uridine monophosphate (UMP), adenosine, guanosine, uridine, cytidine, thymidine, cordycepin, adenine, guanine, uracil, hypoxanthine, cytosine, and thymine (Yang et al., 2010).The AMP, GMP, UMP, thymidine monophosphate (TMP), cytinine monophosphate (CMP), adenosine diphosphate (ADP) and adenosine triphosphate (ATP), in *Cordyceps sinensis,* whereas TMP was not found in *Cordyceps militaris* (Zhou et al., 2019).

Nucleosides and nucleotides found in *Cordyceps,* such as cordycepin (3'-deoxyadenosine), uridine, and adenosine, possess distinct structural motifs (Meng et al., 2015; Lee et al., 2022). Cordycepin, for instance, lacks the

3′-hydroxyl group found in adenosine, which contributes to its antiviral and anticancer activities. The presence of these compounds with altered nucleic acid components underpins their biological effects. The biosynthesis of nucleosides and nucleotides involves complex enzymatic pathways that start with precursor molecules derived from metabolic pathways. Cordycepin, for instance, is synthesized by modifying adenosine through specific enzymatic reactions.

The bioactive secondary metabolites found in *Cordyceps* have drawn considerable scientific interest due to their potential pharmaceutical applications. The identification and extraction of these compounds are essential steps in harnessing their therapeutic potential. Various analytical techniques and extraction methods are employed to isolate and characterize the diverse array of bioactive molecules present in *Cordyceps*.

4. Extraction and Identification of Metabolites

Various methods of extraction and identification of metabolites of *Cordyceps* are given in a flow chart (Fig. 2).

4.1 Extraction

4.1.1 Solvent Extraction

Solvent extraction is a common method to isolate bioactive secondary metabolites from *Cordyceps*. Different solvents, such as ethanol, methanol, and water, are used based on the polarity of the compounds being targeted. This method is efficient in extracting a broad range of compounds, including polysaccharides, nucleosides, and terpenoids (Lu et al., 2013).

4.1.2 Supercritical Fluid Extraction

Supercritical Fluid Extraction (SFE) employs supercritical fluids (such as carbon dioxide) to selectively extract bioactive compounds from *Cordyceps*. This method is advantageous as it evades the use of organic solvents, ensuring higher purity of extracted compounds and minimizing environmental impact (Ling et al., 2009).

4.1.3 Solid-Phase Extraction

Solid-Phase Extraction involves passing a sample through a solid-phase material to selectively retain specific compounds, followed by elution for compound recovery. This technique is valuable for concentrating and purifying bioactive compounds from complex mixtures in *Cordyceps* (Wada et al., 2017).

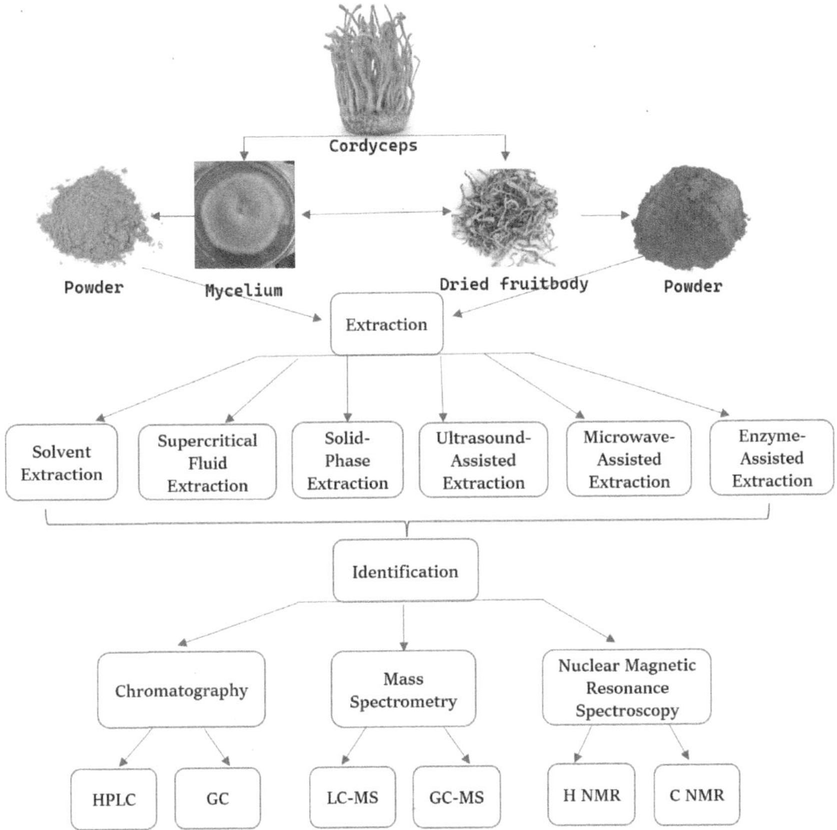

Fig. 2. Representation of extraction and characterization of secondary metabolites from *Cordyceps*.

4.1.4 *Ultrasound-Assisted Extraction and Microwave-Assisted Extraction*

Ultrasound-Assisted Extraction (UAE) and Microwave-Assisted Extraction (MAE) utilize mechanical vibrations or microwave radiation to enhance the extraction process. These methods accelerate the release of bioactive compounds from *Cordyceps*, reducing extraction time while maintaining compound integrity (Zhu et al., 2016).

4.1.5 *Enzyme-Assisted Extraction*

Enzyme-assisted extraction involves using enzymes to degrade cell walls and facilitate the release of bioactive compounds. This method is particularly effective for extracting polysaccharides and proteins from *Cordyceps* (Yang et al., 2017).

4.2 Identification

4.2.1 Chromatography Techniques

High-performance liquid chromatography (HPLC) and gas chromatography (GC) are widely used methods for the separation and quantification of bioactive compounds in *Cordyceps*. These techniques rely on the differential affinities of compounds for stationary and mobile phases, enabling their separation based on chemical properties (Fig. 2). Coupling chromatography with mass spectrometry (MS) allows for precise identification of individual compounds by analysing their molecular masses and fragmentation patterns (Tikhomiroff et al., 2002).

4.2.2 Mass Spectrometry

Mass spectrometry plays a pivotal role in identifying bioactive compounds by providing information about their molecular weight and structural characteristics. Several techniques (e.g., liquid chromatography-mass spectrometry and gas chromatography-mass spectrometry) are often used for profiling the complex mixture of metabolites in *Cordyceps* (Rabha et al., 2023).

4.2.3 Nuclear Magnetic Resonance Spectroscopy

Nuclear Magnetic Resonance (NMR) Spectroscopy aids in elucidating the chemical structure of bioactive compounds. It provides information about molecular connectivity and can be used to confirm the identity of compounds isolated from *Cordyceps* (Wei et al., 2022).

The extraction and identification of bioactive secondary metabolites from *Cordyceps* involve a combination of sophisticated analytical techniques and extraction methods. These processes facilitate the isolation, characterization , and quantification of a diverse array of compounds with potential pharmaceutical applications. The systematic exploration of these methods contributes to the ongoing research aimed at unlocking the full therapeutic potential of *Cordyceps*-derived bioactive secondary metabolites. Figure 3 represents some of the known secondary metabolites with their structural depiction of *Cordyceps*.

5. Pharmacological Activities

5.1 Immunomodulatory Effects

Cordyceps, a genus of fungi renowned for its historical use in traditional medicine, has gained significant attention for its potential immunomodulatory effects. The bioactive secondary metabolites present in *Cordyceps* possess the ability to influence the immune system in various ways, making them a subject of interest in both traditional remedies and

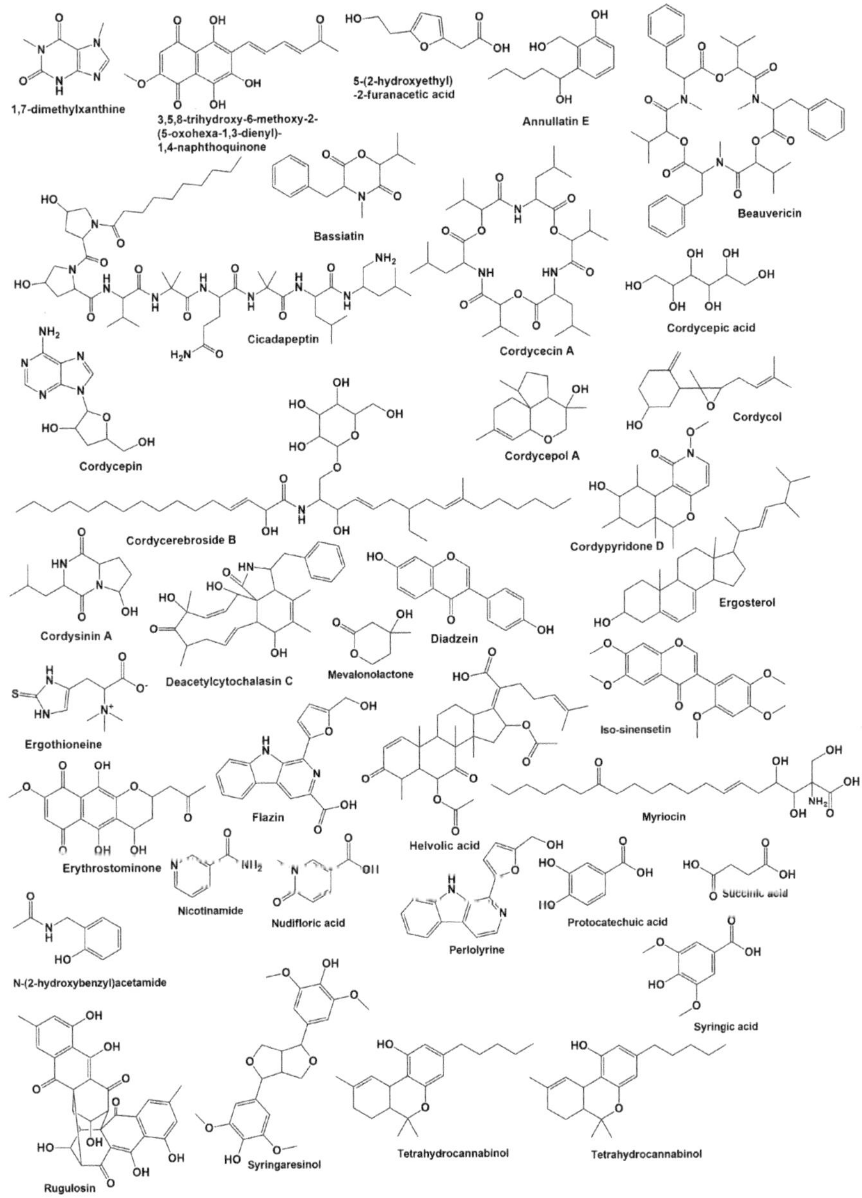

Fig. 3. Chemical structures of key secondary metabolites found in *Cordyceps* species.

modern pharmacology. Understanding the immunomodulatory effects of *Cordyceps*-derived compounds holds promise for developing novel therapeutic approaches (Lan et al., 2022; Jiao et al., 2023).

Cordyceps demonstrates impressive immunomodulatory effects by both enhancing and regulating immune responses. It has the remarkable ability to stimulate macrophages, key immune cells responsible for engulfing pathogens and producing immune signaling molecules called cytokines (Gao et al., 2023). This activation of macrophages, driven by a new galactoglucamannan from *Cordyceps*, enhances their phagocytic activity and immune mediator release, bolstering the body's defense against infections. Additionally, *Cordyceps* metabolites have been reported to enhance the activity of natural killer (NK) cells, which is crucial for detecting and destroying infected or cancerous cells (Bi et al., 2018). Activated NK cells contribute to improved surveillance and targeting of abnormal cells within the body. Furthermore, *Cordyceps* compounds can influence the production and secretion of cytokines, acting as signaling molecules to regulate immune responses. These compounds may promote the production of anti-inflammatory cytokines, balancing the immune system and preventing excessive inflammation (Jiao et al., 2023).

Cordyceps' immunomodulation extends to specific medical conditions. It shows promise in managing autoimmune disorders by modulating the immune response to reduce the attack on self-tissues, with its ability to regulate immune cells and cytokine production (Lan et al., 2022; Jiao et al., 2023). Conditions such as rheumatoid arthritis and multiple sclerosis could benefit from these properties. Moreover, *Cordyceps* metabolites have been explored for their potential in cancer immunotherapy. By enhancing immune responses and promoting the recognition of cancer cells by the immune system, these compounds could augment the body's natural defence mechanisms against tumors (Qi et al., 2020). The diverse array of bioactive secondary metabolites in *Cordyceps* contributes to its multifaceted immunomodulatory effects, often working synergistically to create a comprehensive approach to immune regulation. The immunomodulatory potential of *Cordyceps* holds promise in the ongoing quest to harness natural compounds for therapeutic applications in various health conditions.

5.2 *Antioxidant and Anti-inflammatory Properties*

Cordyceps, a genus of fungi with a rich history in traditional medicine, has garnered attention for its bioactive secondary metabolites that exhibit potent antioxidant and anti-inflammatory properties. These compounds have the potential to play a pivotal role in protecting cells and tissues from oxidative damage and mitigating chronic inflammation, making *Cordyceps* an attractive candidate for both traditional and modern therapeutic applications.

Cordyceps exhibits notable antioxidant properties attributed to a diverse array of bioactive secondary metabolites. These compounds, including polysaccharides and phenolic compounds, have the ability to scavenge free radicals and reactive oxygen species (ROS), effectively neutralizing harmful molecules that contribute to oxidative stress (Oh et al., 2014). By preventing cellular damage, DNA mutations, and lipid peroxidation, *Cordyceps* compounds play a crucial role in protecting overall cellular health. Furthermore, *Cordyceps* metabolites have been shown to upregulate the activity of endogenous antioxidant enzymes (e.g., superoxide dismutase and catalase), which act as the body's natural defense mechanisms against oxidative stress (Zhang et al., 2021). Additionally, some *Cordyceps* compounds can chelate metal ions like iron and copper, known to catalyze the generation of free radicals. By sequestering these ions, *Cordyceps* reduces the oxidative burden on cells and tissues, contributing to long-term cellular well-being (Casasnovas et al., 2013).

The multifaceted antioxidant effects of *Cordyceps* are complemented by its anti-inflammatory properties. *Cordyceps* metabolites have been shown to inhibit pro-inflammatory cytokines: interleukin-1 beta (IL-1β), interleukin-6 (IL-6), and tumor necrosis factor-alpha (TNF-α) (Li et al., 2020; Das et al., 2021). This anti-inflammatory action helps regulate the immune response and prevent chronic inflammation, a key contributor to various health issues. *Cordyceps* compounds also modulate immune cell activities, such as macrophages and neutrophils, by regulating their inflammatory responses, thereby promoting a balanced immune reaction and preventing excessive inflammation that can lead to tissue damage (Phull et al., 2022). Furthermore, some *Cordyceps*-derived compounds can suppress the activation of the nuclear factor kappa B (NF-κB) pathway, a key regulator of inflammation (Bi et al., 2018). By inhibiting this pathway, *Cordyceps* helps downregulate the expression of pro-inflammatory genes. These combined antioxidant and anti-inflammatory effects have prompted investigations into potential therapeutic applications of *Cordyceps*, including neuroprotection, cardiovascular health, and anti-aging, where its properties may contribute to combating oxidative stress, inflammation, and age-related cellular damage (Jędrejko et al., 2021; Lin et al., 2021; Cai et al., 2022).

5.3 *Anti-Tumor and Anti-Cancer Activities*

Cordyceps, a genus of fungi with a long history of use in traditional medicine, has attracted significant scientific interest for its potential anti-tumor and anti-cancer activities. The bioactive secondary metabolites found in *Cordyceps* have demonstrated promising effects in modulating various aspects of cancer development, making it a subject of extensive research in the field of oncology (Qi et al., 2020).

The anti-cancer properties of *Cordyceps* metabolites are multifaceted. These compounds induce apoptosis, a programmed cell death process, in tumor cells, effectively limiting cancer cell proliferation and uncontrolled growth (Bai et al., 2018). Furthermore, they disrupt the cell cycle of tumor cells, causing them to arrest at specific phases, impeding their progression and division (Lee et al., 2022). *Cordyceps* compounds also possess anti-angiogenic effects, inhibiting the formation of new blood vessels that supply nutrients to tumors, ultimately curtailing their growth and metastasis (Tuli et al., 2014). These metabolites not only impact tumor cells but also enhance immune responses, including the activation of natural killer (NK) cells and cytotoxic T cells, which are crucial in recognizing and attacking cancer cells, contributing to the body's natural cancer surveillance and defense mechanisms (Bi et al., 2018).

Moreover, *Cordyceps* compounds have demonstrated anti-metastatic effects, inhibiting the spread of cancer cells to other parts of the body (Jin et al., 2018). They can also modulate various signaling pathways involved in cancer development: the mitogen-activated protein kinase (MAPK) and phosphatidylinositol 3-kinase (PI3K) pathways, which are often dysregulated in cancer cells (Jayakumar et al., 2014). The antioxidant and anti-inflammatory properties of *Cordyceps* compounds further contribute to their anti-cancer effects by reducing oxidative stress and inflammation, creating an environment less favorable for cancer development and progression. These properties, combined with the synergistic effects of various bioactive secondary metabolites in *Cordyceps*, make it a comprehensive approach to combating cancer. In terms of therapeutic applications, *Cordyceps* compounds could be explored as adjunctive cancer therapies alongside conventional treatments like chemotherapy and radiation, enhancing immune responses and mitigating side effects (Qi et al., 2020). Additionally, they may serve as preventive measures for high-risk individuals, helping to delay or inhibit cancer initiation and development (Qi et al., 2020).

5.4 *Anti-Microbial and Anti-Viral Potential*

Cordyceps, a genus of fungi with a rich traditional history, has gained scientific attention for its bioactive secondary metabolites that exhibit remarkable anti-microbial and anti-viral properties. The diverse array of compounds found in *Cordyceps* holds promise in combating microbial infections and viral diseases, making it an intriguing area of research for both traditional medicine and modern pharmacology.

Cordyceps metabolites exhibit potent anti-microbial effects against a range of pathogens. They inhibit bacterial growth, prevent the formation of biofilms, and disrupt bacterial cell membranes (Gyawali et al., 2014). These properties make *Cordyceps* compounds promising candidates for the development of novel antibacterial agents, providing potential solutions to combat bacterial infections. Furthermore, *Cordyceps* compounds are not limited to antibacterial activity; they also display the ability to inhibit the growth of other fungi. This feature is particularly relevant for addressing fungal infections in both humans and plants, offering new possibilities for fungal infection control (Gyawali et al., 2014).

In addition to their anti-microbial effects, *Cordyceps* metabolites demonstrate anti-viral properties that hold promise for managing viral infections. These compounds interfere with viral replication processes, preventing viruses from multiplying and spreading, which is particularly relevant for RNA viruses like influenza and HIV (Li et al., 2021). *Cordyceps* compounds also have the capacity to enhance immune responses, including the activation of natural killer (NK) cells and T cells, which bolster the body's ability to recognize and eliminate viral infections (Lee et al., 2020). Considering these attributes, *Cordyceps* compounds, especially those with anti-viral properties, could be explored for their potential in managing respiratory infections, such as influenza and the common cold. Furthermore, they might find applications in addressing emerging viral diseases like Zika, Ebola, and coronaviruses, given their potential to inhibit viral replication and modulate immune responses, offering hope in the face of evolving viral threats (Li et al. 2021).

5.4 *Neuroprotective and Cognitive-Enhancing Effects*

Cordyceps, a genus of fungi with a history deeply rooted in traditional medicine, has garnered scientific interest for its bioactive secondary metabolites that exhibit potential neuroprotective and cognition-enhancing effects. These compounds hold promise for addressing neurodegenerative disorders and promoting cognitive health, making *Cordyceps* an intriguing subject of research at the intersection of ancient wisdom and modern neuroscience.

Cordyceps metabolites possess a range of neuroprotective effects that make them promising candidates for addressing neurodegenerative diseases. Their potent antioxidant properties enable the neutralization of harmful free radicals and oxidative stress, which play a significant role in neuronal damage and degeneration (He et al., 2021). Additionally, *Cordyceps*-derived compounds exhibit anti-inflammatory activity, mitigating the chronic inflammation associated with neurodegeneration and thus creating a conducive environment for neuron survival (He et al., 2021). Moreover, these compounds may modulate the production and release of neurotrophic factors like brain-derived neurotrophic factor (BDNF), which are essential for promoting neuron growth, synaptic plasticity, and overall brain health (He et al., 2021). These multifaceted mechanisms highlight *Cordyceps'* potential as a neuroprotective agent.

Cordyceps metabolites also demonstrate cognition-enhancing effects that hold promise for improving memory, learning, and cognitive function. They promote synaptic plasticity, a fundamental process for synaptic adaptability and strengthening (Jiao et al., 2023). Furthermore, *Cordyceps* compounds may modulate acetylcholine levels, a neurotransmitter crucial for memory and cognitive processes. Optimizing acetylcholine signaling can lead to improved cognitive function (Rai et al., 2021). In addition, the ability of *Cordyceps* metabolites to inhibit the aggregation of amyloid-beta peptides, which form plaques in the brains of individuals with Alzheimer's disease, offers potential for slowing cognitive decline (Lan et al., 2019). *Cordyceps'* influence on neurotransmitter modulation, activation of neurotrophic factors, and anti-inflammatory effects contributes to its cognitive-enhancing properties. This opens up the possibility of its use in managing conditions such as Alzheimer's disease and mitigating cognitive decline in aging individuals (He et al., 2019; Zhu et al., 2020).

5.5 *Cardiovascular Health Benefits*

Cordyceps, a genus of fungi with a rich history in traditional medicine, has gained scientific attention for its bioactive secondary metabolites that offer potential cardiovascular health benefits. These compounds have demonstrated properties that can support heart health and mitigate risk factors associated with cardiovascular diseases, making *Cordyceps* a promising subject of research at the intersection of traditional knowledge and modern cardiology.

Cordyceps metabolites have demonstrated remarkable effects on cardiovascular health, including cholesterol regulation. These compounds can lower cholesterol levels by inhibiting cholesterol synthesis and enhancing its excretion (Hu et al., 2019). Notably, *Cordyceps* compounds such as CM1, a polysaccharide derived from *Cordyceps* militaris, exhibit the ability to reduce total cholesterol and LDL ("bad") cholesterol levels. Beyond cholesterol management, *Cordyceps* metabolites promote vasodilation, relaxing blood vessel walls and leading to improved blood flow (Lu et al., 2014b). This vasodilatory effect is a key contributor to reduced blood pressure, contributing to overall cardiovascular well-being. Moreover, *Cordyceps* compounds possess the potential to improve glucose control, particularly through their impact on insulin sensitivity and resistance, which may help prevent type 2 diabetes and its associated cardiovascular complications (Wang et al., 2023). Additionally, *Cordyceps* compounds can reduce triglyceride levels, another vital risk factor for cardiovascular diseases, by enhancing lipid metabolism (Ashraf et al., 2020). These multifaceted mechanisms contribute to *Cordyceps'* potential in managing conditions associated with cardiovascular health.

Cordyceps compounds also exhibit significant therapeutic potential in addressing cardiovascular conditions such as atherosclerosis and hypertension. The inhibition of HMG-CoA reductase, an enzyme responsible for cholesterol synthesis, by certain *Cordyceps* metabolites leads to reduced

LDL cholesterol levels (Ashraf et al., 2020). Additionally, these compounds enhance the production of nitric oxide, a molecule that promotes blood vessel relaxation and improves blood flow (Ashraf et al., 2020). This dual effect on cholesterol and blood pressure regulation positions *Cordyceps* as a valuable candidate for managing atherosclerosis, a condition characterized by plaque buildup in blood vessels, as it can regulate cholesterol, blood pressure, and inflammation, potentially slowing the progression of this condition. Moreover, the vasodilatory and blood pressure-lowering properties of *Cordyceps* metabolites make them promising for managing hypertension, offering a comprehensive approach to cardiovascular health (Ashraf et al., 2020). *Cordyceps'* multi-faceted impact on cardiovascular parameters signifies its potential as a holistic intervention for individuals with cardiovascular concerns.

6. Challenges and Future Prospects

Effective utilization of bioactive secondary metabolites from *Cordyceps* in pharmaceutical and healthcare applications demands a comprehensive approach to standardization and quality control. To guarantee product consistency and patient safety, robust measures must be implemented. Taxonomic authentication is fundamental, and advanced techniques like DNA barcoding and molecular analysis can confirm the precise species used, thereby preventing the inadvertent inclusion of adulterants (Moon et al., 2018). In addition to species verification, phytochemical profiling involving techniques like high-performance liquid chromatography (HPLC) and mass spectrometry (MS) plays a pivotal role in identifying and quantifying specific bioactive compounds, thus ensuring product uniformity (Donno et al., 2020).

Quantifying the active compounds necessitates the establishment of marker compounds indicative of bioactivity, allowing for consistent potency assessment. Reference standards calibrated for known purity and potency further contribute to the accurate quantification of active compounds, facilitating batch-to-batch consistency (Yang et al., 2017). Quality control parameters encompass criteria such as purity, potency, and the absence of microbial contaminants, heavy metals, toxins, and residual solvents. These criteria ensure that products adhere to safety standards and deliver the desired therapeutic effects while safeguarding consumers against potential risks (Liu et al., 2015; Dikpati et al., 2020; Gallo et al., 2020). Furthermore, adherence to Good Manufacturing Practices (GMP) guidelines encompasses facility maintenance, process control, and quality management systems, guaranteeing that every facet of production adheres to rigorous standards (WHO, 2003).

Finally, regulatory compliance is imperative, necessitating registration and approval for *Cordyceps* products intended for pharmaceutical use. Stringent labelling requirements are essential, including accurate ingredient listings, dosage instructions, and evidence-based health claims. Additionally, investment in research and development activities, such as bioassays for bioactivity evaluation and optimization of formulations to enhance bioavailability and therapeutic effects, can elevate the overall quality of *Cordyceps*-based pharmaceuticals and healthcare products. These multifaceted measures collectively safeguard the consistent quality and safety of *Cordyceps* products on the market, ultimately benefiting patient health and well-being.

7. Safety and Toxicity Considerations

Exploring the pharmaceutical potential of bioactive secondary metabolites from *Cordyceps* involves not only uncovering their therapeutic benefits but also ensuring their safety for human consumption. Robust safety and toxicity assessments are essential to mitigate any potential adverse effects and ensure the well-being of individuals using *Cordyceps*-derived products.

Despite *Cordyceps'* extensive history of use in traditional medicine, it is vital to recognize that natural products are not inherently risk-free. The safety of *Cordyceps* products is contingent upon various factors, including the composition and concentration of bioactive compounds, which can differ among species and sources. Therefore, comprehensive safety assessments are imperative to mitigate potential risks associated with their consumption (Chen et al., 2020; Ashraf et al., 2020). This involves a series of steps, beginning with the evaluation of acute toxicity through high-dose exposure studies, providing essential insights into initial safety margins. Sub-chronic and chronic studies extend this assessment to observe the compound's effects with repeated administration, unveiling potential cumulative impacts over time. Genotoxicity studies examine whether the compound has the propensity to damage genetic material, a critical consideration for preventing mutations and cancer. Furthermore, reproductive and developmental toxicity studies assess the impact on fertility, pregnancy, and the developing fetus ensuring a holistic evaluation of safety (Liu et al., 2019; Ashraf et al., 2020).

Several potential safety concerns must be addressed, including allergic reactions in individuals sensitive to specific *Cordyceps* compounds. Moreover, the potential for drug interactions should not be underestimated, as *Cordyceps* associated bioactive compounds may interact with various medications, affecting their efficacy or leading to adverse effects. Extremely high doses of *Cordyceps* compounds can result in toxicity, emphasizing the importance of adhering to proper dosage recommendations. Contaminants stemming from substandard sourcing, manufacturing, or handling practices, such as microbes, heavy metals, or pesticides, can introduce health risks. Consequently, stringent quality control and standardization procedures are necessary to ensure the purity of products and consistent dosages, mitigating the risk of unintended effects arising from variations in composition (Wang et al., 2023). Certain risk groups, including pregnant and nursing women, individuals with immune disorders, and those taking medications, should exercise caution when considering the use of *Cordyceps* products. A holistic approach is paramount, recognizing individual differences in response due to genetic, health, and lifestyle factors. Promoting balance and moderation in the use of *Cordyceps* products is key to harnessing their potential benefits while minimizing potential risks (Cetto, 2015).

8. Preclinical and Clinical Development

As research into the pharmaceutical potential of bioactive secondary metabolites from *Cordyceps* advances, numerous challenges must be addressed in both preclinical and clinical development. These challenges span scientific, regulatory, and practical aspects, highlighting the complexity of translating promising findings into safe and effective therapeutic solutions.

The path to harnessing the potential of *Cordyceps* compounds in modern healthcare involves a series of critical steps and considerations. A paramount priority is the identification and standardization of compounds, recognizing that chemical composition can vary among *Cordyceps* species and sources, influencing both potency and safety (Holliday and Cleaver, 2008; Panda and Luyten, 2022). Comprehensive phytochemical profiling, coupled with the development of marker compounds, is essential for precise identification and quantification. Preclinical efficacy and safety evaluations pose their own complexities, necessitating careful considerations regarding dose determination, appropriate animal models, and relevant endpoints. Well-designed preclinical studies with robust methodologies, suitable animal models, and meticulous dosage selection are pivotal in providing valuable insights into the potential benefits and risks.

Understanding bioavailability and pharmacokinetics is another pivotal aspect of *Cordyceps* compound development. An in-depth comprehension of how these compounds are absorbed, distributed, metabolized, and eliminated in the body is fundamental for determining optimal dosages and treatment regimens. This entails the execution of pharmacokinetic studies to assess bioavailability, metabolism, and elimination. Additionally, elucidating the intricate mechanisms of action and potential synergistic interactions among multiple compounds requires multidisciplinary approaches, including molecular biology, cell biology, and systems biology.

Crafting rigorous clinical trial designs that thoroughly evaluate safety, efficacy, and optimal dosages while accommodating individual variability is a formidable task (Liu et al., 2019; Panda and Luyten, 2022). This involves careful planning, randomization, blinding, and the incorporation of appropriate control groups. Recruitment and retention of a diverse patient population for clinical trials are challenging endeavors that can impact the generalizability of study results. Engaging with patient communities, employing innovative recruitment strategies, and addressing participant concerns are strategies to improve recruitment and retention rates. Ethical considerations remain at the forefront of clinical trial conduct, necessitating adherence to ethical guidelines, securing informed consent, and conducting trials under ethical oversight. Navigating regulatory pathways for traditional herbal medicines, dietary supplements, and pharmaceuticals can be intricate, demanding collaboration with regulatory agencies and the provision of robust scientific evidence. Post-market surveillance mechanisms are crucial for detecting rare or delayed adverse effects that may not surface in clinical trials, requiring ongoing collaboration with healthcare professionals and regulatory agencies. Bridging the gap between traditional usage of *Cordyceps* and modern pharmaceutical development entails addressing cultural perceptions and expectations. Engaging with communities, fostering dialogue, and acknowledging cultural traditions are vital steps in building acceptance (Liu et al., 2019).

9. Future Directions and Emerging Research Areas

The exploration of the pharmaceutical potential of bioactive secondary metabolites from *Cordyceps* is an evolving field with exciting prospects. As research advances, new directions and emerging areas of study are shaping the trajectory of *Cordyceps*-based therapeutics, expanding our understanding of their applications and mechanisms.

The frontier of *Cordyceps* research holds immense promise and is ripe for innovative approaches to unlock its full potential. Personalized medicine is a focal point, with the aim of tailoring *Cordyceps* interventions to individual characteristics, from genetics to lifestyle. This involves genetic profiling to identify responders, the discovery of biomarkers for outcome prediction, and the creation of personalized treatment regimens. Embracing multi-omics approaches integrating genomics, proteomics, metabolomics, and transcriptomics is essential for gaining a holistic understanding of *Cordyceps'* effects on various biological levels (Chaubey et al., 2019). This systematic investigation elucidates how *Cordyceps* compounds interact with cellular pathways and influence molecular networks, offering insights into their mechanisms of action. Innovative nanotechnology and targeted delivery systems play a pivotal role in enhancing the bioavailability and therapeutic efficacy of *Cordyceps* compounds (Wasser, 2010). Designing nanoparticles for the encapsulation of *Cordyceps* compounds facilitates controlled release, improved absorption, and targeted delivery to specific tissues. The dynamic interplay between *Cordyceps* compounds and gut microbiota is an emerging area of exploration, offering the potential to unveil their influence on overall health. Investigating the modulation of gut microbiota composition and activity by *Cordyceps* holds relevance for systemic health. The versatile nature of *Cordyceps* extends to its potential role in managing neuroinflammation and neurodegenerative diseases, with its anti-inflammatory and neuroprotective properties garnering attention (Kontogiannatos et al., 2021). Unravelling how *Cordyceps* compounds target neuroinflammatory pathways and impact neurodegenerative processes is essential in this context. Furthermore, immune-modulating effects of *Cordyceps*, coupled with its potential to address immune disorders with metabolic underpinnings, necessitate comprehensive investigations into the intricate connections between its effects on immune cells and metabolic pathways. As the potential of *Cordyceps* as a complementary therapy comes to the forefront, research delves into identifying synergistic interactions between *Cordyceps* and conventional therapies and optimizing combination regimens. Epigenetics and the examination of long-term effects are key aspects, involving a deeper understanding of how *Cordyceps* compounds influence gene expression patterns and epigenetic modifications over time. *Cordyceps* also steps onto the stage in the context of aging and longevity, where its antioxidant and anti-inflammatory properties may contribute to healthy aging (Chaubey et al., 2019). Research explores the effects of *Cordyceps* on aging-related pathways, cellular senescence, and age-associated diseases. As *Cordyceps* becomes a subject of intense interest, environmental sustainability and cultivation practices are of paramount importance. It is important to ensure a consistent supply of *Cordyceps* while protecting natural habitats calls for research into eco-friendly cultivation techniques and assessing their impact on bioactive compound content (Zhang et al., 2021).

Conclusion

The exploration of the pharmaceutical potential of bioactive secondary metabolites from *Cordyceps* represents a journey that intertwines ancient wisdom with modern scientific rigor. The diverse array of compounds found within this genus of fungi offers a multitude of health benefits, spanning from immunomodulation and cardiovascular support to neuroprotection and anti-microbial effects. The cumulative body of research underscores the potential impact of *Cordyceps*-based interventions on healthcare and disclose a range of exciting implications for both preventive and therapeutic approaches.

However, harnessing *Cordyceps'* benefits is quite challenging. Ethical considerations loom large, requiring a delicate balance between cultural acceptance, patient preferences, and scientific evidence. Integrating *Cordyceps* into modern healthcare practices necessitates careful navigation of these complexities to ensure responsible and equitable incorporation. By addressing these concerns thoughtfully, the journey towards leveraging *Cordyceps'* potential can pave the way for a more inclusive and effective approach to healthcare.

Acknowledgment

The authors are thankful to the Department of Biotechnology (DBT) for sponsoring project # BT/PR29526/FcB/125/16/2018. at Pt. Ravishankar Shukla University, Raipur, Chhattisgarh, India.

References

Asai, T., Yamamoto, T., Chung, Y.M., Chang, F.R., Wu et al. (2012). Aromatic polyketide glycosides from an entomopathogenic fungus, *Cordyceps indigotica*. Tetrahedron. Lett., 53(3): 277–280.

Ashraf, S.A., Elkhalifa, A.E.O., Siddiqui, A.J., Patel, M., Awadelkareem et al. (2020). Cordycepin for health and wellbeing: a potent bioactive metabolite of an entomopathogenic medicinal fungus *Cordyceps* with its nutraceutical and therapeutic potential. Molecules, 25(12): 2735.

Bai, K.C. and Sheu, F. (2018). A novel protein from edible fungi *Cordyceps militaris* that induces apoptosis. J. Food Drug Anal., 26(1): 21–30.

Bi, S., Jing, Y., Zhou, Q., Hu, X., Zhu, J. et al. (2018). Structural elucidation and immunostimulatory activity of a new polysaccharide from *Cordyceps militaris*. Food Funct., 9(1): 279–293.

Cai, H., Li, J., Gu, B., Xiao, Y., Chen, R. et al. (2018). Extracts of *Cordyceps sinensis* inhibit breast cancer cell metastasis via down-regulation of metastasis-related cytokines expression. J. Ethnopharmacol., 214: 106–112.

Cai, Y., Feng, Z., Jia, Q., Guo, J., Zhang, P. et al. (2022). *Cordyceps cicadae* ameliorates renal hypertensive injury and fibrosis through the regulation of SIRT1-mediated autophagy. Front. Pharmacol., 12: 801094.

Casasnovas, R., Frau, J., Ortega-Castro, J., Donoso, J. and Muñoz, F. et al. (2013). C–H activation in pyridoxal-5′-phosphate and pyridoxamine-5′-phosphate Schiff bases: effect of metal chelation. A computational study. J. Phys. Chem. B., 117(8): 2339–2347.

Cetto, A.A. (2015). Diabetes and metabolic disorders: an ethnopharmacological perspective. In: Heinrich M, Jäger AK, (Eds.) Ethnopharmacology. pp 227–238. John Wiley & Sons; Hoboken, New Jersey, USA.

Chang, H.S., Cheng, M.J., Wu, M.D., Chan, H.Y., Hsieh, S.Y. et al. (2017). Secondary metabolites produced by an endophytic fungus *Cordyceps ninchukispora* from the seeds of Beilschmiedia erythrophloia Hayata. Phytochem. Lett., 22: 179–184.

Chaubey, R., Singh, J., Baig, M.M. and Kumar, A. (2019). Recent advancement and the way forward for *Cordyceps*. In: Yadav, A.N., Singh, S., Mishra, S. and Gupta, A. (Eds.), Recent Advancement in White Biotechnology Through Fungi: Volume 2: Perspective for Value-Added Products and Environments, pp. 441–474, Springer Cham.

Chen, B., Sun, Y., Luo, F. and Wang, C. (2020). Bioactive metabolites and potential mycotoxins produced by *Cordyceps* fungi: a review of safety. Toxins, 12(6): 410.

Chirivi, J., Danies, G., Sierra, R., Schauer, N., Trenkamp, S. et al. (2017). Metabolomic profile and nucleoside composition of *Cordyceps nidus* sp. nov.(Cordycipitaceae): a new source of active compounds. PLoS One, 12(6): e0179428.

Chiu, C.P., Hwang, T.L., Chan, Y., El-Shazly, M., Wu, T.Y. et al. (2016). Research and development of *Cordyceps* in Taiwan. Food Sci. Hum. Wellness, 5(4): 177–185.

Das, G., Shin, H.S., Leyva-Gómez, G., Prado-Audelo, M.L.D., Cortes et al. (2021). *Cordyceps* spp.: A review on its immune-stimulatory and other biological potentials. Front. Pharmacol., 11: 2250.

Dikpati, A., Mohammadi, F., Greffard, K., Quéant, C., Arnaud, P. et al. (2020). Residual solvents in nanomedicine and lipid-based drug delivery systems: A case study to better understand processes. Pharm. Res., 37: 1–11.

Doan, U. V., Mendez Rojas, B. and Kirby, R. (2017). Unintentional ingestion of *Cordyceps* fungus-infected cicada nymphs causing ibotenic acid poisoning in Southern Vietnam. Clin. Toxicol., 55(8): 893–896.

Donno, D., Mellano, M.G., Gamba, G., Riondato, I. and Beccaro, G.L. et al. (2020). Analytical strategies for fingerprinting of antioxidants, nutritional substances, and bioactive compounds in foodstuffs based on high performance liquid chromatography–mass spectrometry: An overview. Foods, 9(12): 1734.

Gallo, M., Ferrara, L., Calogero, A., Montesano, D. and Naviglio, D. et al. (2020). Relationships between food and diseases: What to know to ensure food safety. Food Res. Int. 137: 109414.

Gao, F., Luo, L. and Zhang, L. (2023). A New Galactoglucomannan from the mycelium of the medicinal parasitic fungus *Cordyceps cicadae* and its immunomodulatory activity *In Vitro* and *In Vivo*. Molecules, 28(9): 3867.

Gyawali, R. and Ibrahim, S.A. (2014). Natural products as antimicrobial agents. Food Control, 46: 412–429.

Haritakun, R., Sappan, M., Suvannakad, R., Tasanathai, K. and Isaka, M. et al. (2010). An antimycobacterial cyclodepsipeptide from the entomopathogenic fungus *Ophiocordyceps communis* BCC 16475. J. Nat. Prod., 73(1): 75–78.

He, M.T., Lee, A.Y., Kim, J.H., Park, C.H., Shin et al. (2019). Protective role of *Cordyceps militaris* in Aβ 1–42-induced Alzheimer's disease *in vivo*. Food Sci. Biotechnol., 28: 865–872.

He, M.T., Park, C.H. and Cho, E.J. (2021). Caterpillar medicinal mushroom, *Cordyceps militaris* (Ascomycota), Attenuates Aβ 1– 42– induced amyloidogenesis and inflammatory response by suppressing amyloid precursor protein progression and p38 MAPK/JNK activation. Int. J. Med. Mushrooms, 23(11).

Heo, J.C., Nam, S.H., Nam, D.Y., Kim, J.G., Lee et al. (2010). Anti-asthmatic activities in mycelial extract and culture filtrate of *Cordyceps sphecocephala* J201. Int. J. Mol. Med., 26(3): 351–356.

Holliday, J. and Cleaver, M.P. (2008). Medicinal value of the caterpillar fungi species of the genus *Cordyceps* (Fr.) Link (Ascomycetes). A review. Int. J. Med. Mushrooms, 10(3).

Hu, S., Wang, J., Li, F., Hou, P., Yin, J., Yang, Z. et al. (2019). Structural characterisation and cholesterol efflux improving capacity of the novel polysaccharides from *Cordyceps militaris*. Int. J. Biol. Macromol., 131: 264–272.

Isaka, M., Tanticharoen, M., Kongsaeree, P. and Thebtaranonth, Y. (2001). Structures of cordypyridones A– D, antimalarial N-hydroxy-and N-methoxy-2-pyridones from the insect pathogenic fungus *Cordyceps nipponica*. J. Org. Chem., 66(14): 4803–4808.

Jansen, D.J. and Shenvi, R.A. (2014). Synthesis of medicinally relevant terpenes: reducing the cost and time of drug discovery. Future Med. Chem., 6(10): 1127–1148.

Jayakumar, T., Chiu, C.C., Wang, S.H., Chou, D.S., Huang, Y.K. et al. (2014). Anti-cancer effects of CME-1, a novel polysaccharide, purified from the mycelia of *Cordyceps sinensis* against B16-F10 melanoma cells. J. Cancer Res. Ther., 10(1): 43–49.

Jędrejko, K.J., Lazur, J. and Muszyńska, B. (2021). *Cordyceps militaris*: An overview of its chemical constituents in relation to biological activity. Foods, 10(11): 2634.

Jessen, H.J. and Gademann, K. (2010). 4-Hydroxy-2-pyridone alkaloids: structures and synthetic approaches. Nat. Prod. Rep., 27(8): 1168–1185.

Jiang, H., Liu, K., Meng, S. and Chu, Z. (2000). Chemical constituents of the dry sorophore of *Cordyceps militaris*. Acta Pharm. Sin., 35(9): 663–668.

Jiao, L., Yu, Z., Zhong, X., Yao, W., Xing, L. et al. (2023). Cordycepin improved neuronal synaptic plasticity through CREB-induced NGF upregulation driven by MG-M2 polarization: a microglia-neuron symphony in AD. Biomed. Pharmacother., 157: 114054.

Jin, Y., Meng, X., Qiu, Z., Su, Y., Yu, P. et al. (2018). Anti-tumor and anti-metastatic roles of cordycepin, one bioactive compound of *Cordyceps militaris*. Saudi J. Biol. Sci., 25(5): 991–995.

Kittakoop, P., Punya, J., Kongsaeree, P., Lertwerawat, Y., Jintasirikul, A. et al. (1999). Bioactive naphthoquinones from *Cordyceps unilateralis*. Phytochem., 52(3): 453–457.

Kontogiannatos, D., Koutrotsios, G., Xekalaki, S. and Zervakis, G.I. (2021). Biomass and cordycepin production by the medicinal mushroom *Cordyceps militaris*—A review of various aspects and recent trends towards the exploitation of a valuable fungus. J. Fungi., 7(11): 986.

Krasnoff, S.B., Reátegui, R.F., Wagenaar, M.M., Gloer, J.B. and Gibson, D.M. et al. (2005). Cicadapeptins I and II: New Aib-Containing peptides from the entomopathogenic fungus *Cordyceps heteropoda* . J. Nat. Prod., 68(1): 50–55.

Lan, N.T., Vu, K.B., Ngoc, M.K.D., Tran, P.T., Hiep, D.M et al. (2019). Prediction of AChE-ligand affinity using the umbrella sampling simulation. J. Mol. Graph. Model., 93: 107441.

Lan, Y.H., Lu, Y.S., Wu, J.Y., Lee, H.T., Srinophakun, P. et al. (2022). *Cordyceps militaris* reduces oxidative stress and regulates immune T cells to inhibit metastatic melanoma invasion.

Lee, C.T., Huang, K.S., Shaw, J.F., Chen, J.R., Kuo, W.S. et al. (2020). Trends in the immunomodulatory effects of *Cordyceps militaris*: total extracts, polysaccharides and cordycepin. Front. Pharmacol., 11: 575704.

Lee, Y.P., Huang, W.R., Wu, W.S., Wu, Y.H., Ho, S.Y. et al. (2022). Cordycepin enhances radiosensitivity to induce apoptosis through cell cycle arrest, caspase pathway and ER stress in MA-10 mouse Leydig tumor cells. Am. J. Cancer Res., 12(8): 3601.

Li, L.Q., Song, A.X., Yin, J.Y., Siu, K.C., Wong, W.T et al. (2020). Anti-inflammation activity of exopolysaccharides produced by a medicinal fungus *Cordyceps sinensis* Cs-HK1 in cell and animal models. Int. J. Biol. Macromol., 149: 1042–1050.

Li, R.F., Zhou, X.B., Zhou, H.X., Yang, Z.F., Jiang, H.M. et al. (2021). Novel fatty acid in *Cordyceps* suppresses influenza A (H1N1) virus-induced proinflammatory response through regulating innate signaling pathways. ACS Omega, 6(2): 1505–1515.

Lin, P., Yin, F., Shen, N., Liu, N., Zhang, B. et al. (2021). Integrated bioinformatics analysis of the anti-atherosclerotic mechanisms of the polysaccharide CM1 from *Cordyceps militaris*. Int. J. Biol. Macromol., 193: 1274–1285.

Ling, J.Y., Zhang, G.Y., Lin, J.Q., Cui, Z.J. and Zhang, C.K. et al. (2009). Supercritical fluid extraction of cordycepin and adenosine from *Cordyceps kyushuensis* and purification by high-speed counter-current chromatography. Sep. Purif. Technol., 66(3): 625–629.

Liu, H.J., Hu, H.B., Chu, C., Li, Q. and Li, P. et al. (2011). Morphological and microscopic identification studies of *Cordyceps* and its counterfeits. Acta Pharm. Sin. B., 1(3): 189–195.

Liu, R.M., Dai, R., Luo, Y. and Xiao, J.H. (2019). Glucose-lowering and hypolipidemic activities of polysaccharides from *Cordyceps taii* in streptozotocin-induced diabetic mice. BMC Complement Altern Med., 19: 1–10.

Liu, S.H., Chuang, W.C., Lam, W., Jiang, Z. and Cheng, Y.C. et al. (2015a). Safety surveillance of traditional Chinese medicine: current and future. Drug Saf., 38: 117–128.

Liu, Y., Wang, J., Wang, W., Zhang, H., Zhang, X. et al. (2015b). The chemical constituents and pharmacological actions of *Cordyceps sinensis*. Evid. Based Complement. Alternat. Med., 2015:575063.

Liu, Z.Y., Liang, Z.Q., Whalley, A.J.S., Liu, A.Y. and Yao, Y.J. et al. (2001). A new species of *Beauveria*, the anamorph of *Cordyceps sobolifera*. Fungal Divers., 7: 61–70.

Lu, R.L., Bao, G.H., Hu, F.L., Huang, B., Li, C.R. et al. (2014a). Comparison of cytotoxic extracts from fruiting bodies, infected insects and cultured mycelia of *Cordyceps formosana*. Food Chem., 145: 1066–1071.

Lu, R.L., Luo, F.F., Hu, F.L., Huang, B., Li, C. R et al. (2013). Identification and production of a novel natural pigment, cordycepoid A, from *Cordyceps bifusispora*. Appl. Microbiol. Biotechnol., 97: 6241–6249.

Lu, W.J., Chang, N.C., Jayakumar, T., Liao, J.C., Lin, M.J. et al. (2014b). *Ex vivo* and *in vivo* studies of CME-1, a novel polysaccharide purified from the mycelia of *Cordyceps sinensis* that inhibits human platelet activation by activating adenylate cyclase/cyclic AMP. Thromb. Res., 134(6): 1301–1310.

Meng, Z., Kang, J., Wen, T., Lei, B. and Hyde, K.D. et al. (2015). Cordycepin and N6-(2-hydroxyethyl)-adenosine from *Cordyceps pruinosa* and their interaction with human serum albumin. PloS One, 10(3): e0121669.

Mongkolsamrit, S., Noisripoom, W., Tasanathai, K., Khonsanit, A., Thanakitpipattana, D. et al., (2020). Molecular phylogeny and morphology reveal cryptic species in *Blackwellomyces* and *Cordyceps* (Cordycipitaceae) from Thailand. Mycol. Prog., 19: 957–983.

Moon, B.C., Kim, W.J., Park, I., Sung, G.H. and Noh, P. et al. (2018). Establishment of a PCR assay for the detection and discrimination of authentic *Cordyceps* and adulterant species in food and herbal medicines. Molecules, 23(8): 1932.

Oh, J.Y., Baek, Y.M., Kim, S.W., Hwang, H.J., Hwang, H.S. et al. (2008). Apoptosis of human hepatocarcinoma (HepG2) and neuroblastoma (SKN-SH) cells induced by polysaccharides-peptide complexes produced by submerged mycelial culture of an entomopathogenic fungus *Cordyceps sphecocephala*. J. Microbiol. Biotechnol., 18(3): 512–519.

Oh, T.J., Hyun, S.H., Lee, S.G., Chun, Y.J., Sung , G.H. et al. (2014). NMR and GC-MS based metabolic profiling and free-radical scavenging activities of *Cordyceps pruinosa* mycelia cultivated under different media and light conditions. PLoS One, 9(3): e90823.

Olatunji, O.J., Tang, J., Tola, A., Auberon, F., Oluwaniyi, O. et al. (2018). The genus *Cordyceps*: An extensive review of its traditional uses, phytochemistry and pharmacology. Fitoterapia, 129: 293–316.

Panda, S.K. and Luyten, W. (2022). Medicinal mushrooms: Clinical perspective and challenges. Drug Discov. Today, 27(2): 636–651.

Phull, A.R., Ahmed, M. and Park, H.J. (2022). Cordyceps militaris as a bio functional food source: pharmacological potential, anti-inflammatory actions and related molecular mechanisms. Microorganisms, 10(2): 405.

Qi, W., Zhou, X., Wang, J., Zhang, K., Zhou, Y., Chen, S. et al. (2020). *Cordyceps sinensis* polysaccharide inhibits colon cancer cells growth by inducing apoptosis and autophagy flux blockage via mTOR signaling. Carbohydr. Polym., 237: 116113.

Rabha, J., Chetri, B.K., Das, S. and Jha, D.K. (2023). *In-vitro* and *in-silico* evaluation of antimicrobial and antibiofilm secondary metabolites of a novel fungal endophyte, *Albophoma* sp. BAPR5. S. Afr. J. Bot., 158: 347–368. phy A, 955(1): 87–93.

Rai, S.N., Mishra, D., Singh, P., Vamanu, E. and Singh, M.P. et al. (2021). Therapeutic applications of mushrooms and their biomolecules along with a glimpse of *in silico* approach in neurodegenerative diseases. Biomed. Pharmacother., 137: 111377.

Steinhardt, J.B. (2018). Mycelium is the Message: open science, ecological values, and alternative futures with do-it-yourself mycologists. University of California, Santa Barbara.

Suh, W., Nam, G., Yang, W.S., Sung, G.H., Shim, S.H. et al. (2017). Chemical constituents identified from fruit body of *Cordyceps bassiana* and their anti-inflammatory activity. Biomol. Ther., 25(2): 165.

Sun, Y., Zhao, Z., Feng, Q., Xu, Q., Lü, L. et al. (2013). Unusual spirodecane sesquiterpenes and a fumagillol analogue from *Cordyceps ophioglossoides*. Helv. Chim. Acta, 96(1): 76–84.

Sung, G.H., Hywel-Jones, N.L., Sung, J.M., Luangsa-Ard, J.J., Shrestha, B. and Spatafora, J.W. (2007). Phylogenetic classification of *Cordyceps* and the clavicipitaceous fungi. Stud. Mycol., 57: 5–59.

Tikhomiroff, C. and Jolicoeur, M. (2002). Screening of *Catharanthus roseus* secondary metabolites by high-performance liquid chromatography. J. Chromatogr. A 955, 1: 87–93.

Tuli, H. S., Sandhu, S. S., Sharma, A.K. and Gandhi, P.U.N.E.E.T. (2014). Anti-angiogenic activity of the extracted fermentation broth of an entomopathogenic fungus, *Cordyceps militaris* 3936. Int. J. Pharm. Sci., 6(7): 581–583.

Umeyama, A., Takahashi, K., Grudniewska, A., Shimizu, M., Hayashi, S. et al. (2014). *In vitro* antitrypanosomal activity of the cyclodepsipeptides, cardinalisamides A–C, from the insect pathogenic fungus *Cordyceps cardinalis* NBRC 103832. J. Antibiot., 67(2): 163–166.

Varughese, T., Rios, N., Higginbotham, S., Arnold, A.E., Coley, P.D. et al. 2012. Antifungal depsidone metabolites from *Cordyceps dipterigena*, an endophytic fungus antagonistic to the phytopathogen Gibberella fujikuroi. Tetrahedron Lett., 53(13): 1624–1626.

Wada, T., Sumardika, I. W., Saito, S., Ruma, I.M.W., Kondo et al. (2017). Identification of a novel component leading to anti-tumor activity besides the major ingredient cordycepin in *Cordyceps militaris* extract. J. Chromatogr. B, 1061: 209–219.

Wang, H., Chen, Y., Wang, L., Liu, Q., Yang, S. et al. (2023a). Advancing herbal medicine: enhancing product quality and safety through robust quality control practices. Front. Pharmacol., 14.

Wang, J., Zhang, D.M., Jia, J.F., Peng, Q.L., Tian et al. (2014). Cyclodepsipeptides from the ascocarps and insect-body portions of fungus *Cordyceps cicadae*. Fitoterapia, 97: 23–27.

Wang, Y.W., Hong, T.W., Tai, Y.L., Wang, Y.J., Tsai, S.H. et al. (2015). Evaluation of an epitypified *Ophioocrdyceps formosana* (*Cordyceps* sl) for its pharmacological potential. Evid. Based Complement. Alternat. Med., 2015: 189891, 1.

Wang, Y., Ni, Z., Li, J., Shao, Y., Yong, Y. et al. (2023b). *Cordyceps cicadae* polysaccharides alleviate hyperglycemia by regulating gut microbiota and its metabolites in high-fat diet/streptozocin-induced diabetic mice. Front. Nutr., 10: 1203430.

Wasser, S.P. (2010). Medicinal mushroom science: history, current status, future trends, and unsolved problems. Int. J. Med. Mushrooms, 12(1).

Wei, J., Zhou, X., Dong, M., Yang, L., Zhao, C. et al. (2022). Metabolites and novel compounds with anti-microbial or antiaging activities from *Cordyceps fumosorosea*. AMB Express, 12(1): 1–14.

Wen, T.C., Xiao, Y.P., Han, Y.F., Huang, S.K., Zha, L.S. et al. (2017). Multigene phylogeny and morphology reveal that the Chinese medicinal mushroom *Cordyceps gunnii* is *Metacordyceps neogunnii* sp. nov. Phytotaxa, 302(1): 27–39.

Wongsa, P., Tasanatai, K., Watts, P. and Hywel-Jones, N. (2005). Isolation and *in vitro* cultivation of the insect pathogenic fungus *Cordyceps unilateralis*. Mycol. Res., 109(8): 936–940.

World Health Organization. (2003). WHO guidelines on good agricultural and collection practices [GACP] for medicinal plants. World Health Organization.

Xiao, J.H., Xiao, D.M., Xiong, Q., Liang, Z.Q. and Zhong, J.J. et al. (2009). Optimum extraction and high-throughput detection of cordycepic acid from medicinal macrofungi *Cordyceps jiangxiensis*, *Cordyceps taii* and *Cordyceps gunnii*. J. Food Agric. Environ, 7(3&4): 328–33.

Yamada, H., Kawaguchi, N., Ohmori, T., Takeshita, Y., Taneya, S.I. et al. (1984). Structure and Antitumor Activity of an alkali-soluble polysaccharide from *Cordyceps ophioglossoides*. Carbohydr. Res., 125(1): 107–115.

Yan, W., Li, T., Lao, J., Song, B. and Shen, Y. (2013). Anti-fatigue property of *Cordyceps guangdongensis* and the underlying mechanisms. Pharm. Biol., 51(5): 614–620.

Yang, F.Q., Li, D.Q., Feng, K., Hu, D.J. and Li, S.P. et al. (2010). Determination of nucleotides, nucleosides and their transformation products in *Cordyceps* by ion-pairing reversed-phase liquid chromatography–mass spectrometry. J. Chromatogr. A, 1217(34): 5501–5510.

Yang, H., Wu, Z., He, D., Zhou, H. and Yang, H et al. (2017). Enzyme-assisted extraction and Pb 2+ biosorption of polysaccharide from *Cordyceps militaris*. J. Polym. Environ., 25: 1033–1043.

Yang, W., Zhang, Y., Wu, W., Huang, L., Guo, D. et al. (2017). Approaches to establish Q-markers for the quality standards of traditional Chinese medicines. Acta Pharm. Sin. B, 7(4): 439–446.

Yu, K.W., Kim, K.M. and Suh, H.J. (2003). Pharmacological activities of stromata of *Cordyceps scarabaecola*. Phytother. Res., 17(3): 244–249.

Zhang, F.L., Yang, X.F., Wang, D., Lei, S.R., Guo, L.A. et al. (2020). A simple and effective method to discern the true commercial Chinese *Cordyceps* from counterfeits. Sci. Rep., 10(1): 2974.

Zhang, J., Wen, C., Duan, Y., Zhang, H. and Ma, H. et al. (2019a). Advance in Cordyceps militaris (Linn) Link polysaccharides: Isolation, structure, and bioactivities: A review. Int. J. Biol. Macromol., 132: 906–914.

Zhang, L., Fasoyin, O.E., Molnár, I. and Xu, Y. (2020b). Secondary metabolites from hypocrealean entomopathogenic fungi: Novel bioactive compounds. Nat. Prod. Rep., 37(9): 1181–1206.

Zhang, X., Li, J., Yang, B., Leng, Q., Li, J. et al. (2021). Alleviation of liver dysfunction, oxidative stress, and inflammation underlines the protective effects of polysaccharides from Cordyceps cicadae on high sugar/high fat diet-induced metabolic syndrome in rats. Chem. Biodiversity, 18(5): e2100065.

Zhang, Y., Liu, S., Liu, H., Liu, X. and Che, Y. et al. (2009). Cycloaspeptides F and G, cyclic pentapeptides from a *Cordyceps*-colonizing isolate of *Isaria farinosa*. J. Nat. Prod., 72(7): 1364–1367.

Zheng, L., Hao, L., Ma, H., Tian, C., Li, T. et al. (2014). Production and in vivo antioxidant activity of Zn, Ge, Se-enriched mycelia by *Cordyceps sinensis* SU-01. Curr. Microbiol., 69: 270–276.

Zhou, D.D., Zhang, H., Zhang, Q., Qian, Z.M., Li, W.J. et al. (2019). Preparation of titanium ion functionalized polydopamine coated ferroferric oxide core-shell magnetic particles for selective extraction of nucleotides from *Cordyceps* and *Lentinus edodes*. J. Chromatogr. A, 1591: 24–32.

Zhu, Y., Yu, X., Ge, Q., Li, J., Wang, D. et al. (2020). Antioxidant and anti-aging activities of polysaccharides from *Cordyceps cicadae*. Int. J. Biol. Macromol., 157: 394–400.

Zhu, Z.Y., Dong, F., Liu, X., Lv, Q., Liu, F. et al. (2016). Effects of extraction methods on the yield, chemical structure and anti-tumor activity of polysaccharides from *Cordyceps gunnii* mycelia. Carbohydr. Polym., 140: 461–471.

8

Bioactive Compounds of *Cordyceps* and their Potential Pharmaceutical Avenues

Koh Gui Jen,[1] *Emilyn Yapp Wye Yi,*[1] *Sabrina Khor Xin Yi*[1] and
Tang Yin Quan[1,2,*]

1. Introduction

Over the years, *Cordyceps* has gained widespread recognition and has become a well-known dietary supplement, renowned for its potential health benefits. For centuries, researchers have carefully examined the medicinal properties of *Cordyceps* in the context of Traditional Chinese Medicine (TCM). TCM has utilised *Cordyceps* to address a range of ailments, including hyperglycaemia, cancer, tumour disorders, respiratory diseases, and to enhance cellular aerobic capacity while strengthening the immunity of cells (Das et al., 2020). Among all the *Cordyceps* species, *Cordyceps sinensis* stands out as the most utilized in TCM, where it is utilized as a tonic food and herbal remedy to aid conditions such as lung and kidney dysfunction, impotence, and fatigue (Prasain, 2013). Furthermore, the combined effects of *Cordyceps* along with other bioactive compounds have also been documented, whereby some traditional healers recommended a

[1] School of Biosciences, Faculty of Health and Medical Sciences Taylor's University, Subang Jaya 47500, Malaysia.

[2] Digital Health and Medical Advancement Impact Lab, Taylor's University, 47500 Subang Jaya, Selangor Darul Ehsan, Malaysia.

* Corresponding author: yinquan.tang@taylors.edu.my

therapeutic approach for cancer involving a combination of *Cordyceps* spp., ginseng root, and taxus leaves (Das et al., 2020). Other common species used for treatments includes *C. pruinosa Petch* for stomach inflammation, *C. bassiana* for dermatitis and eczema, and *C. gunnii* (Berk.) Berk. for immunomodulatory effects like memory enhancement and delaying senescence, have been utilized in traditional Chinese medicine (Wu et al., 2015).

Cordyceps genus has also captured the attention and interest of researchers due to its potential in immune system stimulation. Multiple pre-clinical studies reported the pharmacological effects of *Cordyceps*, including antioxidant, anti-diabetic, anti-inflammatory, anticancer and immunoregulation (Chen et al., 2013; Qi et al., 2013; Wang et al., 2010; Wong et al., 2011). These effects are closely associated with its bioactive compounds present, which includes polysaccharides, essential amino acids, proteins, phenolic compounds, carotenoids, vitamins (B1, B2, B12, E and K) and minerals (Al, Ca, Cr, Fe, Ga, K, Mg, Mn, Na, Ni, Pi, Zr, Si, Sr, Ti, V, Zn, and Se), polyamines, cyclic peptides, nucleosides, and sterols (Ashraf et al., 2020, Prasain, 2013). Among all the compounds, cordycepin is the major bioactive compound associated with various pharmaceutical effects, including apoptotic induction, regulation of cell cycle, and the inhibition of purine and nucleic acid biosynthesis (Krishna et al., 2023). University of Jena (Friedrich-Schiller-Universität Jena. This is because cordycepin resembles nucleoside adenosine structurally. Adenosine is made up of a purine base (adenine) attached to a ribose sugar and a phosphate group, whereas cordycepin lacks one of the hydroxyl groups at position three of the ribose sugar ring. Hence, the chemical constituents of *Cordyceps* and their bioactive potentials, which contribute to its pharmaceutical effect, and the safety and efficacy of its consumption will be reviewed in this chapter.

2. Bioactive Constituents

The chemical constituents of *Cordyceps* contribute substantially to its biological function and pharmaceutical benefit. Major chemical compounds that carry significant bioactive potential will be elucidated in this section, which includes cordycepin and cordycepic acid (CA), polysaccharides, proteins, nucleotide and nucleoside, as well as sterols (Table 1).

2.1 Cordycepin and Cordycepic Acid

Cordycepin and cordycepic acid (CA) are therapeutic bioactive molecules discovered abundantly in *Cordyceps*, especially *C. militaris* (Huang et al., 2003). Cordycepin was initially obtained by fermenting the broth of *Cordyceps militaris*, a therapeutic fungus (Cunningham et al., 1950).

Table 1. Bioactive constituents of *Cordyceps* and their pharmacological effects.

Bioactive constituent	Pharmacological effect/ Bioactivity	Reference
Nucleosides		
Cordycepin	Anticancer, anti-inflammatory, antimicrobial, inhibit platelet aggregation	(Ashraf et al., 2020; Liu et al., 2020; Panya et al., 2021)
Polysaccharides		
CS-F10	Hypoglycemic and hypolipidemic effect	(Kiho et al., 1993)
CS-F30	Hypoglycemic and hypolipidemic effect	(Liu et al., 2020)
CPS-2	Immunomodulatory	(Wang et al., 2010)
CME-1	Inhibiting human platelet aggregation	(Wang et al., 2010)
CS-PS	Immunomodulatory	(Zhang et al., 2011)
Cordysinocan	Immunomodulatory	(Cheung et al., 2009)
EPS	Immunomodulatory, antitumour	(Sheng et al., 2011)
W-CBP50, W-CBP50I, W-CBP50II	Antioxidant	(Chen et al., 2013)
Proteins		
CSDNase	Nucleolytic properties	(Ye et al., 2004)
CSP	Fibrinolytic properties	(Li et al., 2007)
Cordymin	Anti-diabetic, anti-inflammatory, anti-fungal	(Qi et al., 2013; Wong et al., 2011)
Cordycedipeptide A	Cytotoxic activity	(Jia et al., 2009)
Cordyceamides A and B	Cytotoxic activity	(Jia et al., 2009)
Nucleotide and Nucleosides		
Cordycepic acid	Treat liver fibrosis, cerebral edema, respiratory disease, regulate plasma osmotic pressure	(Anderson et al., 2009; LinandLi, 2011; Ouyang et al., 2013)
3′-Deoxyinosine	Cytotoxicity	(Qiu et al., 2017)
involving (2-amino-N-((2S,3S,4R,5R)-5-(6-amino-9H-purin-9-yl)-4-hydroxy-2-(hydroxymethyl)–tetrahydrofuran-3-yl)-6-ureido-hexanamide)	Anti-tumour	(Xiang et al., 2021)

Table 1. contd. ...

Table 1. contd. ...

Bioactive constituent	Pharmacological effect/ Bioactivity	Reference
Nucleotide and Nucleosides		
(2-amino-N-((2S,3S,4R,5R)-5-(6-amino-9H-purin-9-yl)-4-hydroxy-2-(hydroxymethyl)tetrahydrofuran-3-yl)-6-guanidinohexanamide)	Anti-tumour	(Xiang et al., 2021)
Sterols		
Ergosterol	Bone development, anti-inflammatory, cytotoxic activity, anti-microbial	(LinandLi, 2011; Rao et al., 2010; Zheng et al., 2013)
β-Sitosterol	Anticancer, reduce cholesterol absorption	(Awad et al., 2000; Ostlund, 2007)
H1-A	Immunoregulation	(Yang et al., 2003)

Cordycepin ($C_{10}H_{13}N_5O_3$) has a molecular weight of 251.24 Da. In *Ophiocordyceps sinensis*, its content varies from 0.006 mg/g to 6.36 mg/g, while in cultivated *C. militaris*, it can attain a level of up to 2.28 mg/g (Liu et al., 2022). As aforementioned, cordycepin is an analogue of adenosine derivatives. At the third position in the chain the carbon of ribose sugar, cordycepin is distinguished from adenosine nucleoside by the absence of one oxygen molecule. The comparison of their chemical structures is shown in Figure 1. Thus, its metabolism and pharmacokinetic profile is also similar to adenosine as the enzyme is unable to differentiate between adenosine and cordycepin. Hence, it will be quickly degraded by adenosine deaminase, converting it into an inert metabolite, 3'-deoxyhypoxanthinosine *in vivo* (Adamson et al., 1977).

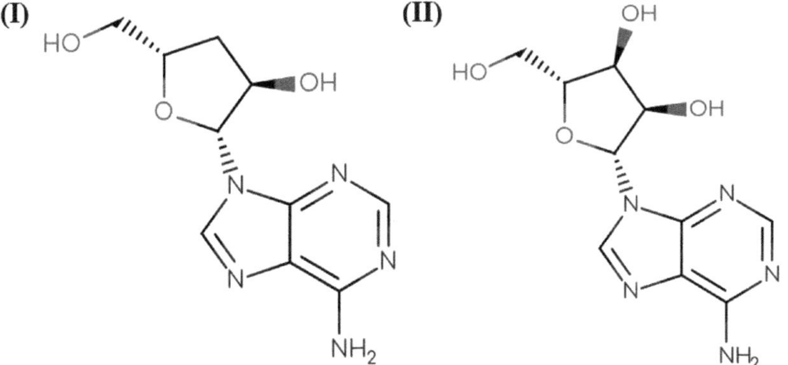

Fig. 1. Chemical structures of (I) Cordycepin and (II) Adenosine.

This similarity contributes to its diverse role in cellular molecular activities, such as intracellular targets, apoptosis, nucleic acid, and cell cycle (Das et al., 2020). In the context of bacteria and viruses, cordycepin can also incorporate DNA structures into their RNA as a result of this structural resemblance. This will potentially impede the microorganisms' ability to grow as nucleic acids and won't synthesised and modified (Łysakowska et al., 2023). This is because when RNA is being synthesised, the incorporation of cordycepin takes place instead of nucleoside (Chen et al., 2008). Consequently, nitrogenous bases (A, U, G, and C) are unable to be incorporated, thus terminating the process of transcription. It was found that cordycepin can inhibit cell attachment, reduce focal adhesion, and reduce the length of poly-A tail in mRNA (Wong et al., 2010). The shortening of mRNA will cause mRNA degradation, hence resulting in downregulation of genes (Passmore and Coller, 2022). On the other hand, interfering with cell attachment may prevent the spread of cancer cells to other tissues, as the cells are unable to adhere to new locations. Reducing focal adhesion can disrupt the ability of cells to interact with their environment, which is beneficial in cases such as preventing cancer cells from migrating and invading (Martin et al., 2013). In addition, the mTOR signaling pathway can be potentially disrupted by cordycepin by triggering the activation of AMP-activated kinase (AMPK), ultimately resulting in the suppression of growth, proliferation, and translation (Wong et al., 2010). For example, at EC_{50} of 26.94 μM, cordycepin can successfully inhibit the replication of the dengue virus and reduce DENV protein (Panya et al., 2021). Another recent *in vitro* research done using anti-SARS-CoV-2 assays revealed that cordycepin is more effective than remdesivir in preventing the emerging SARS-CoV-2 drug-resistant strains from replicating, with an EC_{50} of 2 μM concentration (Rabie, 2022).

Cordycepin may facilitate apoptosis through several pathways such as mitogen-activated protein kinase (MAPK), cysteine–aspartic proteases, and glycogen synthase kinase (GSK)-3β, which is controlled by death receptors, putative adenosine receptors and epidermal growth factor receptors (EGFR) (Khan and Tania, 2020). Furthermore, cordycepin exerts regulatory control over various signalling proteins, including hedgehog, glioblastoma protein (GLI), DNA-dependent protein kinase (DNA-PK), and ERK, which induces cancer cells apoptosis (Liu et al., 2020). Cordycepin additionally triggers apoptosis in the breast cancer cell line (MDA-MB-231) through the stimulation of mitochondrial translocation of Bax and releasing cytochrome C, which ensues by the activation of caspases-3 and caspases-9 (Choi et al., 2011). Likewise, apoptosis was demonstrated via the activation of active caspase-3 and the cleavage of Poly (ADP-ribose) polymerase (PARP) in leukemia cells, neuroblastoma, and melanoma (Baik et al., 2012). Studies have also shown that at certain checkpoints, cordycepin can arrest the cell cycle in human oral squamous cancer cell line (OEC-MI), hence decreasing the percentage of cells in the G1 phase (Wu et al., 2007).

Additionally, the expression on pro-inflammatory cytokines and chemokines can be reduced by cordycepin. For example, TNF-α, IL-1β, and IL-6. This causes an increase in T lymphocytes and B lymphocytes secretions (Ashraf et al., 2020). Through upregulation of cytokines, chemokines and their respective receptors, the cancer cells are suppressed. In LPS-induced microglia cells, cordycepin also reduces the amount of inflammatory mediators being released, which includes PGE2, NO, IL-1β, and TNF-α (Jeong et al., 2010). This reduction was achieved by impeding the degradation of IκB-α, consequently leading to the inactivation of NF-κB. Several studies have also shown that cordycepin can upregulate the expression of IL-10 protein in human peripheral blood mononuclear cells (Tuli et al., 2013). Resultantly, this upregulation will hinder proinflammatory and inflammatory cytokines from secreting.

Moreover, cordycepin also shows suppressive effects on tumour cells by preventing platelet aggregation (Łysakowska et al., 2023). This is because during the process of hematogenous metastasis, cancer cells can trigger platelets in order to survive in the bloodstream (Tuli et al., 2013). Platelets act by shielding tumour cells from destruction of cytotoxic lymphocytes and natural killer (NK) cells, thus enhancing their chances of survival. In this case, cordycepin can inhibit adenosine diphosphate (ADP), which is a potent inducer of platelet aggregation during tumour cell-induced platelet aggregation (Yoshikawa et al., 2009). As a result, cancer cells exhibit reduced hematogenic metastatic potential. Another study has suggested that cordycepin leads to a reduction in intracellular calcium levels and thromboxane A2, subsequently resulting in an increase in intracellular cAMP and cGMP concentrations (Cho et al., 2006). As a result, collagen-induced platelet aggregation is inhibited effectively. Besides that, cordycepin was found to show an inhibitory effect on NF-κB reporter gene in TNF-induced HeLa cells, which occurs within a concentration range of 3 to 100 μM (Sun et al., 2017).

On the other hand, cordycepic acid (CA), also known as mannitol, has a structure of 1,3,4,5-tetrahydroxycyclohexane-1-carboxylic acid which is shown in Figure 2 (Chatterjee et al., 1957). CA was found in *Cordyceps* in significant amounts (Liu et al., 2015). Quinic acid has a structure of 1,3,4,5-tetrahydroxy-cyclohexanoic

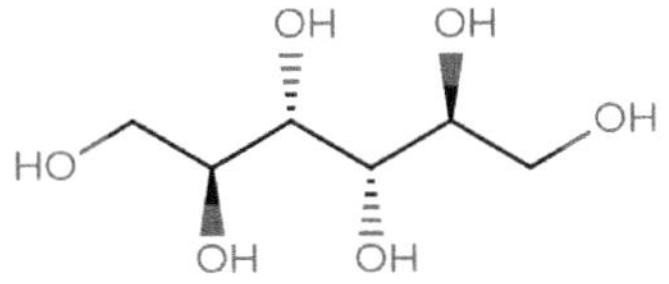

Fig. 2. Chemical structure of cordycepic acid.

acid, making it an isomer of CA. Although the CA content in *C. sinensis* varies in different growing stages, the approximate content is 7 to 29% (Das et al., 2020).

Inhaling powdered mannitol is shown to offer therapeutic benefits to cystic fibrosis and bronchiectasis, which is possible through airway rehydration by clearing up mucociliary (Ilowite et al., 2008). In asthma diagnosis, mannitol has been long used to test for airway hyper-responsiveness (Anderson et al., 2009). Consequently, these pharmacological characteristics of mannitol make it suitable for the treatment of respiratory conditions such as chronic bronchitis and asthma, hypertension, renal dysfunction and renal failure.

In the case of traumatic brain injury, stroke, subarachnoid hemorrhage, or acute renal failure, CA has osmotic activity that can treat refractory intracranial hypertension and accumulation of fluid in the brain (Lin and Li, 2011). In most organs, mannitol is quickly distributed throughout the extracellular compartment. However, in a healthy brain with an intact blood-brain barrier (BBB), mannitol remains within the intravascular compartment (Wise and Chater, 1961). This segregation sets up an osmotic gradient, causing water to shift from the intracellular and interstitial brain compartments into the vasculature (Wise and Chater, 1961). As a result, this osmotic process decreases the net intracranial volume.

CA is also effective in the treatment of liver fibrosis in hepatic stellate cells (HSCs) (Lee and Friedman, 2011). This is possible because CA downregulates the inflammatory phenotype induced by lipopolysaccharides and alleviates the fibrogenic response induced by TGFβ1 in cultured HSCs (Ouyang et al., 2013).

2.2 *Polysaccharides*

Polysaccharides are present in a high proportion in *Cordyceps*, occupying 3 to 8% of the total dry weight (Lin and Li, 2011). This includes cyclofurans, which is a five-carbon sugar in a cyclic ring, cross-connected beta-mannan polymers, heteropolysaccharides beta-glucans, and complex polysaccharides (Das et al., 2020). Due to their excellent water solubility, active polysaccharides can be extracted by water retention with alcohol precipitation. Polysaccharides of *Cordyceps* exist mainly in two forms, which are extracellular polysaccharide (EPS) and intracellular polysaccharide (IPS). EPS is obtained from the fermentation broth of submerged *Cordyceps* spp., whereas IPS is obtained from the cultured mycelium and fruiting bodies of *Cordyceps* (Liu et al., 2022). To obtain crude EPS, ethanol can be added into the fermentation broth. Then, the deproteinized and decolorized polysaccharide is purified by chromatography and the dialysis concentrate is collected to obtain the desired EPS (Sharma et al., 2018). Table 2 shows the

Table 2. Polysaccharides extracted from *Cordyceps\sinensis*.

Polysaccharide	Composition	Molecular Weight (Da)	Bioactivities	Reference
CS-F10	Galactose, glucose, mannose	15,000	Hypoglycemic and hypolipidemic effect	(Kiho et al., 1993)
CS-F30	Galactose, glucose, mannose	45,000	Hypoglycemic and hypolipidemic effect	(Wong et al., 2010)
CPS-2	α-(1–4)-D-glucose, α-(1–3)-D-mannose	4.39×10^4	Immunomodulatory	(Wang et al., 2010)
CME-1	Mannose, galactose	27.6	Inhibiting human platelet aggregation	(Wang et al., 2010)
CS-PS	Mannose, rhamnose, arabinose, xylose, glucose, galactose	12,000	Immunomodulatory	(Zhang et al., 2011)
Cordysinocan	Glucose, mannose, galactose	82,000	Immunomodulatory	(Cheung et al., 2009)
EPS	Mannose, glucose, galactose	1.04×10^5	Immunomodulatory and antitumour effects	(Sheng et al., 2011)

example of polysaccharides extracted from *C. sinensis,* which can be done through methods such as ion-exchange and size chromatography.

Nevertheless, only β-(1→3) glucan, proteopolysaccharide, and galactosaminoglycan show antitumor activity (Xiao et al., 2002). These polysaccharides are derived from *C. cicadae* S.Z. Shing, *C. ophioglossioides* and *Cordyceps* spp. According to several studies, body resistance can be enhanced through the combination of these polysaccharides with chemotherapeutic drugs, as they present a synergistic effect (Xiao et al., 2002; Yang et al., 2005). The polysaccharides are often used in formulating drugs, as they have several medicinal effects such as maintaining blood sugar level, antimetastatic, anti-influenza, immunoprotective and antioxidative effects (Kiho et al., 1993). *In vivo* studies on genetically diabetic mice showed that polysaccharides of *Cordyceps* have potent hypoglycemic activity when introduced through intraperitoneal injection (Lo et al., 2004). This phenomenon was also seen with the oral administration of the neutral polysaccharide (CS-F30) derived from mycelium of *C. sinensis* to normal mice, as it reduces their blood glucose levels (Kiho et al., 1993).

The polysaccharides of *Cordyceps,* particularly *C. sinensis,* also show immense immunomodulatory effects. Interestingly, the pharmacological activities of these polysaccharides are closely related to its molecular weight. The higher the molecular weight, the greater the water solubility, thus the better the biological activity (Zhao et al., 2009). The polysaccharides that have an average molecular weight of 83kDA can significantly zz boost the level of serums IgG, IgG1, and IgG2 (Wu et al., 2006). They are shown to improve phagocytosis of macrophages and monocytes, besides increasing the mass of thymus and spleen in mice that are administered with dexamethasone (Prasain, 2013). For example, Cordysinocan cultured from *C. sinensis* was found to stimulate T-lymphocytes and macrophages (Cheung et al., 2009). Meanwhile, polysaccharides derived from mycelia of *Cordyceps* that have a molecular weight of 210 kDa show a strong antioxidative effect (Li et al., 2014). For instance, when the cultured rat pheochromocytoma PC12 cells were pretreated with CSP-1, strong protection against hydrogen peroxide (H_2O_2)-induced damage was demonstrated (Li et al., 2003). Akaki et al. (2009) also showed that the soluble polysaccharides from *C. sinensis* fractions that resemble 1,3-beta-D-glucan with 1,6 branched chains can enhance the production of TNF-α.

As described earlier, polysaccharides of *Cordyceps* show strong antioxidant activity. Antioxidant extracts from plants and fungi have been extensively investigated and used to combat oxidation. Oxidation plays a significant role in the onset of chronic diseases such as cancer, nephritis, stroke, and arteriosclerosis (Zhang et al., 2019). From the fruiting bodies of *C. militaris,* Three distinct polysaccharides W-CBP50, W-CBP50I, and W-CBP50II were extracted from the bodies of C.militaris (Chen et al., 2013). Among these, W-CBP50II displayed significant free radical scavenging activity against 1,1-diphenyl-2-picrylhydrazyl (DPPH). Conversely, both W-CBP50 and W-CBP50I exhibited potent scavenging abilities. Other than DPPH, this action also targets hydroxyl and superoxide radicals. In addition, some research reported that polysaccharides extracted from cultured *C. militaris* also show antioxidative activities such as the ability to chelate ferrous ions and reduce ferric ions (Chen et al., 2014).

Moreover, polysaccharides of *Cordyceps* can also deal with oxidative stress, which is a condition linked to an aberrant immune response and implicated in various diseases due to the production of excessive free radicals. This is because oxidative stress gives rise to reactive oxygen species (ROS) and reactive nitrogen species (RNS), resulting in adverse consequences such as peroxidation of lipids, oxidation of proteins, and damage of DNA (Zhang et al., 2019). Eventually, this process may elicit changes structurally and functionally, or apoptosis. *C. militaris* polysaccharides (CMP) can prevent ROS scavenging activity caused by Cyclophosphamide, which is possible through increasing the antioxidative activity of superoxide dismutase (SOD), catalase (CAT), and glutathione peroxidase (GSH-Px) (Wang et al., 2012). A similar study also confers the protection mechanism of CMP, whereby it was found that middle and high doses of CMP successfully inhibited malondialdehyde (MDA) in organs such as kidney, heart, and liver. MDA serves as a reliable indicator of lipid peroxidation (Liu et al., 2016).

Lastly, *Cordyceps* polysaccharides also show anti-inflammatory effects. Inflammation is often the root cause of various diseases such as rheumatoid arthritis, Alzheimer's disease, cancer, and cardiovascular disease. It can also damage body tissues by causing defects in genes and disrupting immune regulation (Zhang et al., 2019). It was found that polysaccharides of *C. milituris* can reduce the secretion of NO, TNF-α and IL-6 triggered by LPS. This provides evidence supporting their notable ability in inhibiting inflammatory mediators, emphasizing their anti-inflammatory efficacy (Jo et al., 2010).

2.3 *Proteins*

Cordyceps spp contains all essential amino acids, peptides, polyamines, and proteins (Das et al., 2020). They also contain some less common cyclic dipeptides such as cyclo-[Leu-Pro], cyclo-[Gly-Pro], cyclo-[Val-Pro], cyclo-[Thr-Leu], and cyclo-[Ala-Leu]. In addition, polyamines like 1,3-diamino propane, cadaverine, spermidine, spermine, and putrescine are also present (Mishra and Upadhyay, 2011). In addition, some nitrogenous compounds such as putrescine and putrescine were also found in *Cordyceps* (Mizuno, 1999).

A large portion of proteins found in *Cordyceps* are enzymes, which include both intracellular proteases and extracellular proteases. One of the components found in *C. sinensis* is CSDNase, which is an acid deoxyribonuclease (DNase) (Ye et al., 2004). It is a single-chained DNase with a molecular mass of 34kDa. However, its structure requires further studies. DNases can be classified into two major classes, namely DNase I and DNase II. In this case, CSDNase is a type of DNase II, which acts on dsDNA and ssDNA. Being an endocellular enzyme, CSDNase exhibits elevated levels during the phase when fungal mycelium is still growing. Additionally, CSDNase can hydrolyze DNA to produce3-phosphate and 5-OH termini, which demonstrated its nucleolytic properties similar to DNases (Ye et al., 2004). Along with CSDNase, another serine protease, CSP, was extracted from *C. sinensis*. CSP was obtained by isolating it from its purified culture supernatant. It has a molecular weight of 31kD (Liu et al., 2015). CSP is a fibrinolytic enzyme consisting of a singular polypeptide chain, and its active site contains a free cysteine residue. It is capable of hydrolysing human serum albumin (HSA) and bovine serum albumin (BSA). As a plasmin-link protease, CSP can also cleave the Aα chain and α-chain of fibrinogen and fibrin respectively (Li et al., 2007). Therefore, CSP holds promise in the field of cardiovascular pharmacology, presenting new opportunities for protein engineering in developing new thrombolytic agents.

The specific biological activity of protein is also determined by factors such as the composition and sequence of amino acids, as well as the length of the polypeptide chain. Cordymin is a low molecular weight peptide which was found in *Ophiocordyceps sinensis* (Berk.) G.H. Sung, J.M. Sung, Hywel-Jones and Spatafora. In a study using alloxan-induced hyperglycaemic rats, cordymin showed a protective effect in reducing their blood glucose level (Qi et al., 2013). When a cordymin dose ranging from 50 to 100 mg per kg of the rats' body weight was administered, it was found that the glycated haemoglobin (HbA11C) levels dropped after a period of 5 weeks. Besides that, the body weight of the rats and the oxidative stress caused by high glucose level also significantly decreased. This effect is possible due to the action of cordymin that directly reduces the action of ALP and TRAP. Indirectly, it also restores β cells and reduction of glucose level in blood. Resultantly, this caused oxidative stress level to diminish in diabetic rats (Ouyang et al., 2013). Other than regulating blood glucose, cordymin is also anti-inflammatory, anti-nociceptive (Qian et al., 2012) and anti-fungal (Wong et al., 2011).

Cyclic dipeptides from the protein of *Cordyceps* also shows antimicrobial and anti-mutagenic ability. For instance, cyclo-(Leu-Pro) and cyclo-(Phe-Pro) can effectively inhibit the vancomycin-resistant enterococci (VRE) and pathogenic yeasts from growing (Rhee, 2004). Furthermore, a separate experiment also reported that cyclic dipeptides can inhibit aflatoxin production (Yan et al., 2004). This antimicrobial activity possessed by *Cordyceps* protein was further observed, where it was found that when

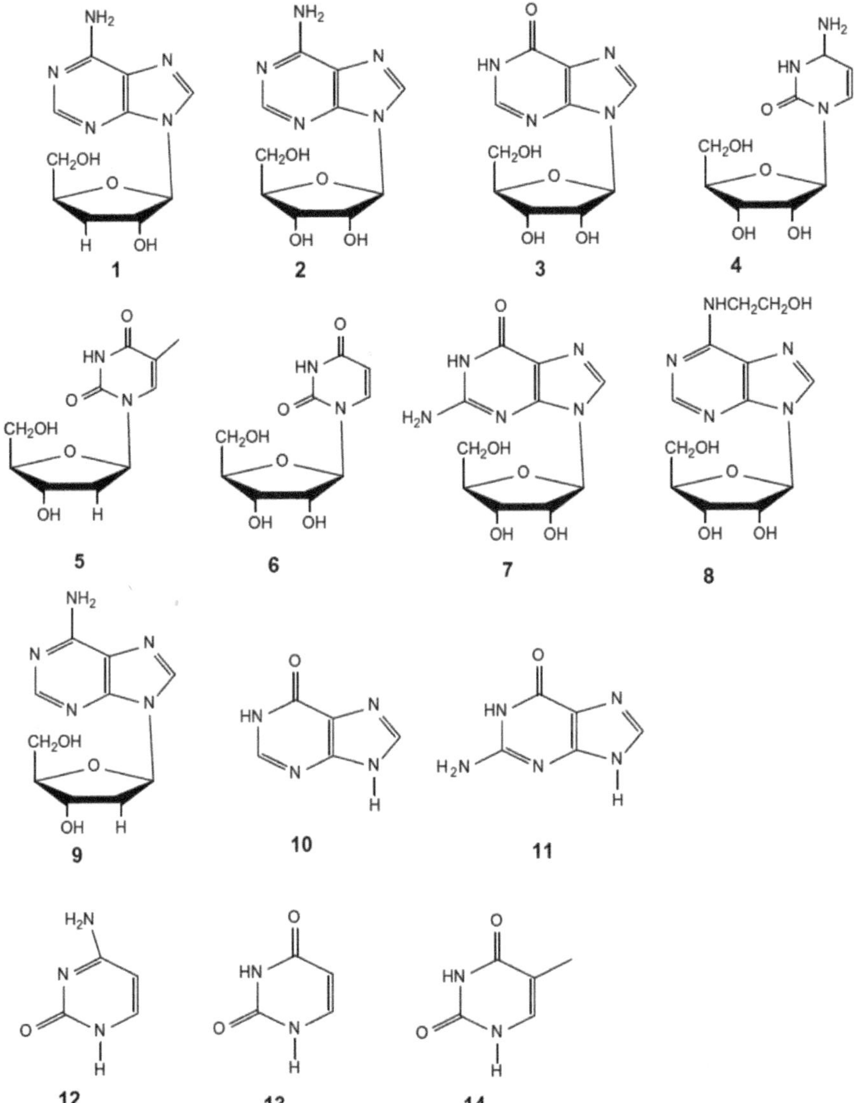

Fig. 3. Chemical structures of Nucleosides from *Cordyceps*: 1, cordycepin; 2, adenosine; 3, inosine; 4, uridine; 5, thymidine; 6, cytidine; 7, guanosine; 8, 6-hydroxyethyladenosine; 9, 20 -deoxyadenosine; 10, hypoxanthine; 11, guanine; 12, cytosine; 13, uracil; 14, thymine (Prasain, 2013).

dealing with Gram-positive and Gram-negative bacteria, these proteins presented a 60% anti-microbial activity and enhanced overall temperature stability without agglutinating blood cells (Zheng et al., 2006). Among the various types of proteins present, Cordycedipeptide A, Cordyceamides A and B, were also found to exhibit cytotoxic effects against cell lines such as L-929, A375, and HeLa cells (Jia et al., 2009).

2.4 Nucleotide and Nucleoside

Nucleosides constitute a significant portion of the components found in *Cordyceps*. Nucleosides are glycosylamines composed of a nitrogenous base (either a purine or pyrimidine) attached to a ribose or deoxysugar. More than 10 nucleosides such as cordycepin, adenosine, inosine, uridine, thymidine, cytidine, guanosine, 6-hydroxyethyladenosine, 20-deoxyadenosine, hypoxanthine, guanine, cytosine, uracil, and thymine have been identified in *Cordyceps*. Figure 3 illustrates the chemical structures of the nucleosides.

High-performance liquid chromatography (HPLC), coupled with diode array detection and liquid chromatography-mass spectrometry (LC-MS), has been extensively employed in various studies to quantitatively analyze nucleosides in both natural and cultured *C. sinensis* (Prasain, 2013). Adenosine receptors, namely A1, A2A, A2B, and A3, are found widespread throughout body organs such as the brain, lungs, heart, liver, and kidneys. These receptors play crucial roles in functions mediated by the central nervous system. For example, sleep regulation, respiratory control, immune responses, cardiovascular performance, as well as liver and kidney functions (Lo et al., 2013). Remarkably, the pharmacological advantages shown by *Cordyceps* are interrelated with the distribution of adenosine receptors, which includes anticancer properties, anti-aging effects, antithrombotic action, antiarrhythmic potential, antihypertensive effects, immune system modulation, and protective attributes for the kidneys, liver, and lungs (Lo et al., 2013). Given the ability of nucleosides to modulate human physiological functions through adenosine receptors, this highlights the therapeutic potential of *Cordyceps* nucleosides in addressing these diseases.

Nucleosides of *Cordyceps* are shown to have immunomodulatory properties. The ratio of nucleosides differs in natural and cultured *Cordyceps,* which plays a role in its bioactivities. In natural *C. sinensis*, uridine, inosine, and guanosine are present in a ratio of 8:11:5 respectively. Although this ratio has been observed to have no impact on LPS-stimulated cells, it can enhance the release of NO, TNF-α, and IL-1 from resting macrophages in mice (Lo et al., 2013). Contrarily, in cultured *C. sinensis,* guanosine, adenosine, and uridine are present in a ratio of 9:7:11 respectively. This ratio has been shown to effectively prevent LPS-stimulated cells from releasing cytokines, at the same time stimulate resting macrophages to produce NO, as well as cytokines such as TNF-α and IL-1 (Lo et al., 2013). Overall, these findings suggest that different components lead to distinct immune responses, which is also dependent on the ratios of these nucleosides.

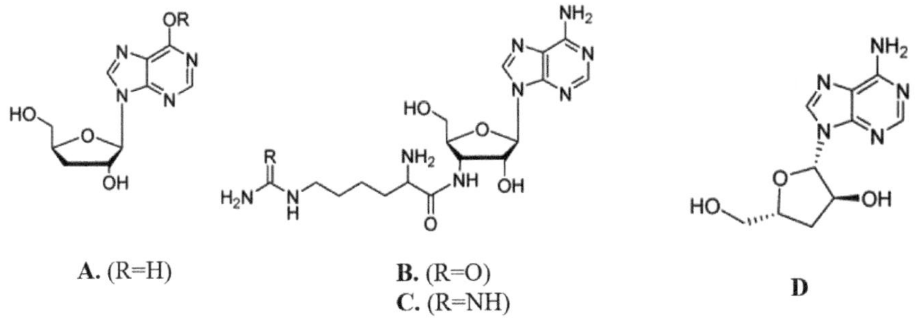

A. (R=H)   **B.** (R=O)
                     C. (R=NH)   **D**

Fig. 4. Chemical structures: A, 3'-Deoxyinosine; B, (2-amino-N-((2S,3S,4R,5R)-5-(6-amino-9H-purin-9-yl)-4-hydroxy-2-(hydroxymethyl)–tetrahydrofuran-3-yl)-6-ureido-hexanamide); C, (2-amino-N-((2S,3S,4R,5R)-5-(6-amino-9H-purin-9-yl)-4-hydroxy-2-(hydroxymethyl) tetrahydrofuran-3-yl)-6-guanidinohexanamide); D; 3'-Deoxyadenosine (Qu et al., 2022).

An isolated nucleoside, 3'-Deoxyinosine, derived from *C. militaris*, demonstrated weak cytotoxicity against cancer cells A549 and PANC-1 (Qiu et al., 2017). However, when administered at a concentration of 30 µM, it exhibited a remarkable cytotoxic effect against cancer cells MCF-7. Furthermore, it can also inhibit breast cancer cells from metastasizing and proliferating when administered in combination with doxorubicin (Xin et al., 2021). Utilizing the Chou-Talalay method, the most potent synergistic effect was observed when 3'-deoxyinosine and doxorubicin were co-administered at a dosage of 80 µmol/L and 1 µmol/L respectively (Xin et al., 2021). This combination resulted in an index (CI) value of 0.665 and achieved a cell inhibition rate of 60.31% ± 1.06% (Xin et al., 2021). Anti-tumour activity was also observed in the derivatives of *C. militaris*, which involved (2-amino-N-((2S,3S,4R,5R)-5-(6-amino-9H-purin-9-yl)-4-hydroxy-2-(hydroxymethyl)–tetrahydrofuran-3-yl)-6-ureido-hexanamide) and (2-amino-N-((2S,3S,4R,5R)-5-(6-amino-9H-purin-9-yl)-4-hydroxy-2-(hydroxymethyl)tetrahydrofuran-3-yl)-6-guanidinohexanamide) (Fig. 4). Both nucleosides exhibited suppressive effects on the HepG2 liver cancer cells at the 72-hour mark, with IC_{50} values of 0.09 µM and 0.51 µM respectively (Xiang et al., 2021).

2.5 Sterols

Numerous sterols have been identified in *Cordyceps*, such as cholesterol, ergosterol peroxide, β-sitosterol, ergosteryl-3-0-β-D-glucopyranoside, daucosterol, Δ^3ergosterol, campesterol, 22,23-dihydroergosteryl-3-O-β-D-glucopyranoside, H1-A, cholesteryl palmitate, cereisterol, and dihydrobrassicasterol. Other than pressurized liquid extraction (PLE), the determination of sterols in wild *C. sinensis* can be done through trimethylsilyl (TMS) derivatization and Gas chromatography–mass spectrometry (GC-MS) analysis (Yang et al., 2009).

Ergosterol is present in *Cordyceps*, both in free form and combined forms. In both natural and cultured *Cordyceps*, the content of the free form ergosterol is notably high (Yang et al., 2009). However, the amount of ergosterol in *Cordyceps* varies according to their stage of growth. For instance, the mycelium contained 1.44 mg/g of ergosterol, while the fruiting bodies had a higher content of 10.68 mg/g (Li et al., 2001). When *Cordyceps* receive exposure to UV radiation, ergosterol undergoes photolysis and transform into vitamin D_2 (Urbain et al., 2011). Being a precursor of vitamin D_2, ergosterol is a crucial for human bone development (Lin and Li, 2011).

In an *in vitro* experiment, it was demonstrated that ergosterol extracted from *C. militaris* effectively inhibits inflammatory mediators, specifically IL-12, NO, and TNF-α. The ergosterol extract is associated with their antiproliferative effects on colon tumour cells (Rao et al., 2010). Besides that, ergosterol showed cytotoxic effects against liver cancer cells and acute promyelocytic leukaemia (APML) cancer cells. Their study also showed that ergosterol delivers a moderate level of antimicrobial activity against the bacteria *P. aeruginosa* and *E. aerogenes,* as well as the fungus *C. albicans* (Zheng et al., 2013). Esterified ergosterol peroxide from *C. sinensis* has a higher potency in the inhibition of tumour cells proliferation as compared to aglycon (Bok et al., 1999). Meanwhile, ergosterol peroxide extracted from *C. ciacadae* was found to inhibit phytohaemagglutinin-induced T-cell proliferation (Kuo et al., 1996). This resulted in arresting cell cycle, whereby activated T cells from G1 were unable to progress into the S phase. Thus, early gene transcripts were suppressed, especially in cyclin E, interferon, and interleukins. Other than that, key phytosterols like β-sitosterol, campesterol, and stigmasterol, exhibit the ability to reduce cholesterol absorption while having limited absorption themselves (Ostlund, 2007). Among phytosterols, β-Sitosterol demonstrates a protective effect against cancers, particularly in the context of prostate, breast, and colon cancer (Awad et al., 2000).

Another distinct sterol from *Cordyceps*, named H1-A, exhibits structural similarity to testosterone and dehydroepiandrosterone. It shares similarities with ergosterol, but it lacks the binding ability to the glucocorticosteroid receptor (Liu et al., 2015). H1-A has the capability to inhibit activated human mesangial cells. It also alleviates a chronic kidney disease, known as Immunoglobulin A (IgA) nephropathy or Berger's disease, leading to both clinical and histological improvements (Lin et al., 1999). This compound had demonstrated potential effectiveness in treating specific autoimmune diseases (Yang et al., 2003). This is because H1-A can regulate the signalling pathways in mesangial cells, thus regulating the balance between cell growth and programmed cell death of the cells. Therefore, H1-A holds promise as a potential treatment for managing autoimmune disorders, given its capacity to modulate proteins involved in signal transduction, such as Bcl-XL and Bcl-2 (Yang et al., 2003).

3. Mechanism and Pharmacological Significance

3.1 *Diabetes*

Diabetes mellitus (DM) is characterized by chronic hyperglycaemia or the persistent elevation of blood glucose levels, attributed to either impaired insulin secretion or reduced insulin sensitivity (Dong et al., 2014). Individuals with diabetes may experience a range of metabolic disturbances, culminating in complications that detrimentally affect the kidneys, nerves, cardiovascular system, and various other tissues and organs, potentially leading to fatality (Liu et al., 2023). Noninsulin-dependent diabetes mellitus (NIDDM), also termed Type 2 diabetes mellitus (T2DM), arises from the resistance to insulin and represents the most prevalent manifestation of diabetes (Dong et al., 2014). Its global incidence stands at approximately 537 million individuals as of 2021 (Sun et al., 2022), with projections anticipating a surge to 643 million by 2030 and further to 738 million by 2045 (International Diabetes Federation, 2023). Presently, in the context of urban living, diabetes poses a formidable challenge, for which currently there is no entirely effective therapeutic regimen. While many approaches are utilized to manage diabetes by restoring normal blood glucose and lipid levels, lowering blood pressure, and enhancing microcirculation, they do not provide a definitive cure (Levterova et al., 2013). Furthermore, pharmaceutical products used for diabetes treatment often exhibit various adverse effects, with their efficacy occasionally being a subject of debate (Ashraf et al., 2020). For instance, Pioglitazone can lead to hepatocellular-cholestatic liver injury, while metformin may result in gastrointestinal issues such as diarrhoea, nausea, or vomiting (Andújar-Plata et al., 2012). Consequently, alternative medicinal strategies, such as traditional herbal medicine containing bioactive antidiabetic compounds have been researched (Ashraf et al., 2020).

The precise mechanism underlying the use of *Cordyceps*, particularly cordycepin, as a potential treatment for managing diabetes still requires further research efforts. Although several investigations have explored plausible pathways, a comprehensive understanding of the molecular and physiological mechanisms involved is essential for the development of targeted therapies (Ashraf et al., 2020). These studies have suggested that cordycepin hinders the generation of nitric oxide (NO) generation and pro-inflammatory cytokines including TNF-α, IL-1β, and IL-6, in macrophages activated by lipopolysaccharides (LPS). This inhibition occurs via the protein synthesis suppression of pro-inflammatory mediators. Consequently, the gene expression associated with type 2 diabetes regulation, specifically PPARλ and 11β-HSD1, reduces. As presented in Figure 5(A), an increase in cordycepin concentration also decreases co-stimulatory molecule expression, including ICAM-1 and B7-1/-2 (Shin et

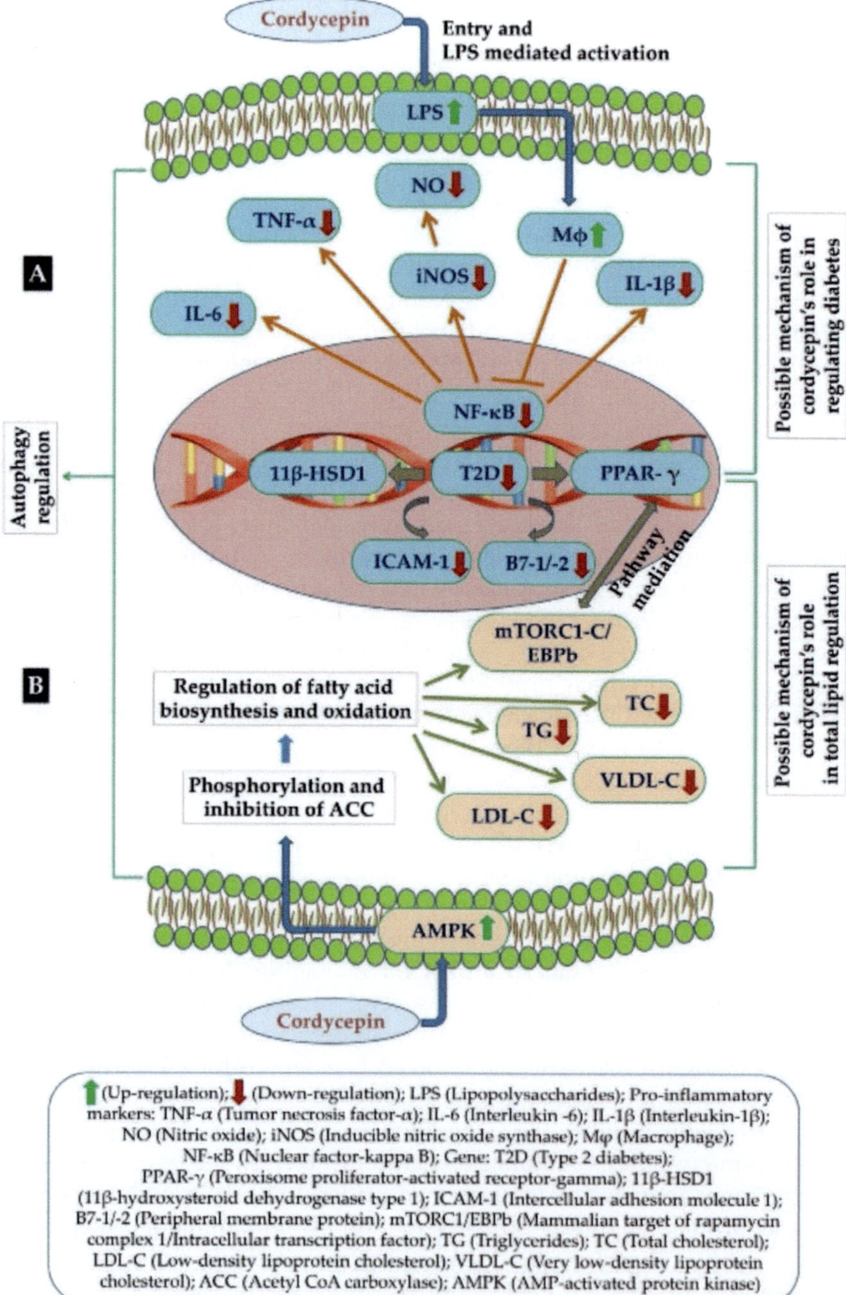

Fig. 5. Possible mechanisms of anti-diabetic activity (A) and regulation of fat metabolism in hyperlipidaemia (B) of cordycepin (Ashraf et al., 2020).

al., 2009). Moreover, it has been observed that cordycepin also mitigates gene expression implicated in diabetes regulation by deactivating NF-κB-dependent inflammatory responses (Ji et al., 2009; Shweta et al., 2023). The antidiabetic efficacy of cordycepin was demonstrated through a diabetic mouse model induced with alloxan. The glucose tolerance test results later indicated vast improvements, following an effective dosage of cordycepin (Ma et al., 2015).

Cordycepin is also a pivotal factor in mitigating insulin resistance, with the capacity to enhance insulin sensitivity by monitoring serum insulin level and the assessment of the insulin resistance index homeostasis model (Niu et al., 2010). For instance, β-sitosterol is a potent antioxidant and plant sterol known for its anti-hyperlipidaemic properties in *C. sinensis*, exhibiting abilities in reducing levels of lipid, reinstating sensitivity to insulin, activating insulin receptors and stimulating GLUT4 in the adipose tissue, thereby exerting anti-hyperglycaemic effects in rats with diabetes (Ponnulakshmi et al., 2019).

Multiple bioactive compounds from *Cordyceps* were identified as contributors to the anti-hyperglycaemic properties. The elevated fibre content within *C. sinensis*, especially within its fruiting body, has the potential to enhance cellular glucose uptake and mitigate insulin resistance, potentially through its effects on glucose absorption mechanisms (Lo et al., 2004). Additionally, polysaccharides have been identified as critical bioactive components in anti-hyperglycaemic effects, with extracts enriched in polysaccharides demonstrating the capacity to reduce glycaemic levels by 60% to 70% (Zhang et al., 2006). Notably, a study highlighted the significant glucose-reducing effects of a polysaccharide derived from *C. sinensis* hot water extracts identified as CS-F10, which achieved this outcome by augmenting glucokinase activity in liver tissues of both normoglycemic and hyperglycaemic mice (Kiho et al., 1999). Furthermore, cordycepin has exhibited the potential to stimulate insulin secretion and cholinergic activation. Studies have indicated that both *C. sinensis* mycelia and the liquid fermentation products can substantially enhance glucose responses from oral glucose tolerance tests (OGTT) and elevate insulin levels in diabetic rats induced with nicotinamide as well as streptozotocin (STZ) (Lo et al., 2006).

3.2 *Cardiovascular Diseases*

Cardiovascular disease (CVD) (Hyperlipidaemia) stands out as a predominant factor of mortality in both advanced and emerging nations. Projections indicate that CVD-related annual deaths will reach 23.6 million by the year 2030 (Prevention, 2011). CVD is influenced by a spectrum of risk factors that can manifest at any age, including early in life (Hart et al., 2022). These risk factors encompass factors such as excessive tobacco consumption, alcoholism, sedentary lifestyles, and unhealthy dietary practices. Among these risk factors, elevations in lipid levels are recognised as a prominent contributor (Le, 2008). Hyperlipidaemia is classified as disruptions in lipid metabolism, involving elevated levels of low-density lipoprotein cholesterol (LDL-C), increased triglycerides (TG), elevated total cholesterol (TC), and decreased high-density lipoprotein cholesterol (HDL-C) (Kopin and Lowenstein, 2017). The plasma concentrations of total cholesterol and LDL-C are pivotal contributors to CVD, coronary heart disease (CHD) in particular (Kannel, 1995). This significance arises as LDL conveys approximately 60% to 70% of cholesterol in the blood and transports cholesterol to tissues on the periphery from the liver. Consequently, high LDL-C levels pose a significant threat as they can accumulate and initiate the formation of atherosclerotic plaques within arterial walls (Elshourbagy et al., 2014), ultimately leading to atherosclerotic CVD, which has been recognised as another major global cause of high mortality (Zhao et al., 2019). However, the degree of increased CVD risk largely depends on the specific pattern as well as the underlying causes.

Drugs with cholesterol-lowering abilities, such as statins, represent the primary pharmaceutical approach to prevent CVD by targeting 3-hydroxy-3-methylglutaryl coenzyme A reductase (HMGCR) inhibition. While statins are typically well-received, increasing evidence indicates their tendency to elicit muscular discomfort and various adverse effects (Irvine, 2020; Shahbaz et al., 2018). Furthermore, it is noteworthy that the existing therapeutic strategies, primarily centred on lipid-lowering drugs, do not entirely eliminate the risk of CVD. For this reason, it is necessary to explore alternative approaches, particularly the utilization of natural compounds found in food as they may offer reduced toxicity levels, in the pursuit of CVD prevention and management.

Cordycepin might have a role in preventing the intracellular accumulation of lipids by engaging with the $\gamma 1$ subunit and activating AMP-activated protein kinase (AMPK) (Wu et al., 2014a). AMPK is crucial in sensing cellular energy and contributes mechanistically to the regulation of fat metabolism . Furthermore, AMPK activation reduces fatty acid levels through phosphorylating and suppressing acetyl-CoA carboxylase (ACC), thus regulating both the biosynthesis and oxidation of fatty acids, as seen in Fig. 5(B). Additionally, the activation of AMPK has been shown to reduce overall cholesterol and TG levels via the inhibition of glycerol-3-phosphate acyltransferase (GPAT) and HMG CoA reductase activity, which are key enzymes responsible for the synthesis of total cholesterol and TG, respectively (Atkinson et al., 2003). Therefore, observations have shown that AMPK regulation is a potential remedy for individuals struggling with obesity and overweight conditions associated with hyperlipidaemia (Hardie et al., 2012).

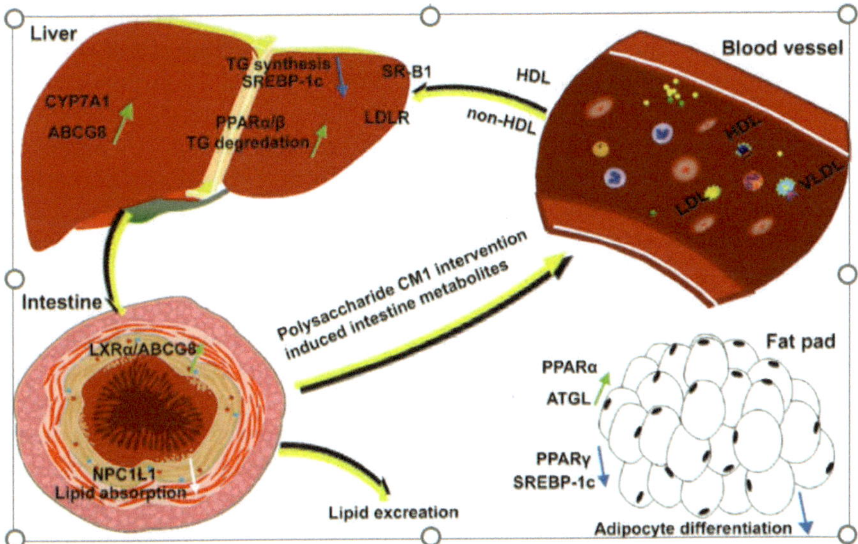

Fig. 6. Mechanism of action pf polysaccharide CM1 in LDLR(+/−) hamsters (Yu et al., 2021).

In accordance with prior research, cordycepin has demonstrated significant efficacy in lowering lipid levels. This effect is attributed to its structural resemblance to adenosine, an AMPK activator. Likewise, cordycepin administration exhibited notable reductions in LDL-C, TC and TG, making it a promising nutraceutical candidate for mitigating hyperlipidaemia induced by high-fat diets (Aymerich et al., 2006, Guo et al., 2010). Alternatively, cordycepin was also examined for its potential role in the regulation of autophagy and lipid metabolism. It was observed that cordycepin had displayed efficacy in mitigating the build-up of hepatic lipid elicited by palmitic acid (PA) through the initiation of autophagy. The underlying mechanism behind this efficacy may involve the PKA/mTOR pathway (Ashraf et al., 2020). Furthermore, it was noted that cordycepin exhibited significant efficacy in decreasing intracellular concentrations of overall lipids, TC, cholesterol and TG, LDL-C, very low-density lipoprotein cholesterol (VLDL-C), and ratios of LDL-C/HDL-C and TC/HDL-C. Consequently, these lipid-reducing properties of cordycepin may hold promise for addressing hyperlipidaemia, as suggested by prior research.

Polysaccharide CM1 sourced from *C. militaris* exhibited notable lipid-lowering effects in LDLR (+/−) hamsters, which are known to possess a lipid profile resembling that of humans (Guo et al., 2018, Wu et al., 2019; Xia et al., 2017) (Fig. 6). Polysaccharide CM1 is primarily composed of (1→4)-β-D-glucopyranosyl and (1→2)-α-D-mannopyranosyl units, with branching occurring at the O-6 positions of (1→2,6)-α-D-mannopyranosyl residues linked to (1→2)-β-D-galactopyranosyl, (1→2)-α-D-mannopyranosyl, or methyl groups (Hu et al., 2019).

Firstly, CM1 elevates CYP7A1 and ABCG5/8 levels, suggesting the possible enhancement of bile acid synthesis and lipid excretion, respectively. This contributes to potential cholesterol conversion and excretion (Yu et al., 2021). Secondly, CM1 directly inhibits the protein levels of NPC1L1, thus indicating its ability to inhibit the absorption of NPC1L1-mediated cholesterol in the small intestine. CM1 can also decrease TG concentration in plasma by suppressing the formation of TG and facilitating fatty acid degradation (Yu et al., 2021). This reduction in TG synthesis is achieved through the inhibition of SREBP-1c transcription. Research has presented that SREBP-1c regulates the transcription of genes involved in lipid synthesis, thereby influencing TG metabolism (Moslehi and Hamidi-Zad, 2018). Consequently, this cascade of effects may ultimately lead to a decrease in cholesterol absorption and synthesis (Yu et al., 2021). Thirdly, CM1 has the potential to lower TG levels by upregulating lipoprotein lipase (LPL) and peroxisome proliferator-activated receptor-alpha (PPARα) expression in the liver. PPARα and peroxisome proliferator-activated receptor-beta (PPARβ) are known as major inducers of hepatic fatty acid oxidation, whereas peroxisome proliferator-activated receptor-gamma (PPARγ) functions significantly in adipocyte differentiation (Lin et al., 2021; Poulsen et al., 2012). This suggests the role of CM1 in reducing TG levels through the increase of LPL-mediated lipoprotein degradation in the liver and enhancing PPARα-mediated β-oxidation, thereby promoting the breakdown of fatty acids and contributing to a reduction of the fat pad index (Yu et al., 2021). Lastly, CM1 intervention results in a decreased rate of adipocyte differentiation, potentially achieved through the modulation of various molecules in the epididymal fat (Yu et al., 2021). CM1's inhibitory effect on adipose tissue in LDLR(+/−) hamsters can be partly linked to the reduction of PPARγ, given that PPARγ acts as a significant activator of adipocyte differentiation (Poulsen et al., 2012). Moreover, CM1 also inhibits TG formation via DGAT1 and DGAT2 downregulation, hindering differentiation of adipocyte as well (Liu et al., 2012).

3.3 Anti-Inflammatory Effects

Inflammation serves as a fundamental component of the body's defence mechanism, functioning as a favorable host response in combating infections as well as repairing tissue damage, to eventually restore normal tissue structure and function. Typical inflammatory responses are self-regulating by nature, however, prolonged inflammation contributes significantly to the progression of numerous inflammatory diseases (Dinarello, 2000; Jo et al., 2010), and contributes to the onset of cancer (Li et al., 2014; Wu et al., 2014b). Inflammation can manifest in either acute or chronic forms (Michels da Silva et al., 2019). For instance, acute and chronic lung inflammations have both been documented in multiple pulmonary disorders, including but not limited to asthma, acute respiratory distress syndrome, cystic fibrosis (CF), and chronic obstructive pulmonary disease (COPD) (Aghasafari et al., 2019).

Significantly, cordycepin was documented in the repression of intestinal inflammation in a mouse model with acute colitis and the suppression of microglial activation via the inhibition of pro-inflammatory molecules such as tumour necrosis factor alpha (TNFα) (Han et al., 2011, Jeong et al., 2010). Cordycepin has also demonstrated efficacy in a murine model of bronchial asthma, enhancing mucociliary clearance and respiratory tract lubrication, which are both particularly pertinent to respiratory disorders like COPD and asthma where excessive secretion is a concern (Hsu et al., 2008). Furthermore, studies indicate that cordycepin can diminish airway remodelling in COPD-diagnosed rats by curbing inflammation of the respiratory tract and modulating the TGF-β1/Smad signalling pathway. This suggests that cordycepin might hold promise in addressing disorders like COPD (Antoniou et al., 2007; Yang et al., 2018). Cordycepin extracts were found to also alleviate lung fibrosis by decreasing transforming growth factor beta-1 (TGF-β1) expression as well as promoting the degradation of collagen (Li et al., 2006; Wang et al., 2007; Xu et al., 2011).

Moreover, cordycepin exhibits anti-proliferative and anti-inflammatory properties by stimulating AMPK activity, inhibiting protein kinase B (AKT) phosphorylation by mTOR, and frequently diminishing ERK phosphorylation by MEK. This observed phenomenon can be attributed to the established involvement of mTOR/AKT and AMPK in the modulation of inflammation and their role in connecting metabolic changes with the inflammatory response within macrophages, as substantiated by previous research (Dickson and Lehmann, 2019; Kang and Kumanogoh, 2020).

Smiderle et al (2011) found that isolated β-(1→3)-D-glucan found in *C. militaris* polysaccharide extracts, exhibit the highest anti-inflammatory activity. It is widely recognized that toll-like receptor-4 (TLR-4) serves as the major signal transducer for lipopolysaccharides (LPS), whereas TLR-2 can be considered as a receptor with limited affinity for this bacterial endotoxin. Given these observations, it is plausible that β-(1→3)-D-glucan and LPS compete for TLR-4 binding, thereby suppressing the LPS-induced pro-inflammatory gene stimulation (Smiderle et al., 2011).

Consequently, considering the cumulative evidence and data from these studies, cordycepin possesses substantial potential as a potent bioactive anti-inflammatory component.

3.4 Immunomodulatory Effects

Immune modulation, commonly termed as the regulation of the immune system, is achieved through the use of various chemical agents that influence immune responses and immune system functioning by either stimulating the production of antibodies or inhibiting the activity of white blood cells. *Cordyceps* spp. contain a variety of compounds with immunomodulatory properties and are recognized by TLRs and C-type lectin receptors (CLRs) when initiating APC immunomodulation and hypo responsiveness. Notably, these bioactive components not only induce changes in TLR and CLR expression in APCs, however they also specifically regulate their intracellular signaling pathways (Das et al., 2020).

Firstly, the active constituents found in the various *Cordyceps* spp., specifically *C. cicadae, C.militaris, C. sinensis,* and *C.sobolifera* are involved in the transmission of TLR-4 signalling to the MAPK pathway, as well as the activation of ERK1/2, thus contributing to the promotion of regulatory T cell (Treg) and T helper 2 (Th2) cell induction (Das et al., 2020). Moreover, the concurrent presence of dendritic cell-specific intercellular adhesion molecule-3-grabbing non-integrin (DC-SIGN) alongside TLR-4 facilitates the activation of these active *Cordyceps* constituents, initiating previously uncharacterized intracellular pathways that are capable of cross-inhibiting MyD88 and NF-κB activation.

Additionally, in order to reprogram host-immunity, *Cordyceps* spp. employs a strategy that involves the degradation of intracellular molecules vital to the host. Specifically, the polysaccharide constituents found within *Cordyceps* spp. have the capacity to degrade endosomal receptors such as TLR2, TLR3, TLR4, and TLR6, as well as host mRNA, thereby promoting and facilitating the activation of Treg and Th2. This process is associated with the engagement of CLRs as reported in previous studies (Das et al., 2020).

Cordycepin has also demonstrated its capacity to induce the secretion of cytokines in resting peripheral blood mononuclear cells (PBMCs) and impacting the proliferation of PBMC as well as THP-1, the transcription factors within a human acute monocytic leukemia cell line. Furthermore, it has been reported that cordycepin can modulate the functionality of human immune cells in *in vitro* settings (Cheng et al., 2019; Zhou et al., 2008). Additionally, isolated bioactive constituents derived from *C. militaris* have exhibited noteworthy immunomodulatory effects through enhancing the survival of mice infected with lupus and decreasing the production of anti-ds-DNA (Shashidhar et al., 2013; Shin et al., 2009). Besides, research conducted by Zhang & Xia (1990) notes of the immunosuppressive role of *C. sinensis* in a rat model with a heterotopic heart allograft, thereby extending its survival period. Similarly, this is seen in a study conducted by Zhu & Yu (1990), where the survival period of mouse skin allografts was prolonged with *C.sinensis* (Zhang and Xia, 1990; Zhu andYu, 1990).

In light of these findings, *Cordyceps* has proven to be a promising immunomodulatory agent, and can be specifically utilized in managing autoimmune disorders and preventing post-organ transplant rejection (Taylor et al., 2005). An expanding number of research also suggests the dual modulatory function of cordycepin, including both inhibitory as well as stimulatory impacts on immunity, by having the ability to regulate both the adaptive and innate immunity (Kitamura et al., 2011; Ng and Wang, 2005; Zhao et al., 2019).

3.5 *Anti-Osteoporosis Effects*

Osteoporosis is characterized by diminished bone mineral density (BMD) and the deterioration of both structural and biomechanical attributes of bone tissue (Ashraf et al., 2020). This reduction in bone mass, coupled with heightened bone fragility, heightens the susceptibility of individuals with osteoporosis to fractures (Khosla and Hofbauer, 2017). This condition predominantly manifests in the elderly, particularly post-menopausal women and individuals who have undergone prolonged steroid therapy (Barnsley et al., 2021).

Cordycepin's anti-osteoporotic potential was investigated via ovariectomized osteopenic rats, revealing its efficacy in mitigating bone loss within the experimental model (Ashraf et al., 2020). The mechanistic approach employed in this study presented limited contributions in both *in vitro* and *in vivo* tartrate-resistant acid phosphatase and alkaline phosphatase enzymes. Furthermore, results demonstrated that the oral administration of cordycepin led to an elevation in osteocalcin (OC) levels, indicative of enhanced bone formation, while simultaneously lowering the levels of C-terminal cross-linked telopeptide of type 1 collagen (CTX), indicating reduced bone resorption. Additionally, it was observed that oxidative stress levels in ovariectomized rats were restored by cordycepin as well (Zhang et al., 2015).

Furthermore, it was noted that cordycepin exhibits the ability to inhibit processes of osteoclast differentiation induced by RANKL. This is accompanied by a down-regulation in mRNA expression of key genes associated with osteoclastogenesis, including matrix metalloproteinase-9 (MMP-9), cathepsin K, tartrate-resistant alkaline phosphatase (TRAP), and nuclear factor of activated T-cells, cytoplasmic 1 (Kim et al., 2016). Notably, IRF-8, a member of the IRF family, exhibits a specific expression in immune cells and plays a regulatory role in myeloid cell development (TamuraandOzato, 2002). It has also been elucidated that IRF-8 serves as a crucial inhibitory regulator in osteoclastogenesis, contributing to the maintenance of bone health (Zhao et al., 2009). The influence of cordycepin treatment results in an upregulation of IRF-8 expression, accompanied with a reduced NFATc1 expression. This observation indicates that cordycepin can also effectively inhibit osteoclastogenesis by activating IFR-8, consequently leading to the downregulation of NFATc1 (Dou et al., 2016).

Furthermore, it is important to note that cordycepin's efficacy in the context of osteosarcoma treatment has been substantiated through a

Table 3. Details of compounds with potential toxic effect.

Compound	Toxic effect	Reference
Cordycepin	Alter male fertility, gastrointestinal and bone marrow toxicity	(Qin et al., 2019)
Pentostatin	Diarrhea, nausea, renal and neurological toxicity	(Margolis and Grever, 2000)
HEA	Insecticidal	(Fang et al., 2016)
2-pyridone alkaloid tenellin-like compounds	Cell cytotoxicity	(Schmidt et al., 2003)
Bibenzoquinone oosporein,	Insecticidal, antibiotic activity	(Fan et al., 2017; Schmidt et al., 2003)
Beauveriolide	Cytotoxicity, cell apoptosis	(Lu et al., 2017; Mallebrera et al., 2016)

combination of *in vitro* experimentation and *in vivo* trials in mice (Jędrejko et al., 2022). The research findings conclude that cordycepin elicits apoptosis, while repressing osteosarcoma cell proliferation. Through *in vitro* testing, results have established that cordycepin downregulates key marker expressions, such as Bcl-2, MMP-2, and MMP-9, while upregulating PARP and proapoptotic protein Bax expressions. Moreover, cordycepin instigates the AMPK pathway activation while simultaneously suppressing the AKT/mTOR signaling pathway, thus further preventing osteosarcoma cell growth (Li et al., 2021).

Consequently, these findings offer valuable insights into cordycepin's potential as a bioactive compound for addressing osteoporosis. Notably, it also exhibits promise in mitigating bone loss induced by oestrogen deficiency (Ashraf et al., 2020) or postmenopausal osteoporosis (Dou et al., 2016).

4. Safety and Efficacy

In general, *Cordyceps* is considered to be a safe and non-toxic natural ingredient for human consumption. Nevertheless, safety concerns still exist and has been a debate till date. Documented side effects, including but not limited to diarrhea, abdominal distension, headache, dry mouth, throat discomfort, nausea, and other allergic reactions, contribute to the ongoing discussions surrounding the safety of the treatment. (Shashidhar et al., 2013; Tuli et al., 2013; Yu et al., 2019). Certain *Cordyceps* compounds, such as cordycepin, exhibits dose-dependent cytotoxicity and neurotoxicity in animals as well as humans despite owning many positive pharmacological potentials (Table 3). Furthermore, the possibility of *Cordyceps* fungi to produce mycotoxin remains a concern. Moreover, the fungi genome encodes biosynthetic gene clusters (BGCs) that have the potential to produce unidentified secondary metabolites. In order to identify and address the safety concerns of *Cordyceps*, a comparative analysis method can be used to deduce metabolites from conserved BGCs. The purpose of this analytic methodology is to compare the genome information of *Cordyceps* fungi with BGCs involved in the production of mycotoxins (Chen et al., 2020).

Despite having various beneficial bioactivities, cordycepin remains a subject of ongoing research. As mentioned earlier, cordycepin shows properties such as anticancer, anti-inflammatory, antimicrobial, and inhibition platelet aggregation. However, it has been reported that cordycepin may alter male fertility as it can stimulate Leydig cells to produce testosterone in a mouse model (Qin et al., 2019). Other than cordycepin, through the strategy of protector–protégé, *C. militaris* is able to produce pentostatin using the same BGC employed for cordycepin production (Xia et al., 2017). Pentostatin is a drug isolated from *Streptomyces antibioticus*, which is commonly used to treat leukemia. This drug can inhibit adenosine deaminase irreversibly. Besides that, it also carries dosage- dependent and schedule-dependent side effects such as diarrhea, nausea, as well as renal and neurological toxicities (Margolis and Grever, 2000). If combined, cordycepin and pentostatin may lead to adverse effects such as toxicity in bone marrow and gastrointestinal tract, which is shown in an experiment using a dog model (Rodman et al., 1997).

Another analog of adenosine, N6-(2-Hydroxyethyl)-adenosine (HEA), was discovered in several *Cordyceps* spp. such as *C. cicadae* and *C. militaris* (Zheng et al., 2018). HEA has a protective effect on kidneys and shows anticancer properties. Nevertheless, HEA can target the adenosine receptor (AdoR) found in insects, leading to insecticidal results (Fang et al., 2016). AdoRs of a different family is also found in humans. It is a type of G-protein coupled receptors that act as potential drug targets. Henceforth, the role of HEA of AdoRs in mammals, regardless of being agonists or antagonists, requires further research and clarification before consuming it in the long-term.

Furthermore, *C. militaris* and *C. cicadae* produces metabolites such as 2-pyridone alkaloid tenellin-like compounds and bibenzoquinone oosporein, which is shown in Figure 7. The 2-pyridone is involved in

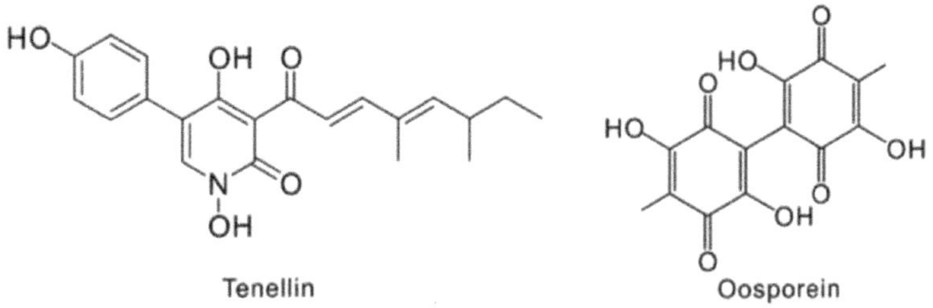

Fig. 7. Chemical structures of 2-pyridone alkaloid tenellin-like compounds and bibenzoquinone oosporein.

iron homeostasis and demonstrates both neuritogenic activity and cell cytotoxicity (Schmidt et al., 2003). On the other hand, oosporein promotes fungal infection of insect hosts via insecticidal and antibiotic activities (Fan et al., 2017).

Besides that, cyclodepsipeptides such as beauveriolide I and III, as well as beauvericin has also been discovered in *C. militaris* and *C. cicadae* respectively (Lu et al., 2017). Beauveriolides have shown anti-aging properties, the ability to reduce beta-amyloid levels, and anti-atherogenic effect without any apparent side effects. These are achieved by inhibiting lipid droplet accumulation in macrophages (Ohshiro et al., 2017). Meanwhile, beauvericin can cause cell apoptosis and cytotoxicity depending on the dosage (Mallebrera et al., 2016). This is due to the enhanced permeability of ions through the membrane as a result of its ionophoric activity.

According to conservation analysis of BGCs, it was found that *Cordyceps* fungi may produce mycotoxin analogues of PR-toxin, which may cause adverse effects such as kidney damage, liver damage, infertility, and abortions in cattle (Chen et al., 2020).

In addition, *C. cicadae* contains a highly conserved gene cluster when compared to BGC responsible in producing trichothecene (Tri) mycotoxins in *Fusarium graminearum,* a fungus commonly found in wheat. Trichothecenes encompass a large group of compounds that are chemically related, which includes type A mycotoxins T-2, NX-2, and HT-2, as well as type B deoxynivalenol (DON), nivalenol toxins, and acetylated deoxynivalenol. These toxins can potentially cause neurotoxic, immunosuppressive, and nephrotoxic effects in mammals as it can impede the synthesis of proteins (Woloshuk and Shim, 2013). Collectively, there is a strong possibility that *C. cicadae* could produce mycotoxins with similarities to trichothecenes.

Due to safety concerns, consumption by individuals suffering from systemic lupus erythematosus (SLE), rheumatoid arthritis, and multiple sclerosis are not advisable (Ashraf et al., 2020). Additional clinical, experimental, and epidemiological data are necessary to identify potential molecular targets and establish correlations between *Cordyceps* and the diseases, as well as to perform detailed assessments to unveil the unknown BGCs This will help to determine the optimal dosage to facilitate its use in the medicinal field and daily consumption.

6. Current Trend and Global Prospects

The global demand for natural remedies has surged in response to the growing trend of non-communicable diseases such as cardiovascular disease, chronic respiratory disease, diabetes, and stroke. This drives the demand of *Cordyceps* as a preventative medical product, specifically the well-studied *C. sinensis*. Within the *C. sinensis* market, manufacturers are focusing on creating versatile products suitable for various applications, including pharmaceuticals, food products, and dietary supplements. The emergence of major markets, such as China, has triggered increased investments in research and development across the Asia Pacific region. According to a report by Data Bridge Market Research, the global *C. sinensis* market is projected to skyrocket from USD 1.07 billion to USD 2.44 billion from 2022 to 2030, along with a compound annual growth rate (CAGR) of 5.01% (Research, 2023).

Cordyceps products can be categorized into two types, namely dry and wet products, and their applications can range from therapeutic to general healthcare. Due to increased chronic illnesses and healthcare initiatives, a strong presence of major industry players, as well as increased research endeavours in North America, it is anticipated that this region will dominate the global *C. sinensis* market (Research, 2023). Meanwhile, from year 2023 to 2030, the Asia-Pacific region is expected to experience large growth, which is driven by government initiatives to raise awareness, a surge in research activities, access to new markets, a vast population base, and a surging need for better healthcare. Notable players in this competitive market are The Lubrizol Corporation U.S., Nutra Green Biotechnology Co., Ltd. (China), Xi'an Saina Biological Technology Co., Ltd (China), Naturalin Bio-Resources Co., Ltd. (China), and Kshipra Biotech Private Limited (India) (Research, 2023).

Conclusion

There has been a growing inclination and confidence towards the utilization of naturally occurring food-based products in the area of managing and treating chronic diseases in recent decades, wth *Cordyceps* being one of the popular compounds that is highly consumed and extensively studied. In short, *Cordyceps* spp. contains a rich source of diverse bioactive constituents, some of which have exhibited therapeutic and pharmacological potential in preclinical investigations, both *in vitro* and *in vivo*. Key constituents such as cordycepin, cordycepic acid (CA), polysaccharides, proteins, nucleotides, nucleosides, and sterols hold significant therapeutic potential. These active constituents exert their therapeutic effects via regulation of multiple critical cellular signaling pathways, primarily by influencing inflammatory and oxidative stress processes. Looking ahead, there is a pressing need for further chemical investigations to discover the unidentified molecules within *Cordyceps* spp. Additionally, expanded preclinical research is also crucial in identifying and understanding the most promising biological activities of these compounds and to comprehend the potential synergies that can or may exist among its many constituents. Nevertheless, comprehensive trial-based and clinical investigations are also necessary for the identification of precise action mechanisms underpinning the role of cordycepin, pentostatin, HEA, and other compounds, as well as to evaluate its effectiveness and safety, both as a standalone treatment and in combination with other therapies. In conclusion, future research endeavors are essential for the further discovery of bioactive compounds inherent to the *Cordyceps* genus and their therapeutic potential. This exploration is of utmost importance as they could play significant roles in preventing and treating a range of health issues stemming from metabolic disorders and infections.

References

Adamson, R.H., Zaharevitz, D.W. and Johns, D.G. (1977). Enhancement of the biological activity of adenosine analogs by the adenosine deaminase inhibitor 2'-deoxycoformycin. Pharmacology, 15: 84-89. doi: 10.1159/000136666.

Aghasafari, P., George, U. and Pidaparti, R. (2019). A review of inflammatory mechanism in airway diseases. Inflamm Res, 68: 59-74. doi: 10.1007/s00011-018-1191-2.

Akaki, J., Matsui, Y., Kojima, H., Nakajima, S., Kamei, K. and Tamesada, M. (2009). Structural analysis of monocyte activation constituents in cultured mycelia of *Cordyceps sinensis*. Fitoterapia, 80(3): 182–187. doi: 10.1016/j.fitote.2009.01.007

Anderson, S.D., Charlton, B., Weiler, J.M., Nichols, S., Spector et al. (2009). Comparison of mannitol and methacholine to predict exercise-induced bronchoconstriction and a clinical diagnosis of asthma. Respir Res, 10: 4. doi: 10.1186/1465-9921-10-4.

Andújar-Plata, P., Pi-Sunyer, X. and Laferrère, B. (2012). Metformin effects revisited. Diabetes Res Clin Pract, 95: 1-9. doi: 10.1016/j.diabres.2011.09.022.

Antoniou, K.M., Pataka, A., Bouros, D. and Siafakas, N.M. (2007). Pathogenetic pathways and novel pharmacotherapeutic targets in idiopathic pulmonary fibrosis. Pulm Pharmacol Ther, 20: 453-461. doi: 10.1016/j.pupt.2006.01.002.

Ashraf, S.A., Elkhalifa, A.E.O., Siddiqui, A.J., Patel, M., Awadelkareem, A.M. et al. (2020). Cordycepin for Health and Wellbeing: A potent bioactive metabolite of an entomopathogenic *Cordyceps* medicinal fungus and its nutraceutical and therapeutic potential. Molecules, 25(12): 2735; https://doi.org/10.3390/molecules25122735.

Atkinson, L.L., Kozak, R., Kelly, S.E., Onay Besikci, A., Russell, J.C. et al. (2003). Potential mechanisms and consequences of cardiac triacylglycerol accumulation in insulin-resistant rats. Am J Physiol Endocrinol Metab, 284: E923-930. doi: 10.1152/ajpendo.00360.2002.

Awad, A.B., Chan, K.C., Downie, A.C. and Fink, C.S. (2000). Peanuts as a source of beta-sitosterol, a sterol with anticancer properties. Nutr. Cancer, 36: 238–241. doi: 10.1207/s15327914nc3602_14.

Aymerich, I., Foufelle, F., Ferré, P., Casado, F.J. and Pastor-Anglada, M. et al. (2006). Extracellular adenosine activates AMP-dependent protein kinase (AMPK). J. Cell Sci., 119: 1612–1621. doi: 10.1242/jcs.02865.

Baik, J.S., Kwon, H.Y., Kim, K.S., Jeong, Y.K., Cho, Y.S. et al. (2012). Cordycepin induces apoptosis in human neuroblastoma SK-N-BE(2)-C and melanoma SK-MEL-2 cells. Indian J. Biochem. Biophys., 49: 86–91.

Barnsley, J., Buckland, G., Chan, P.E., Ong, A., Ramos, A.S. et al. (2021). Pathophysiology and treatment of osteoporosis: challenges for clinical practice in older people. Aging. Clin. Exp. Res., 33: 759–773. doi: 10.1007/s40520-021-01817-y.

Bok, J.W., Lermer, L., Chilton, J., Klingeman, H.G. and Towers, G.H et al. (1999). Antitumor sterols from the mycelia of *Cordyceps* sinensis. Phytochemistry, 51: 891–898. doi: 10.1016/s0031-9422(99)00128-4.

Chatterjee, R., Srinivasan, K.S. and Maiti, P.C. (1957). *Cordyceps sinensis* (Berkeley) Saccardo: Structure of Cordycepic Acid**Department of Chemistry, Medical College, Calcutta, India. Journal of the American Pharmaceutical Association (Scientific ed.), 46: 114–118. doi: 10.1002/jps.3030460211.

Chen, B., Sun, Y., Luo, F. and Wang, C. (2020). Bioactive Metabolites and Potential Mycotoxins Produced by *Cordyceps* Fungi: A Review of Safety. Toxins (Basel), 25(12): 2735; https://doi.org/10.3390/molecules25122735

Chen, L.S., Stellrecht, C.M. and Gandhi, V. (2008). RNA-directed agent, cordycepin, induces cell death in multiple myeloma cells. Br. J. Haematol., 140: 682–391. doi: 10.1111/j.1365-2141.2007.06955.x.

Chen, R., Jin, C., Li, H., Liu, Z., Lu, J. et al. (2014). Ultrahigh pressure extraction of polysaccharides from *Cordyceps militaris* and evaluation of antioxidant activity. Separation and Purification Technology, 134: 90–99. doi:10.1016/j.seppur.2014.07.017.

Chen, X.-Y., Liang, Y.-R., Xu, F.-L., Wu, Q. and Lin, X.-. et al. (2013). Stereoselective synthesis of spiro[5.5]undecane derivatives via biocatalytic [5+1] double Michael additions. Journal of Molecular Catalysis B: Enzymatic, 97: 18–22. doi:10.1016/j.molcatb.2013.07.012.

Cheng, Y.H., Hsieh, Y.C. and Yu, Y.H. (2019). Effect of *Cordyceps militaris* hot water extract on immunomodulation-associated gene expression in broilers, Gallus gallus. J. Poult. Sci., 56: 128–139. doi:10.2141/jpsa.0180067.

Cheung, J.K., Li, J., Cheung, A.W., Zhu, Y., Zheng, K.Y. et al. (2009). Cordysinocan, a polysaccharide isolated from cultured *Cordyceps*, activates immune responses in cultured T-lymphocytes and macrophages: signaling cascade and induction of cytokines. J. Ethnopharmacol., 124: 61–68. doi:10.1016/j.jep.2009.04.010.

Cho, H.J., Cho, J.Y., Rhee, M.H., Lim, C.R. and Park, H.J. et al. (2006). Cordycepin (3'-deoxyadenosine) inhibits human platelet aggregation induced by U46619, a TXA2 analogue. J. Pharm. Pharmacol., 58: 1677–1682. doi:10.1211/jpp.58.12.0016.

Choi, S., Lim, M.H., Kim, K.M., Jeon, B.H., Song, W.O. et al. (2011). Cordycepin-induced apoptosis and autophagy in breast cancer cells are independent of the estrogen receptor. Toxicol. Appl. Pharmacol., 257: 165–173. doi:10.1016/j.taap.2011.08.030.

Cunningham, K.G., Manson, W., Spring, F.S. and Hutchinson, S.A. (1950). Cordycepin, a metabolic product isolated from cultures of *Cordyceps militaris* (Linn.) Link. Nature, 166: 949–949. doi:10.1038/166949a0.

Das, G., Shin, H.S., Leyva-Gómez, G., Prado-Audelo, M.L.D., Cortes, H. et al. (2020). *Cordyceps* spp.: A review on its immune-stimulatory and other biological potentials. Front. Pharmacol., 11: 602364. doi:10.3389/fphar.2020.602364.

Dickson, K. and Lehmann, C. (2019). Inflammatory response to different toxins in experimental sepsis models. Int. J. Mol. Sci., 20(18): 4341; https://doi.org/10.3390/ijms20184341.

Dinarello, C.A. (2000). Proinflammatory cytokines. Chest., 118: 503–508. doi:10.1378/chest.118.2.503.

Dong, Y., Jing, T., Meng, Q., Liu, C., Hu, S. et al. (2014). Studies on the antidiabetic activities of *Cordyceps militaris* extract in diet-streptozotocin-induced diabetic Sprague-Dawley rats. Biomed. Res. Int., 2014: 160980. doi:10.1155/2014/160980.

Dou, C., Cao, Z., Ding, N., Hou, T., Luo, F. et al. (2016). Cordycepin prevents bone loss through inhibiting osteoclastogenesis by scavenging ROS generation. Nutrients, 8: 231. doi:10.3390/nu8040231.

Elshourbagy, N.A., Meyers, H.V. and Abdel-Meguid, S.S. (2014). Cholesterol: the good, the bad, and the ugly—therapeutic targets for the treatment of dyslipidemia. Med. Princ. Pract, 23: 99–111. doi:10.1159/000356856.

Fan, Y., Liu, X., Keyhani, N.O., Tang, G., Pei, Y. et al. (2017). Regulatory cascade and biological activity of Beauveria bassiana oosporein that limits bacterial growth after host death. Proc. Natl. Acad. Sci. USA, 114: E1578–e1586. doi:10.1073/pnas.1616543114.

Fang, M., Chai, Y., Chen, G., Wang, H. and Huang, B. et al. (2016). N6-(2-Hydroxyethyl)-adenosine exhibits insecticidal activity against *Plutella xylostella* via adenosine receptors. PLoS One, 11: e0162859. doi:10.1371/journal.pone.0162859.

Guo, P., Kai, Q., Gao, J., Lian, Z.Q., Wu, C.M. et al. (2010). Cordycepin prevents hyperlipidemia in hamsters fed a high-fat diet via activation of AMP-activated protein kinase. J. Pharmacol. Sci., 113: 395–403. doi:10.1254/jphs.10041fp.

Guo, X., Gao, M., Wang, Y., Lin, X., Yang, L. et al. (2018). LDL receptor gene-ablated hamsters: a rodent model of familial hypercholesterolemia with dominant inheritance and diet-induced coronary atherosclerosis. EBioMedicine, 27: 214–224. doi:10.1016/j.ebiom.2017.12.013.

Han, E.S., Oh, J.Y. and Park, H.J. (2011). *Cordyceps militaris* extract suppresses dextran sodium sulfate-induced acute colitis in mice and production of inflammatory mediators from macrophages and mast cells. J. Ethnopharmacol., 134: 703–710. doi:10.1016/j.jep.2011.01.022.

Hardie, D.G., Ross, F.A. and Hawley, S.A. (2012). AMP-activated protein kinase: a target for drugs both ancient and modern. Chem. Biol., 19: 1222–1236. doi:10.1016/j.chembiol.2012.08.019.

Hart, T.L., Petersen, K.S. and Kris-Etherton, P.M. 2022 Chapter 12 - Early nutrition and development of cardiovascular disease. pp. 309–325. *In*: Saavedra, J.M. and Dattilo, A.M. (eds.). Early Nutrition and Long-Term Health (Second Edition). Woodhead Publishing.

Hsu, C.-H., Sun, H.-L., Sheu, J.-N., Ku, M.-S., Hu, C.-M. et al. (2008). Effects of the immunomodulatory agent *Cordyceps militaris* on airway inflammation in a mouse asthma model. Pediatrics & Neonatology, 49: 171–178. doi:10.1016/S1875-9572(09)60004-8.

Hu, S., Wang, J., Li, F., Hou, P., Yin, J. et al. (2019). Structural characterisation and cholesterol efflux improving capacity of the novel polysaccharides from *Cordyceps militaris*. Int. J. Biol. Macromol., 131: 264–272. doi:10.1016/j.ijbiomac.2019.03.078.

Huang, L.F., Liang, Y.Z., Guo, F.Q., Zhou, Z.F. and Cheng, B.M. et al. (2003). Simultaneous separation and determination of active components in *Cordyceps sinensis* and *Cordyceps militarris* by LC/ESI-MS. J. Pharm. Biomed. Anal., 33: 1155–1162. doi: 10.1016/s0731-7085(03)00415-1.

Ilowite, J., Spiegler, P. and Chawla, S. (2008). Bronchiectasis: new findings in the pathogenesis and treatment of this disease. Curr. Opin. Infect. Dis, 21: 163–167. doi:10.1097/QCO.0b013e3282f4f237.

International Diabetes Federation, I. 2023 *Facts & figures*. https://idf.org/about-diabetes/diabetes-facts-figures/ [Accessed 2 Oct 2023].

Irvine, N.J. (2020). Anti-HMGCR myopathy: A rare and serious side effect of statins. J Am Board. Fam. Med., 33: 785–788. doi:10.3122/jabfm.2020.05.190450.

Jędrejko, K., Kała, K., Sułkowska-Ziaja, K., Pytko-Polończyk, J. and Muszyńska, B. et al. (2022). Effect of *Cordyceps* spp. and Cordycepin on functions of bones and teeth and related processes: A Review. Molecules, 27(23): 8170; https://doi.org/10.3390/molecules27238170.

Jeong, J.W., Jin, C.Y., Kim, G.Y., Lee, J.D., Park, C. et al. (2010). Anti-inflammatory effects of cordycepin via suppression of inflammatory mediators in BV2 microglial cells. Int Immunopharmacol., 10: 1580–1586. doi:10.1016/j.intimp.2010.09.011.

Ji, D.B., Ye, J., Li, C.L., Wang, Y.H., Zhao, J. et al. (2009). Antiaging effect of *Cordyceps sinensis* extract. Phytother. Res., 23: 116–122. doi:10.1002/ptr.2576.

Jia, J.M., Tao, H.H. and Feng, B.M. (2009). Cordyceamides A and B from the culture liquid of *Cordyceps sinensis* (BERK.) SACC. Chem. Pharm. Bull. (Tokyo), 57: 99–101. doi:10.1248/cpb.57.99.

Jo, W.S., Choi, Y.J., Kim, H.J., Lee, J.Y., Nam, B.H. et al. (2010). The anti-inflammatory effects of water extract from *Cordyceps militaris* in murine macrophage. Mycobiology, 38: 46–51. doi:10.4489/myco.2010.38.1.046.

Kang, S. and Kumanogoh, A. (2020). The spectrum of macrophage activation by immunometabolism. Int Immunol, 32: 467–473. doi:10.1093/intimm/dxaa017.

Kannel, W.B. (1995). Range of serum cholesterol values in the population developing coronary artery disease. The American Journal of Cardiology, 76: 69C–77C. doi:10.1016/S0002-9149(99)80474-3.

Khan, M.A. and Tania, M. (2020). Cordycepin in anticancer research: molecular mechanism of therapeutic effects. Curr. Med. Chem., 27: 983–996. doi:10.2174/0929867325666618100 1105749.

Khosla, S. and Hofbauer, L.C. (2017). Osteoporosis treatment: recent developments and ongoing challenges. Lancet Diabetes Endocrinol., 5: 898–907. doi:10.1016/s2213-8587(17)30188-2.

Kiho, T., Hui, J., Yamane, A. and Ukai, S. (1993). Polysaccharides in fungi. XXXII. Hypoglycemic activity and chemical properties of a polysaccharide from the cultural mycelium of *Cordyceps sinensis*. Biol. Pharm. Bull., 16: 1291–1293. doi:10.1248/bpb.16.1291.

Kiho, T., Ookubo, K., Usui, S., Ukai, S. and Hirano, K. et al. (1999). Structural features and hypoglycemic activity of a polysaccharide (CS-F10) from the cultured mycelium of *Cordyceps sinensis*. Biol. Pharm. Bull., 22: 966–970. doi:10.1248/bpb.22.966.

Kim, J.H., Kim, E.Y., Lee, B., Min, J.H., Song, D.U. et al. (2016). The effects of *Lycii Radicis Cortex* on RANKL-induced osteoclast differentiation and activation in RAW 264.7 cells. Int. J. Mol. Med., 37: 649–658. doi:10.3892/ijmm.2016.2477.

Kitamura, M., Kato, H., Saito, Y., Nakajima, S., Takahashi, S. et al. (2011). Aberrant, differential and bidirectional regulation of the unfolded protein response towards cell survival by 3'-deoxyadenosine. Cell Death Differ., 18: 1876–1888. doi:10.1038/cdd.2011.63.

Kopin, L. and Lowenstein, C. (2017). Dyslipidemia. Ann. Intern. Med., 167: Itc81–itc96. doi: 10.7326/aitc201712050.

Krishna, K.V., Ulhas, R.S. and Malaviya, A. (2023). Bioactive compounds from *Cordyceps* and their therapeutic potential. Crit. Rev. Biotechnol., 1–21. doi:10.1080/07388551.2023.22311 39.

Kuo, Y.C., Tsai, W.J., Shiao, M.S., Chen, C.F. and Lin, C.Y. et al. (1996). *Cordyceps sinensis* as an immunomodulatory agent. Am. J. Chin. Med., 24: 111–125. doi:10.1142/ s0192415x96000165.

Le, N.A. (2008). Hyperlipidemia and cardiovascular disease: cardiovascular update. Curr. Opin. Lipidol., 19: 545–547. doi:10.1097/MOL.0b013e32830f4a57.

Lee, U.E. and Friedman, S.L. (2011). Mechanisms of hepatic fibrogenesis. Best. Pract. Res. Clin. Gastroenterol., 25: 195–206. doi:10.1016/j.bpg.2011.02.005.

Levterova, B.A., Dimitrova, D.D., Levterov, G.E. and Dragova, E.A. (2013). Instruments for disease-specific quality-of-life measurement in patients with type 2 diabetes mellitus—a systematic review. Folia. Med. (Plovdiv), 55: 83–92. doi:10.2478/folmed-2013-0010.

Li, F.H., Liu, P., Xiong, W.G. and Xu, G.F. (2006). [Effects of corydyceps polysaccharide on liver fibrosis induced by DMN in rats]. Zhongguo Zhong Yao Za Zhi., 31: 1968–1971.

Li, H.B., Chen, J.K., Su, Z.X., Jin, Q.L., Deng, L.W. et al. (2021). Cordycepin augments the chemosensitivity of osteosarcoma to cisplatin by activating AMPK and suppressing the AKT signaling pathway. Cancer Cell. Int., 21: 706. doi:10.1186/s12935-021-02411-y.

Li, H.P., Hu, Z., Yuan, J.L., Fan, H.D., Chen, W. et al. (2007). A novel extracellular protease with fibrinolytic activity from the culture supernatant of *Cordyceps sinensis*: purification and characterization. Phytother. Res., 21: 1234–1241. doi:10.1002/ptr.2246.

Li, S., Li, P. and 1, H.J. (2001). RP-HPLC Determination of ergosterol in natural and cultured *Cordyceps*. Chin JMAP. 2001; 18: 297–299.

Li, S.P., Zhao, K.J., Ji, Z.N., Song, Z.H., Dong, T.T. et al. (2003). A polysaccharide isolated from *Cordyceps sinensis*, a traditional Chinese medicine, protects PC12 cells against hydrogen peroxide-induced injury. Life Sci., 73: 2503–2513. doi:10.1016/s0024-3205(03)00652-0.

Li, Y., Zhang, J. and Ma, H. (2014). Chronic inflammation and gallbladder cancer. Cancer Lett, 345: 242–248. doi:10.1016/j.canlet.2013.08.034.

Lin, B.-q. and Li, S.-p. 2011 *Cordyceps* as an Herbal Drug. *In*: Benzie IFF & S, W.-G. (eds.) Herbal Medicine: Biomolecular and Clinical Aspects. 2nd ed., Boca Raton (FL): CRC Press/Taylor & Francis.

Lin, C.Y., Ku, F.M., Kuo, Y.C., Chen, C.F., Chen, W.P. et al. (1999). Inhibition of activated human mesangial cell proliferation by the natural product of *Cordyceps sinensis* (H1-A): an implication for treatment of IgA mesangial nephropathy. J. Lab. Clin. Med., 133: 55–63. doi:10.1053/lc.1999.v133.a94239.

Lin, P., Yin, F., Shen, N., Liu, N., Zhang, B. et al. (2021). Integrated bioinformatics analysis of the anti-atherosclerotic mechanisms of the polysaccharide CM1 from *Cordyceps militaris*. Int. J. Biol. Macromol., 193: 1274–1285. doi: 10.1016/j.ijbiomac.2021.10.175.

Liu, C., Qi, M., Li, L., Yuan, Y., Wu, X. et al. (2020). Natural cordycepin induces apoptosis and suppresses metastasis in breast cancer cells by inhibiting the Hedgehog pathway. Food Funct, 11: 2107-2116. doi: 10.1039/c9fo02879j.

Liu, J.Y., Feng, C.P., Li, X., Chang, M.C., Meng, J.L. et al. (2016). Immunomodulatory and antioxidative activity of *Cordyceps militaris* polysaccharides in mice. Int. J. Biol. Macromol., 86: 594–598. doi:10.1016/j.ijbiomac.2016.02.009.

Liu, Q., Siloto, R.M., Lehner, R., Stone, S.J. and Weselake, R.J. et al. (2012). Acyl-CoA:diacylglycerol acyltransferase: molecular biology, biochemistry and biotechnology. Prog. Lipid. Res., 51: 350–377. doi:10.1016/j.plipres.2012.06.001.

Liu, W., Gao, Y., Zhou, Y., Yu, F., Li, X. et al. (2022). Mechanism of *Cordyceps sinensis* and its extracts in the treatment of diabetic kidney disease: a review. Front Pharmacol., 13: 881835. doi:10.3389/fphar.2022.881835.

Liu, X., Dun, M., Jian, T., Sun, Y., Wang, M. et al. (2023). *Cordyceps militaris* extracts and cordycepin ameliorate type 2 diabetes mellitus by modulating the gut microbiota and metabolites. Front Pharmacol., 14: 1134429. doi: 10.3389/fphar.2023.1134429.

Liu, Y., Wang, J., Wang, W., Zhang, H., Zhang, X. et al. (2015). The Chemical Constituents and Pharmacological Actions of *Cordyceps sinensis*. Evid Based Complement Alternat Med, 2015: 575063. doi:10.1155/2015/575063.

Lo, H.-C., Hsieh, C., Lin, F.-Y. and Hsu, T.-H. (2013). A systematic review of the mysterious caterpillar fungus *Ophiocordyceps sinensis* in DongChongXiaCao (冬蟲夏草 Dōng Chóng Xià Cǎo) and related bioactive ingredients. J. Tradit. Complement. Med. 3: 16–32. doi:10.1016/S2225-4110(16)30164-X.

Lo, H.-C., Tu, S.-T., Lin, K.-C. and Lin, S.-C. (2004). The anti-hyperglycemic activity of the fruiting body of *Cordyceps* in diabetic rats induced by nicotinamide and streptozotocin. Life Sci., 74: 2897–2908. doi: 10.1016/j.lfs.2003.11.003.

Lo, H.C., Hsu, T.H., Tu, S.T. and Lin, K.C. (2006). Anti-hyperglycemic activity of natural and fermented *Cordyceps sinensis* in rats with diabetes induced by nicotinamide and streptozotocin. Am. J. Chin. Med., 34: 819–832. doi:10.1142/s0192415x06004314.

Lu, Y., Luo, F., Cen, K., Xiao, G., Yin, Y. et al. (2017). Omics data reveal the unusual asexual-fruiting nature and secondary metabolic potentials of the medicinal fungus *Cordyceps cicadae*. BMC Genomics, 18: 668. doi:10.1186/s12864-017-4060-4.

Łysakowska, P., Sobota, A. and Wirkijowska, A. (2023). Medicinal mushrooms: their bioactive components, nutritional value and application in functional food production-a review. Molecules, 28(14): 5393; https://doi.org/10.3390/molecules28145393.

Ma, L., Zhang, S. and Du, M. (2015). Cordycepin from *Cordyceps militaris* prevents hyperglycemia in alloxan-induced diabetic mice. Nutr. Res., 35: 431–439. doi:10.1016/j.nutres.2015.04.011.

Mallebrera, B., Juan-Garcia, A., Font, G. and Ruiz, M.J. (2016). Mechanisms of beauvericin toxicity and antioxidant cellular defense. Toxicol Lett, 246: 28–34. doi:10.1016/j.toxlet.2016.01.013.

Margolis, J. and Grever, M.R. (2000). Pentostatin (Nipent): a review of potential toxicity and its management. Semin Oncol., 27: 9–14.

Martin, T.A., Ye, L., Sanders, A.J., Lane, J. and Jiang, W.G et al. (2013). Cancer invasion and metastasis: molecular and cellular perspective. *In*: Jandial, R. (ed.). Metastatic Cancer: Clinical and Biological Perspectives. Landes Bioscience. www.ncbi.nlm.nih.gov/books/NBK164700/.

Michels da Silva, D., Langer, H. and Graf, T. (2019). Inflammatory and molecular pathways in heart failure-ischemia, HFpEF and transthyretin cardiac amyloidosis. Int. J. Mol. Sci., 20(9): 2322. doi:10.3390/ijms20092322.

Mishra, R. and Upadhyay, Y. (2011). *Cordiceps sinensis*: The Chinese Rasayan-Current Research Scenario. Int. J. Res. Pharm. Biomed. Sci., 2(4): 1503–1519.

Mizuno, T. (1999). Medicinal effects and utilization of *Cordyceps* (Fr.) Link (Ascomycetes) and Isaria Fr. (Mitosporic Fungi) Chinese caterpillar fungi, "Tochukaso" (Review). Int. J. Med. Mushrooms, 1: 251–261.

Moslehi, A. and Hamidi-Zad, Z. (2018). Role of SREBPs in liver diseases: a mini-review. J. Clin. Transl. Hepatol., 6: 332–338. doi:10.14218/jcth.2017.00061.

Ng, T.B. and Wang, H.X. (2005). Pharmacological actions of *Cordyceps*, a prized folk medicine. J. Pharm. Pharmacol., 57: 1509–1519. doi:10.1211/jpp.57.12.0001.

Niu, Y.J., Tao, R.Y., Liu, Q., Tian, J.Y., Ye, F. et al. (2010). Improvement on lipid metabolic disorder by 3'-deoxyadenosine in high-fat-diet-induced fatty mice. Am. J. Chin. Med., 38: 1065–1075. doi:10.1142/s0192415x10008470.

Ohshiro, T., Kobayashi, K., Ohba, M., Matsuda, D., Rudel, L.L. et al. (2017). Selective inhibition of sterolO-acyltransferase 1 isozyme by beauveriolide III in intact cells. Sci. Rep., 7: 4163. doi:10.1038/s41598-017-04177-8.

Ostlund, R.E., Jr. (2007). Phytosterols, cholesterol absorption and healthy diets. Lipids, 42: 41–45. doi:10.1007/s11745-006-3001-9.

Ouyang, Y.Y., Zhang, Z., Cao, Y.R., Zhang, Y.Q., Tao, Y.Y. et al. (2013). [Effects of cordyceps acid and cordycepin on the inflammatory and fibrogenic response of hepatic stellate cells]. Zhonghua Gan Zang Bing Za Zhi, 21: 275–278. doi:10.3760/cma.j.is sn.1007-3418.2013.04.009.

Panya, A., Songprakhon, P., Panwong, S., Jantakee, K., Kaewkod, T. et al. (2021). Cordycepin Inhibits Virus Replication in Dengue Virus-Infected Vero Cells. Molecules, 26(11): 3118. doi: 10.3390/molecules26113118.

Passmore, L.A. and Coller, J. (2022). Roles of mRNA poly(A) tails in regulation of eukaryotic gene expression. Nat. Rev. Mol. Cell. Biol., 23: 93–106. doi:10.1038/s41580-021-00417-y.

Ponnulakshmi, R., Shyamaladevi, B., Vijayalakshmi, P. and Selvaraj, J. (2019). *In silico* and *in vivo* analysis to identify the antidiabetic activity of beta sitosterol in adipose tissue of high fat diet and sucrose induced type-2 diabetic experimental rats. Toxicol. Mech. Methods., 29(4): 276–290. doi:10.1080/15376516.2018.1545815.

Poulsen, L., Siersbæk, M. and Mandrup, S. (2012). PPARs: fatty acid sensors controlling metabolism. Semin Cell. Dev. Biol., 23: 631–639. doi:10.1016/j.semcdb.2012.01.003.

Prasain, J.K. (2013). Chapter 13 - Pharmacological Effects of *Cordyceps* and Its Bioactive Compounds. pp. 453–468. *In*: Atta ur, R. (ed.). Studies in Natural Products Chemistry. Elsevier,

Prevention, C.f.D.C.a. (2011). Million hearts: strategies to reduce the prevalence of leading cardiovascular disease risk factors United States, 2011. MMWR Morb. Mortal. Wkly Rep., 60: 1248–1251.

Qi, W., Zhang, Y., Yan, Y.B., Lei, W., Wu, Z.X. et al. (2013). The protective effect of Cordymin, a peptide purified from the medicinal mushroom *Cordyceps sinensis*, on diabetic osteopenia in alloxan-induced diabetic rats. Evid. Based Complement Alternat Med., 2013: doi:985636. 10.1155/2013/985636.

Qian, G.M., Pan, G.F. and Guo, J.Y. (2012). Anti-inflammatory and antinociceptive effects of cordymin, a peptide purified from the medicinal mushroom *Cordyceps sinensis*. Nat. Prod. Res., 26: 2358-2362. doi:10.1080/14786419.2012.658800.

Qin, P., Li, X., Yang, H., Wang, Z.Y. and Lu, D. et al. (2019). Therapeutic potential and biological applications of cordycepin and metabolic mechanisms in cordycepin-producing fungi. Molecules, 24(12): 2231. doi:10.3390/molecules24122231.

Qiu, W., Wu, J., Choi, J.H., Hirai, H., Nishida, H. et al. (2017). Cytotoxic compounds against cancer cells from Bombyx mori inoculated with *Cordyceps militaris*. Biosci Biotechnol Biochem, 81: 1224-1226. doi: 10.1080/09168451.2017.1289075.

Qu, S.L., Li, S.S., Li, D. and Zhao, P.J. (2022). Metabolites and their bioactivities from the genus *Cordyceps*. Microorganisms, 10(8): 1489. doi:10.3390/microorganisms10081489.

Rabie, A.M. (2022). Potent Inhibitory Activities of the adenosine analogue cordycepin on SARS-CoV-2 replication. ACS Omega, 7: 2960–2969. doi:10.1021/acsomega.1c05998.

Rao, Y.K., Fang, S.H., Wu, W.S. and Tzeng, Y.M. (2010). Constituents isolated from *Cordyceps militaris* suppress enhanced inflammatory mediator's production and human cancer cell proliferation. J. Ethnopharmacol., 131: 363–367. doi:10.1016/j.jep.2010.07.020.

Research, D.B.M. 2023 *Global Cordyceps Sinensis Market—Industry Trends and Forecast to 2030.* https://www.databridgemarketresearch.com/reports/global-cordyceps-sinensis-market [Accessed 2 Oct 2023].

Rhee, K.H. (2004). Cyclic dipeptides exhibit synergistic, broad spectrum antimicrobial effects and have anti-mutagenic properties. Int. J. Antimicrob. Agents, 24: 423–427. doi:10.1016/j. ijantimicag.2004.05.005.

Rodman, L.E., Farnell, D.R., Coyne, J.M., Allan, P.W., Hill, D.L. et al. (1997). Toxicity of cordycepin in combination with the adenosine deaminase inhibitor 2'-deoxycoformycin in beagle dogs. Toxicol. Appl. Pharmacol., 147: 39–45. doi:10.1006/taap.1997.8264.

Schmidt, K., Riese, U., Li, Z. and Hamburger, M. (2003). Novel tetramic acids and pyridone alkaloids, militarinones B, C, and D, from the insect pathogenic fungus *Paecilomyces militaris*. J. Nat. Prod., 66: 378–383. doi:10.1021/np020430y.

Shahbaz, A., Mahendhar, R., Fransawy Alkomos, M., Zarghamravanbakhsh, P. and Sachmechi, I. et al. (2018). Drug-induced angioedema: a rare side effect of rosuvastatin. Cureus, 10: e2965. doi:10.7759/cureus.2965.

Sharma, S.K., Gautam, N. and Atri, N.S. (2018). Retraction Note: Optimized extraction, composition, antioxidant and antimicrobial activities of exo and intracellular polysaccharides from submerged culture of *Cordyceps cicadae*. BMC Complement Altern. Med., 18: 276. doi:10.1186/s12906-018-2344-0.

Shashidhar, M.G., Giridhar, P., Udaya Sankar, K. and Manohar, B. (2013). Bioactive principles from *Cordyceps sinensis*: A potent food supplement—A review. J. Funct. Foods, 5: 1013–1030. doi:10.1016/j.jff.2013.04.018.

Sheng, L., Chen, J., Li, J. and Zhang, W. (2011). An exopolysaccharide from cultivated *Cordyceps sinensis* and its effects on cytokine expressions of immunocytes. Appl. Biochem. Biotechnol., 163: 669–678. doi:10.1007/s12010-010-9072-3.

Shin, S., Lee, S., Kwon, J., Moon, S., Lee, S. et al. (2009). Cordycepin suppresses expression of diabetes regulating genes by inhibition of lipopolysaccharide-induced inflammation in macrophages. Immune Netw., 9: 98–105. doi:10.4110/in.2009.9.3.98.

Shweta, Abdullah, S., Komal and Kumar, A. (2023). A brief review on the medicinal uses of *Cordyceps militaris*. Pharmacological Research—Modern Chinese Medicine, 7: 100228. doi:10.1016/j.prmcm.2023.100228.

Smiderle, F.R., Ruthes, A.C., van Arkel, J., Chanput, W., Iacomini, M. et al. (2011). Polysaccharides from *Agaricus bisporus* and *Agaricus brasiliensis* show similarities in their structures and their immunomodulatory effects on human monocytic THP-1 cells. BMC Complement Altern. Med., 11: 58. doi:10.1186/1472-6882-11-58.

Sun, H., Saeedi, P., Karuranga, S., Pinkepank, M., Ogurtsova, K. et al. (2022). IDF Diabetes Atlas: Global, regional and country-level diabetes prevalence estimates for 2021 and projections for 2045. Diabetes Research and Clinical Practice, 183: 109119. doi:10.1016/j. diabres.2021.109119.

Sun, J., Jin, M., Zhou, W., Diao, S., Zhou, Y. et al. (2017). A new ribonucleotide from *Cordyceps militaris*. Nat. Prod. Res., 31: 2537–2543. doi:10.1080/14786419.2017.1323210.

Tamura, T. and Ozato, K. (2002). ICSBP/IRF-8: its regulatory roles in the development of myeloid cells. J. Interferon. Cytokine Res., 22: 145–152. doi:10.1089/107999002753452755.

Taylor, A.L., Watson, C.J. and Bradley, J.A. (2005). Immunosuppressive agents in solid organ transplantation: Mechanisms of action and therapeutic efficacy. Crit. Rev. Oncol. Hematol., 56: 23–46. doi:10.1016/j.critrevonc.2005.03.012.

Tuli, H.S., Sharma, A.K., Sandhu, S.S. and Kashyap, D. (2013). Cordycepin: a bioactive metabolite with therapeutic potential. Life Sci., 93: 863–869. doi:10.1016/j.lfs.2013.09.030.

Urbain, P., Singler, F., Ihorst, G., Biesalski, H.K. and Bertz, H. et al. (2011). Bioavailability of vitamin D_2 from UV-B-irradiated button mushrooms in healthy adults deficient in serum

25-hydroxyvitamin D: a randomized controlled trial. Eur. J. Clin. Nutr., 65: 965–971. doi:10.1038/ejcn.2011.53.

Wang, M., Meng, X.Y., Yang, R.L., Qin, T., Wang, X.Y. et al. (2012). *Cordyceps militaris* polysaccharides can enhance the immunity and antioxidation activity in immunosuppressed mice. Carbohydr. Polym., 89: 461–466. doi: 10.1016/j.carbpol.2012.03.029.

Wang, N.Q., Jiang, L.D., Zhang, X.M. and Li, Z.X. (2007). [Effect of dongchong xiacao capsule on airway inflammation of asthmatic patients]. Zhongguo Zhong Yao Za Zhi, 32: 1566–1568.

Wang, Y., Yin, H., Lv, X., Wang, Y., Gao, H. et al. (2010). Protection of chronic renal failure by a polysaccharide from *Cordyceps sinensis*. Fitoterapia, 81: 397–402. doi: 10.1016/j.fitote.2009.11.008.

Wise, B.L. and Chater, N. (1961). Use of hypertonic mannitol solutions to lower cerebrospinal fluid pressure and decrease brain bulk in man. Surg. Forum, 12: 398–399.

Woloshuk, C.P. and Shim, W.-B. (2013). Aflatoxins, fumonisins, and trichothecenes: a convergence of knowledge. FEMS Microbiology Reviews, 37: 94–109. doi:10.1111/1574-6976.12009 %J FEMS Microbiology Reviews.

Wong, J.H., Ng, T.B., Wang, H., Sze, S.C., Zhang, K.Y. et al. (2011). Cordymin, an antifungal peptide from the medicinal fungus *Cordyceps militaris*. Phytomedicine, 18: 387–392. doi:10.1016/j.phymed.2010.07.010.

Wong, Y.Y., Moon, A., Duffin, R., Barthet-Barateig, A., Meijer, H.A. et al. (2010). Cordycepin inhibits protein synthesis and cell adhesion through effects on signal transduction. J. Biol. Chem., 285: 2610–2621. doi:10.1074/jbc.M109.071159.

Wu, C., Guo, Y., Su, Y., Zhang, X., Luan, H. et al. (2014a). Cordycepin activates AMP-activated protein kinase (AMPK) via interaction with the $\gamma 1$ subunit. J. Cell. Mol. Med., 18: 293–304. doi:doi: 10.1111/jcmm.12187.

Wu, P., Tao, Z., Liu, H., Jiang, G., Ma, C. et al. (2015). Effects of heat on the biological activity of wild *Cordyceps sinensis*. Journal of Traditional Chinese Medical Sciences, 2: 32-38. doi: 10.1016/j.jtcms.2014.12.005.

Wu, W.C., Hsiao, J.R., Lian, Y.Y., Lin, C.Y. and Huang, B.M. et al. (2007). The apoptotic effect of cordycepin on human OEC-M1 oral cancer cell line. Cancer Chemother Pharmacol, 60: 103-111. doi: 10.1007/s00280-006-0354-y.

Wu, Y., Antony, S., Meitzler, J.L. and Doroshow, J.H. (2014b). Molecular mechanisms underlying chronic inflammation-associated cancers. Cancer Lett, 345: 164–173. doi:10.1016/j.canlet.2013.08.014.

Wu, Y., Sun, H., Qin, F., Pan, Y. and Sun, C. (2006). Effect of various extracts and a polysaccharide from the edible mycelia of *Cordyceps sinensis* on cellular and humoral immune response against ovalbumin in mice. Phytother. Res., 20: 646–652. doi:10.1002/ptr.1921.

Wu, Y., Xu, M.J., Cao, Z., Yang, C., Wang, J. et al. (2019). Heterozygous Ldlr-deficient hamster as a model to evaluate the efficacy of PCSK9 antibody in hyperlipidemia and atherosclerosis. Int. J. Mol. Sci., 20(23): 5936. doi:10.3390/ijms20235936.

Xia, Y., Luo, F., Shang, Y., Chen, P., Lu, Y. et al. (2017). Fungal cordycepin biosynthesis is coupled with the production of the safeguard molecule pentostatin. Cell. Chem. Biol., 24: 1479–1489.e1474. doi:10.1016/j.chembiol.2017.09.001.

Xiang, T., Xia, C., Liu, J., Wang, C. and Shen, J. et al. (2021). Separation, structural identification and anti-tumor effects of new compounds from *Cordyceps militaris*[J]. Food Science, 42: 235–242. doi: 10.7506/spkx1002-6630-20191017-168.

Xiao, J.H., Liang, Z.Q. and Liu, A.Y. (2002). [Advances of polysaccharides research and exploitation of anamorph and its related fungi from *Cordyceps*]. Yao Xue Xue Bao, 37: 589–592.

Xin, H., Cheng-Yi, L., Fan-Zheng, X., Wen-Ya, W., Xiao-Ping, W. et al. (2021). Effects of cordycepin combined with doxorubicin on proliferation and metastasis of breast cancer cells., 40: 3012–3022. doi:10.13346/j.mycosystema.210124.

Xu, H., Li, S., Lin, Y., Liu, R., Gu, Y. et al. (2011). [Effectiveness of cultured *Cordyceps sinensis* combined with glucocorticosteroid on pulmonary fibrosis induced by bleomycin in rats]. Zhongguo Zhong Yao Za Zhi, 36: 2265–2270.

Yan, P.S., Song, Y., Sakuno, E., Nakajima, H., Nakagawa, H. et al. (2004). Cyclo(L-leucyl-L-prolyl) produced by *Achromobacter xylosoxidans* inhibits aflatoxin production by *Aspergillus parasiticus*. Appl. Environ. Microbiol., 70: 7466–7473. doi:10.1128/aem.70.12.7466-7473.2004.

Yang, F.Q., Feng, K., Zhao, J. and Li, S.P. (2009). Analysis of sterols and fatty acids in natural and cultured *Cordyceps* by one-step derivatization followed with gas chromatography–mass spectrometry. J. Pharm. Biomed. Anal., 49: 1172–1178. doi:10.1016/j.jpba.2009.02.025.

Yang, J., Zhang, W., Shi, P., Chen, J., Han, X. et al. (2005). Effects of exopolysaccharide fraction (EPSF) from a cultivated *Cordyceps sinensis* fungus on c-Myc, c-Fos, and VEGF expression in B16 melanoma-bearing mice. Pathology—Research and Practice, 201: 745–750. doi:10.1016/j.prp.2005.08.007.

Yang, L., Jiao, X., Wu, J., Zhao, J., Liu, T. et al. (2018). *Cordyceps sinensis* inhibits airway remodeling in rats with chronic obstructive pulmonary disease. Exp. Ther. Med., 15: 2731–2738. doi: 10.3892/etm.2018.5777.

Yang, L.Y., Huang, W.J., Hsieh, H.G. and Lin, C.Y. (2003). H1-A extracted from *Cordyceps sinensis* suppresses the proliferation of human mesangial cells and promotes apoptosis, probably by inhibiting the tyrosine phosphorylation of Bcl-2 and Bcl-XL. J. Lab. Clin. Med., 141: 74–83. doi: 10.1067/mlc.2003.6.

Ye, M., Hu, Z., Fan, Y., He, L., Xia, F. et al. (2004). Purification and characterization of an acid deoxyribonuclease from the cultured mycelia of *Cordyceps sinensis*. J. Biochem. Mol. Biol., 37: 466–473. doi:10.5483/bmbrep.2004.37.4.466.

Yoshikawa, N., Kunitomo, M., Kagota, S., Shinozuka, K. and Nakamura, K. et al. (2009). Inhibitory effect of cordycepin on hematogenic metastasis of B16-F1 mouse melanoma cells accelerated by adenosine-5′-diphosphate. Anticancer Res., 29: 3857–3860.

Yu, W.-Q., Yin, F., Shen, N., Lin, P., Xia, B. et al. (2021). Polysaccharide CM1 from *Cordyceps militaris* hinders adipocyte differentiation and alleviates hyperlipidemia in LDLR(+/−) hamsters. Lipids Health Dis., 20: 178. doi: 10.1186/s12944-021-01606-6.

Yu, X., Mao, Y., Shergis, J.L., Coyle, M.E., Wu, L. et al. (2019). Effectiveness and safety of oral *Cordyceps sinensis* on stable COPD of GOLD stages 2–3: systematic review and meta-Analysis. Evid. Based Complement Alternat Med., 2019: 4903671. doi: 10.1155/2019/4903671.

Zhang, D.W., Deng, H., Qi, W., Zhao, G.Y. and Cao, X.R. et al. (2015). Osteoprotective effect of cordycepin on estrogen deficiency-induced osteoporosis *in vitro* and *in vivo*. Biomed. Res. Int., 2015: 423869. doi: 10.1155/2015/423869.

Zhang, G., Huang, Y., Bian, Y., Wong, J.H., Ng, T.B. et al. (2006). Hypoglycemic activity of the fungi *Cordyceps militaris, Cordyceps sinensis, Tricholoma mongolicum,* and *Omphalia lapidescens* in streptozotocin-induced diabetic rats. Appl. Microbiol. Biotechnol., 72: 1152–1156. doi: 10.1007/s00253-006-0411-9.

Zhang, J., Wang, W.-Y., Du, W.-B. and Cao, X.-B. (2011). Evolution of cooperation among mobile agents with heterogenous view radii. Physica A: Statistical Mechanics and its Applications, 390: 2251–2257. https://doi.org/10.1016/j.physa.2011.02.036.

Zhang, J., Wen, C., Duan, Y., Zhang, H. and Ma, H. et al. (2019). Advance in *Cordyceps militaris* (Linn) Link polysaccharides: Isolation, structure, and bioactivities: A review. Int. J. Biol. Macromol., 132: 906–914. doi:10.1016/j.ijbiomac.2019.04.020.

Zhang, Z. and Xia, S.S. (1990). *Cordyceps Sinensis-I* as an immunosuppressant in heterotopic heart allograft model in rats. J Tongji Med Univ, 10: 100-103. doi: 10.1007/bf02887870.

Zhao, B., Takami, M., Yamada, A., Wang, X., Koga, T. et al. (2009). Interferon regulatory factor-8 regulates bone metabolism by suppressing osteoclastogenesis. Nat. Med., 15: 1066–1071. doi: 10.1038/nm.2007.

Zhao, D., Liu, J., Wang, M., Zhang, X. and Zhou, M. et al. (2019). Epidemiology of cardiovascular disease in China: current features and implications. Nature Reviews Cardiology, 16: 203–212. doi: 10.1038/s41569-018-0119-4.

Zheng, H., Maoqing, Y., Liqiu, X., Wenjuan, T., Liang, L. et al. (2006). Purification and characterization of an antibacterial protein from the cultured mycelia of *Cordyceps sinensis*. Wuhan University Journal of Natural Sciences, 11: 709–714. doi:10.1007/BF02836695.

Zheng, J., Wang, Y., Wang, J., Liu, P., Li, J. et al. (2013). Antimicrobial ergosteroids and pyrrole derivatives from halotolerant *Aspergillus flocculosus* PT05-1 cultured in a hypersaline medium. Extremophiles, 17: 963–971. doi:10.1007/s00792-013-0578-9.

Zheng, R., Zhu, R., Li, X., Li, X., Shen, L. et al. (2018). N6-(2-Hydroxyethyl) adenosine from *Cordyceps cicadae* ameliorates renal interstitial fibrosis and prevents inflammation via TGF-β1/Smad and NF-κB signaling pathway. Front. Physiol., 9: 1229. doi:10.3389/fphys.2018.01229.

Zhou, X., Luo, L., Dressel, W., Shadier, G., Krumbiegel, D. et al. (2008). Cordycepin is an immunoregulatory active ingredient of *Cordyceps sinensis*. Am. J. Chin. Med., 36: 967–980. doi:10.1142/s0192415x08006387.

Zhu, X.Y. and Yu, H.Y. (1990). [Immunosuppressive effect of cultured *Cordyceps sinensis* on cellular immune response]. Zhong Xi Yi Jie He Za Zhi, 10: 485–487, 454.

9

Cosmeceutical Applications of *Cordyceps*

Sarita Sangthong,[1,2,*] *Lutfun Nahar,*[3,4,*] *Phanuphong Chaiwut,*[1,2]
Punyawatt Pintathong[1,2] and *Satyajit D. Sarker*[4]

1. Introduction

Cordyceps is a genus of ascomycetous fungi that consists of 600 species, most of which are parasitic (they often grow on insects). Many of these species have been used mainly as antifatigue, antihyperglycemic, anti-inflammatory, antioxidant, antimicrobial, antitumor and immunomodulating agents. Additionally it has been used as a remedy for kidney and liver diseases in traditional Chinese medicine (TCM) over the past 1500 years (Lin and Li, 2011; Wan et al., 2023; Das et al., 2024; Khan et al., 2024; Pu et al., 2024). Among the *Cordyceps* species, only *C. sinensis* is included as a herbal

[1] School of Cosmetic Science, Mae Fah Luang University, 333 M.1 Thasud, Muang, Chiang Rai, Thailand.

[2] Green Cosmetic Technology Research Group, Mae Fah Luang University, 333 M.1 Thasud, Muang, Chiang Rai, Thailand.

[3] Center for Global Health Research, Saveetha Medical College and Hospital, Saveetha Institute of Medical and Technical Sciences, Saveetha University, Chennai, India.

[4] Centre for Natural Products Discovery, School of Pharmacy and Biomolecular Sciences, Liverpool John Moores University, James Parsons Building, Byrom Street, Liverpool L3 3AF, United Kingdom.

* Corresponding authors: sarita.san@mfu.ac.th; l.nahar@ljmu.ac.uk; lutfunnahar.smc@saveetha.com

medicine in the Chinese Pharmacopoeia. In traditional Chinese medicine, *C. militaris* has emerged as a more feasible substitute, primarily because of its lower production costs of and ease of cultivation (Wong et al., 2011; Pintathong et al., 2021). Because of the potential anti-inflammatory, antioxidant, and antimicrobial properties of the *Cordyceps* species, they have now been used in various cosmeceutical products (Mago et al., 2023; Zhang et al., 2023). It can be noted that cosmeceuticals are cosmetic products that contain active pharmaceutical ingredients aimed at enhancing skincare efficacy (Martin and Glaser, 2011; Pandey et al., 2023). The main difference between a cosmetic product and a cosmeceutical product is the presence of a pharmacologically active ingredient in the latter. This chapter, as the title implies, provides an overview of the potential applications of various *Cordyceps* species that are documented in the published literature available to date.

2. *Cordyceps* Used in Cosmeceuticals

Although the *Cordyceps* genus contains several species, only a few species have been studied so far for their cosmeceutical potential. Table 1 lists the species that could be used as cosmeceuticals based on the information

Table 1. Cosmeceutical Applications of *Cordyceps* species.

Cordyceps	Potential application	Reference
Cordyceps cicadae (Miq,) Massee	Moisturizing and anti-aging cosmeceuticals	Shao et al., 2023
C. militaris (L.) Fr.	Anti-aging cosmeceuticals	Wang et al., 2020
	Anti-dandruff cosmeceuticals	Zhang et al., 2023
	Antioxidant, antimicrobial and anti-inflammatory cosmeceuticals	Ahn and Cho, 2013; Rupa et al., 2020
	Antiwrinkle cosmeceuticals	Prommaban et al., 2022
	Skin care cosmeceutical creams	Kanlayavattanakul and Lourith, 2023
	Skin cell regeneration cosmeceuticals	Upatcha et al., 2023
	Skin-whitening cosmeceuticals	Chien et al., 2009; Hyde et al., 2010
C. sinensis (Berk.) Sacc.	Anti-aging cosmeceuticals	Hyde et al., 2010
	Anticollagenase cosmeceutical	Cheng et al., 2018
	Moisturising cosmeceuticals	Chen et al., 2014
	Skin care cosmeceuticals	Chen et al., 2019
Cordyceps spp.	Antioxidant, antiwrinkle, anti-aging and moisturizing cosmeceuticals	Tang et al., 2019; He et al., 2020; Mago et al., 2023

available in the literature. Mainly, the extracts of *Cordyceps cicade, C. militaris,* and *C. sinensis* have been documented as potential cosmeceutical ingredients. These extracts have been used in cosmeceutical formulations that possess anti-aging, anti-dandruff, antioxidant, antimicrobial, anti-wrinkle, anti-inflammatory, moisturizing and skin-whitening properties (Table 1).

3. *Cordyceps* in Cosmeceutical Applications

The *Cordyceps* species possesses various pharmacological properties, including anti-inflammatory, antimicrobial, antioxidant, and anti-tyrosinase properties, which have been successfully exploited to formulate various skincare cosmeceuticals. Highly branched polysaccharides produced by these fungi have been used as moisturizing cosmeceuticals (Chen et al., 2014). In most cases, the bioactive compound cordycepin, present in *Cordyceps* species, has been implicated as the main contributor to most of the bioactivities relevant to cosmeceutical applications (Qin et al., 2019). However, the compounds of interest are not at all limited to the well-known cordycepin but also to several polyphenols like phenolic acids and flavonoids. Additionally, *Cordyceps*-derived skin nutrients like proteins, peptides, and amino acids are noted as cosmeceutical functional ingredients. For cosmetic and cosmeceutical applications, *Cordyceps* species are prepared in extract form, both crude extract and partially purified extract or even cordycepin-enriched extracts (Kunhorm et al., 2019). This section revisits the bioactive constituents in the *Cordyceps*, especially *Cordyceps militaris*, and provides an insight into their potential cosmeceutical applications.

3.1 *Cordycepin*

Ashraf et al. (2020) reported a group of nucleosides and polysaccharides as the bioactive compounds from *C. militaris*; the nucleosides cordycepin and adenosine were present in *C. militaris,* and their content was higher in *C. militaris* than *C. sinensis.* Additionally, the presence of γ-amino butyric acid, cordycepic acid, ergothioneine, glycolipids (cerebrosides), glycoproteins (lectins), xanthophylls including carotenoids, minerals (magnesium, potassium, selenium, and sulfur), phenolic compounds, sterols, and vitamins was also confirmed.

Cordycepin (3′-deoxyadenosine) (Fig. 1) is known for its nutraceutical and therapeutic potential, including antidiabetic, antihyperlipidemia, antifungal, anti-inflammatory, antimalarial, immunomodulatory, anti-aging, anti-arthritic, anticancer, anti-osteoporotic, antioxidant, antiviral, cardiovascular diseases, cosmeceutical properties, hepatoprotective, and hyposexuality (Qin et al., 2019; Ashraf et al., 2020). It has been

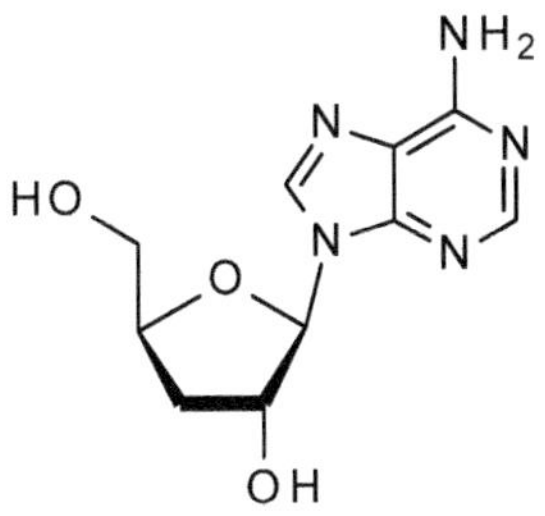

Fig. 1. Bioactive compound cordycepin present in *Cordyceps*.

shown to be the main bioactive compound in the *Cordyceps*, possessing antimicrobial and antioxidant properties against dermal pathogenic bacteria like *Cutibacterium acnes*, *Pseudomonas aeruginosa*, *Staphylococcus aureus*, and methicillin-resistant S. aureus (MRSA). These properties could easily be exploited to develop skincare cosmeceuticals (Eiamthaworn et al., 2022; Upatcha et al., 2023). Skin aging is accelerated by oxidative stress, and compounds that can reduce such stress could be useful as ingredients in anti-aging cosmeceutical products. Cordycepin, purified from *C. militaris* displayed remarkable antioxidant activity and skin regeneration potential (Upatcha et al., 2023).

Cordycepin exhibits antipigmentation and antiphotoaging properties on 2D cell culture platforms (Lee et al., 2009). Aging is caused by external stimulus such as UV radiation, which is responsible for matrix metalloproteinase in dermal fibroblasts to degrade the network of extracellular matrix in the dermis layer. The UV radiation activates the role of the NF-κB transcription factor to translocate to the nucleus and upregulate matrix metalloproteinase genes. Treatment of cordycepin in UVB-exposed dermal fibroblasts suppresses the activation and translocation of NF-κB to the nucleus, leading to a decreased expression of matrix metalloproteinase genes. The biosynthesis pathway of melanogenesis to produce melanin in melanocytes could also be induced by external stimuli such as UV radiation and chemicals. Owing to such stimuli, the signal transduction cascade is mediated by cAMP response element-binding protein (CREB), which results in the activation of microphthalmia-associated transcription factor (MITF). Thus, MITF upregulates genes involved in melanogenesis. In the experiment carried out in mouse melanoma, it is understood that cordycepin activates protein kinase B (Akt) as well as extracellular signal-regulated kinase (ERK)-dependent mechanisms leading to inhibition of MITF activity. It was suggested that because of the above property, cordycepin is useful as a bioactive compound in cosmeceutical products possessing anti-photoaging and anti-pigmentation properties. Besides, cordycepin could be used to treat different dermatologic hyperpigmentation disorders such as freckles and melasma (Kunhorm et al., 2019). It is well known that adenosine (cordycepin is 3′-deoxyadenosine) plays anti-aging roles via the adenosine receptors in various skin layers. Stimulating lipogenesis and adipogenesis in the hypodermis layer and collagen production in the dermis layer rejuvenate the skin composition. Adenosine also promotes the DNA and protein synthesis of dermal cells. This decreases the aging signs of wrinkles, roughness, and sin dryness (Marucci et al., 2022). Previously, cultivated *Cordyceps* was shown to protect the skin from UVB-induced acute photopathy. Thus, it was suggested that *Cordyceps* extract could serve as an ideal compound for topical application to avoid adverse effects of UVB-induced photopathy (He et al., 2020).

Zhang et al. (2022) stated the preparation and evaluation of a nanoformulation containing cordycepin and its functional properties, such as its protective effect on the skin. Its anti-aging effect was demonstrated. No skin irritation was caused by the nanoemulsion, and the cordycepin was found to be stable in it for up to eight months. This nanoformulation was also able to prevent any microbial contamination during storage. Based on this finding, it was stated that this nanoemulsion containing cordycepin conformed to the provisions of the 2015 Safety and Technical Standards for Cosmetics of the Chinese Ministry of Health. It was shown that *Cordyceps militaris* extracts serve as nanoemulsions enhancing stability as well as dermal delivery (Marsup et al., 2020).

Cordycepin, which has been a key component in Chinese and Korean traditional medicine for centuries, was shown to inhibit melanogenesis, revealing its potential as a skin-whitening or depigmentation agent for cosmeceutical use (Jin et al., 2012). This melanogenesis-inhibitory activity of cordycepin was mediated by inhibition of melanin-synthesis-related enzymes [tyrosinase, tyrosinase-related protein-1 (TRP1) and tyrosinase-related protein-2 (TRP2), alpha-MSH, and IBMX]. Additionally, this compound was shown to decrease the phosphorylation of CREB induced by a-MSH and IBMX in B16E10 melanoma cells. Thus, cordycepin could inhibit melanogenesis pathways by activating ERK and AKT pathways and regulating the suppression of downstream pathways, including MITF and tyrosinase (TRP1 and TRP2).

3.2 *Cordymin*

Cordymin is an antifungal peptide (10,906 Da) isolated from *C. militaris* by ion-exchange chromatography (Wong et al., 2011). The *N*-terminal amino acid sequence in this peptide is different from that of earlier reported proteins. Cordymin was shown to inhibit mycelial growth (*Bipolaris maydis, Candida albicans, Mycosphaerella arachidicola,* and *Rhizoctonia solani*) (IC_{50} with concentrations of 50, 10, 80 μM, and 0.75 mM, respectively), which indicated its potential use in cosmeceutical preparations as an antifungal agent. It is important to note that this peptide is thermostable up to 100°C and stable in the pH range of 6 to 13 in the presence of Zn ($^{2+}$) and Mg

($^{2+}$) in a concentration of 10 mM. Furthermore, this peptide did not show any noticeable mitogenic potential up to a concentration of 1 mM in mouse spleen cells as well as mouse macrophage nitric oxide-inducing activity.

3.3 Fatty Acids and Sterols

Several fatty acids, which are generally alkanes bearing one or more acid functionalities, have been identified in *Cordyceps* species to date, and many of them possess bioactive properties that are suitable for developing cosmeceutical formulations (Yang et al., 2009). Docosanoic, lauric, lignoceric, linoleic, myristic, oleic, palmitic, pentadecanoic, and stearic acids are the major fatty acids found in *Cordyceps*, while β-sitosterol, cholesterol, cordycepic acid, campesterol, and ergosterol are the main sterols in these extracts (Ashraf et al., 2020). Some of these compounds have known applications in cosmeceutical formulations.

3.4 Polyphenols

Yu et al. (2006) documented the antioxidant properties of *C. militaris* and *C. sinensis* extracts that could significantly protect lipids and proteins, including low-density lipoprotein (LDL) oxidation, and prevent cell damage. This effect was ascribed to the presence of various polyphenolic compounds; the polyphenolic and flavonoid contents were found to be 60.2 and 0.598 mg/mL in *C. militaris* and 31.8 and 0.616 mg/mL in *C. sinensis*.

3.5 Polysaccharides

Natural polysaccharides are used as moisturizers, conditioners, and rheology-modifiers in cosmeceutical formulations. Simple sugars as well as oligosaccharides can provide hygroscopic impacts owing to their molecules containing polyols that can exert moisturizing and hydrating impacts on the hair and skin, respectively. Oligosaccharides have also served as prebiotics, and are currently being used as active ingredients in skincare formulations. Moreover, some polysaccharides possess biological activities that enable their use as active anti-aging and/or anti-microbial ingredients in cosmeceutical formulations.

Cordyceps species produce highly branched bioactive polysaccharides (Chen et al., 2014; Shao et al., 2023). Yan et al. (2008) opined that galactomannan is one of the major extracellular or intracellular polysaccharides found in *C. militaris*. These polysaccharides are secreted into the culture filtrate through submerged fermentation (Wang et al., 2019). A polysaccharide, named CSP-1, from the water extract of cultured *C. militaris* was separated through ethanol precipitation, deproteination, and gel-filtration chromatography (Yu et al., 2004). The CPS-1 displayed anti-inflammatory activity by suppressing the humoral immunity in mice, but it has no significant impacts on cellular as well as non-specific immunities. It was reported that a polysaccharide present in *Cordyceps* extracts could also possess antioxidant properties (Yu et al., 2006) that could be used in cosmeceutical products. Reutilization of waste materials as a solid substrate fermentation medium was reported for their polysaccharide content (Li et al., 2021). For the first time, *C. militaris* mycelium was reutilized to ferment soybean curd residue to obtain bioactive polysaccharides from the fermented products. The purified polysaccharide CSCPS displayed antioxidant properties. Furthermore, this polysaccharide could activate the proliferation of RAW 264.7 cells, protect cells from doxorubicin-induced and LPS-stimulated cell damage. All these findings revealed that this polysaccharide could potentially be used in cosmeceutical products. The application of polysaccharides from *C. militaris* in cosmeceutical formulations is well documented (Kanlayavattanakul and Lourith, 2023).

4. Cosmeceuticals and Personal care Products

Environmental Working Group (EWG), a non-profit organization in DC, USA, published the function of *C. sinensis* extract as a skin-conditioning agent-humectant antioxidant, emollient, humectant, skin conditioning, and skin protecting (https://www.ewg.org/skindeep/), where the Prospector® (UL solution) search engine of raw materials in cosmetics showed multiple commercialized extracts of *Cordyceps*, either as a single active ingredient or used in combination with others. The main applications of *Cordyceps* in cosmeceuticals appear to be anti-aging and skin conditioners, and as an antimicrobial active ingredient (https://www.ulprospector.com/). The extracts are generally available in both lyophilized (dry) and solution forms. The directly fermented filtrates and the extracted form from the fruiting body, or mycelium, are utilized in cosmetic products. The cosmetic products in the market containing the above-mentioned extracts have been found to vary from the niche market brand, the indie brand, and the mass market product worldwide. Multiple dosage forms of cosmetic products are compatible with *Cordyceps* extracts. A low concentration of 0.1% to 3% is recommended in *Cordyceps* extracts by the Chinese skin care suppliers for product applications targeting skin-aging treatment including the skin lightening effect. A patent was published in 2016 on topical formulations containing extracts of *Cordyceps* spp., methods of formulating the extract and formulations, and methods of use. In one embodiment, *Cordyceps* spp. is *Cordyceps sinensis* (Gopaul et al., 2016). Multiple finished cosmetic products containing *Cordyceps* are available and recognized for their active moisturizing function. Table 2 illustrates some examples of commercial product with *Cordyceps* as active ingredients.

Table 2. Examples of cosmetic products containing *Cordyceps* extract

Product name	Brand, Country	*Cordyceps* ingredient in formula	Product claims
GinZing™ SPF 40 Energy-Boosting Tinted Moisturizer [A]	Origins, NY, USA.	*Cordyceps Sinensis* Extract	Moisturizer hydrates & energizes skin as it protects from skin-damaging UV rays
DR. ANDREW WEIL FOR ORIGINS™ Mega-Mushroom Relief & Resilience Soothing Gel Cream For Eyes [A]	Origins, NY, USA.	*Cordyceps Sinensis* Extract	Hydrates and helps reduce the look of irritation and fatigue for a more radiant appearance.
DR. ANDREW WEIL FOR ORIGINS™Mega-Mushroom Relief & Resilience Hydra Burst Gel Lotion [A]	Origins, NY, USA.	*Cordyceps Sinensis* (Mushroom) Extract	Hydration
DR. ANDREW WEIL FOR ORIGINS™Mega-Mushroom Skin Relief Micellar Cleanser [A]	Origins, NY, USA.	*Cordyceps Sinensis* Extract	Gently removes pollutants to help support skin's resilience against environmental irritants
Ultra Hydrating 12 Hour Facial Moisturizer [B]	ACURE®, FL, USA	*Cordyceps Sinensis* Extract	Intense 12-hour moisturizer for super thirsty skin types. Adaptogens turmeric and spirulina soothe, protect and de-stress.
Ultra Hydrating Eye Cream [C]	ACURE®, FL, USA	*Cordyceps Sinensis* Extract	Target puffiness and dark circles with deep moisturizing adaptogens and stimulating Green Coffee Oil
Cordyceps care makeup remover cream/ *Cordyceps sinensis* cleansing balm [D]	Floralsis®, Zhejiang, China	*Cordyceps Sinensis* Extract	Removes your makeup while leaving clear, tender, and smooth skin
Intensive Serum Foundation SPF 40/30 [E]	Bobbi Brown ®, NY, USA.	*Cordyceps Sinensis* Extract	Radiance-boosting serum foundation
Invisible physical defense mineral sunscreen spf30 [F]	dermalogica®, CA, USA.	*Cordyceps Sinensis* Extract	Invisible, weightless defense that blends easily on skin, featuring only non-nano Zinc Oxide. Bio-active Mushroom Complex helps soothe skin, and reduce UV-induced redness and dryness.

Sources: [A] https://www.origins.com/
[B] https://acure.com/products/ultra-hydrating-12-hour-facial-moisturizer-1
[C] https://acure.com/collections/eye-care/products/ultra-hydrating-eye-cream-1
[D] https://florasis.com/products/cordyceps-care-makeup-remover-cream?_pos
[E] https://www.bobbibrowncosmetics.com/product/14017/93890/makeup/face/foundation/
   intensive-serum-foundation-spf-4030/radiance-boosting-foundation
[F] https://www.dermalogica.com/products/invisible-physical-defense-spf30

Conclusion

The cosmetic industry is in a constant search for novel natural ingredients because of their competitive effectiveness, low toxicity, and low cost. The use of mushrooms in cosmeceutical products has grown over the years and has become a 'trendy' option in modern cosmeceutical formulations, mainly because of their antityrosinase, antihyaluronidase, anticollagenase, antioxidant, anti-inflammatory, antimicrobial, and anti-elastase properties. *Cordyceps* extracts and their main cosmeceutical compounds, like cordycepin and polysaccharides, have certainly become a popular choice among the mushrooms used in cosmeceuticals owing to their versatile and multifunctional characteristics.

Acknowledgements

The authors acknowledge the support of the Reinventing University System 2023 funding support from the Office of the Permanent Secretary of the Ministry of Higher Education, Science, Research and Innovation, Thailand.

References

Ahn, H. and Cho, Y.S. (2013). Antioxidative activity and chemical characteristics of cordycepin-enriched *Cordyceps militaris* JLM0636 powder. J. Life Sci., 23: 249–258.

Ashraf, S.A., Elkhalifa, A.E.O., Siddiqui, A.J., Patel, M., Awadelkareem, A.M. et al. (2020). Cordycepin for health and wellbeing: A potent bioactive metabolite of an entomopathogenic *Cordyceps* medicinal fungus and its nutraceutical and therapeutic potential. Molecules, 25: 2735.

Chen, L., Ge, M.D., Zhu, Y.J., Song, Y., Cheung, P.C.K. et al. (2019). Structure, bioactivity and applications of natural hyperbranched polysaccharides. Carbohydr. Polym., 223: 115076.

Chen, X., Siu, K.C., Cheung, Y.C. and Wu, J.Y. (2014). Structure and properties of a $(1\rightarrow3)$-β-D-glucan from ultrasound-degraded exopolysaccharides of a medicinal fungus. Carbohydr. Polym, 106: 270–275.

Cheng, W.Y., Wei, X.Q, Siu, K.C., Song, A.X. and Wu, J. et al. (2018). The cosmetic skincare benefits of cultivated *Cordyceps sinensis* fungal mycelia. Int. J. Med. Mushrooms, 20: 623–636.

Chien, C.C., Tsai, M.L., Chen, C.C., Chang, S.J. and Tseng, C.H. et al. (2009). Effects of tyrosinase activity by the extracts of *Ganoderma lucidum* and related mushrooms. Micopathologia, 166: 117–120.

Cordyceps sinensis Extract, EWG Skin Deep. 2023. https://www.ewg.org/skindeep/ingredients/701653-cordyceps-sinensis-extract/) (retrieved on October 10, 2023).

Cordyceps. Prospector, UL solution. 2023. https://www.ulprospector.com/en/asia/PersonalCare/Product/search?k=cordyceps (retrieved on October 10, 2023).

Das, J., Fatmi, M.Q., Devi, M., Singh, N.S. and Verma, A.K. et al. (2024). Antitumor activity of cordycepin in murine malignant tumor cell line and *in silico* study. J. Mol. Struct., 1297: 136946.

Eiamthaworn, K., Kaewkod, T., Bovonsombut, S. and Tragoolpua, Y. (2022). Efficacy of *Cordyceps militaris* extracts against some skin pathogenic bacteria and antioxidant activity. J. Fungi., 8(4): 327. 10.3390/jof8040327.

Gopaul, R., Namkoong, J., Ballantyne, A., Knaggs, H., Carter, D. et al. (2016). *Cordyceps* containing topical skin care formulation. Patent: WO2016197144A1, Application filed by Nse Products, Inc., USA.

He, H., Tang, J., Ru, D., Shu, X., Li, W. et al. (2020). Protective effects of *Cordyceps* extract against UVBinduced damage and prediction of application prospects in the topical administration: An experimental validation and network pharmacology study. Biomed. Pharmacother., 121: 109600. 10.1016/j.biopha.2019.109600.

Hyde, K.D., Bahkali, A.H. and Moslem, M.A. (2010). Fungi—An unusual source for cosmetics. Fungal Diversity, 43: 1–9.

Jin, M.L., Park, S.Y., Kim, Y.H., Park, G., Son, H.J. et al (2012). Suppression of α-MSH and IBMX-induced melanogenesis by cordycepin via inhibition of CREB and MITF, and activation of PI3K/Akt and ERK-dependent mechanisms. Int. J. Mol. Med., 29: 119–124.

Kanlayavattanakul, M. and Lourith, N. (2023). *Cordyceps millitaris* polysaccharides: Preparation and topical product application. Fungal. Biol. Biotechnol., 10: 3.

Khan, J., Yadav, S. and Alam, M.A. (2024). Herbal components for the treatment of various kidney disorders. Nat. Prod. J., 14(1): 10.2174/2210315513666230418104500.

Kunhorm, P., Chaicharoenaudomrung, N. and Noisa, P. (2019). Enrichment of cordycepin for cosmeceutical applications: culture systems and strategies. Appl. Microbiol. Biotechnol., 103: 1681–1691.

Lee, Y.R., Noh, E.M., Jeong, E.Y., Yun, S.K., Jeong, Y.J. et al. (2009). Cordycepin inhibits UVB-induced matrix metalloproteinase expression by suppressing the NF-kB pathway in human derma, fibroblasts. Exp. Mol. Med., 41: 548–554.

Li, Y., Ban, L., Meng, S., Huang, L., Sun, N. et al. (2021). Bioactivities of crude polysaccharide extracted from fermented soybean curd residue by *Cordyceps militaris*. Biotechnol. Biotechnol. Equip., 35: 334–345.

Lin, B.-Q. and Li, S.-P. (2011). *Cordyceps* as an herbal drug. pp. 73–105. *In*: Benzie, I.F.F. and Wachtel-Galor, S. (eds.). Herbal Medicine: Biomolecular and Clinical Aspects, 2nd edition. CRC Press, Boca Raton, USA. 10.1201/b10787-6.

Mago, P., Sharma, R., Hafeez, I., Nawaz, I., Joshi, M. et al. (2023). Mushroom based cosmeceuticals: An upcoming biotechnology sector. Biosci. Biotechnol. Res. Asia, 20: 381–394.

Marruci, G., Buccioni, M., Varlaro, V., Volpini, R. and Amenta, F. et al. (2022). The possible role of the nucleoside adenosine in countering skin aging: A review. Biofactors, 48: 1027–1035.

Marsup, P., Yeerong, K., Neimkhum, W., Sirithunyalug, J., Anuchapreeda, S. et al. (2020). Enhancement of chemical stability and dermal delivery of *Cordyceps militaris* extracts by nanaemulsion. Nanomater., 10: 1565.

Martin, K. and Glaser, D.A. (2011). Cosmeceuticals: The new medicine of beauty. Missouri Medicine, 108(1): 60–63.

Pandey, A., Jatana, G.K. and Sonthalia, S. (2023). Cosmeceuticals. StatPearls. https://www.ncbi.nlm.nih.gov/books/NBK544223/.

Pintathong. P., Chomnunti, P., Sangthong, S., Jirarat, A. and Chaiwut, P. et al. (2021). The feasibility of utilizing cultured *Cordyceps militaris* residues in cosmetics: biological activity assessment of their crude extracts. J. Fungi, 7: 973. 10.3390/jof7110973.

Prommaban, A., Sriyab, S., Marsup, P., Neimkhum, W., Sirithunyalug, J. et al. (2022). Comparison of chemical profiles, antioxidation, inhibition of skin extracellular matrix degradation, and anti-tyrosinase activity between mycelium and fruiting body of *Cordyceps militaris* and *Isaria tenuipes*. Pharm. Biol., 60: 225–234.

Pu, F., Li, T., Shen, C., Wang, Y., Tang, C. et al. (2024). Fermented *Ophiocordyceps sinensis* mycelium products for preventing contrast-associated acute kidney injury: a systematic review of randomized controlled trials. Ren. Fail, 46(1): 2300302. 10.1080/0886022X.2023.2300302

Qin, P., Li, X.K., Yang, H., Wang, Z.Y. and Lu, D.X. et al. (2019). Therapeutic potential and biological applications of cordycepin and metabolic mechanisms in cordycepin-producing fungi. Molecules, 24(12): 2231. 10.3390/molecules24122231.

Rupa, E.J., Li, J.F., Arif, M.H., Yaxi, H., Puja, A.M. et al. (2020). *Cordyceps militaris* fungus extracts-mediated nanoemulsion for improvement antioxidant, antimicrobial, and anti-inflammatory activities. Molecules, 25(23): 5733. 10.3390/molecules25235733

Shao, L., Jiang, S.J., Li, Y., Yu, L., Liu, H. et al. (2023). Aqueous extract of *Cordyceps cicadae* (Miq.) promotes hyaluronan synthesis in human skin fibroblasts: A potential moisturizing and anti-aging ingredient. PLOS One, 18(7): e0274479. 10.1371/journal.pone.0274479

Tang, J., Xiong, L.D., Shu, X.H., Chen, W., Li, W.J. et al. (2019). Antioxidant effects of bioactive compounds isolated from *Cordyceps* and their protective effects against UVB-irradiated HaCaT cells. J. Cosmet. Dermatol., 18: 1899–1906.

Upatcha, N., Kaokaen, P., Sorraksa, N., Phonchai, R., Kunhorn, P. et al. (2023). Nanocapsulated *Cordyceps* extract enhances collagen synthesis and skin cell regeneration through antioxidation and autophagy. J. Microencapsul., 40: 303–317.

Wan, J.Q., Zhu, Y.L., Jiang, X., Wang, F.X., Zhou, Z.Y. et al. (2023). Immunomodulation of RAW 264.7 cells by CP80-1, a polysaccharide of *Cordyceps cicadae,* via Dectin-1/Syk/NF-κB signaling pathway. Food Agric. Immunol., 34(1): 2231172. 10.1080/09540105.2023.2231172

Wang, C.C., Wu, J.Y., Chang, C.Y., Yu, S.T. and Liu, Y.C. et al. (2019). Enhanced exopolysaccharide production by *Cordyceps militaris* using repeated batch cultivation. J. Biosci. Bioeng., 127: 499–505.

Wang. S.S., Wang, C.B., Cao, H., Cui, X.W., Guo, H. et al. (2020). Comparing the cosmetic effects of liquid-fermented culture of some medicinal mushrooms including antioxidant, moisturizing, and whitening activities. Int. J. Med. Mushrooms, 22: 693–703.

Wong, J.H., Ng, T.B., Wang, H., Sze, S.C.W., Zhang, K.Y. et al. (2011). Cordymin, an antifungal peptide from the medicinal fungus *Cordyceps militaris*. Phytomedicine, 18: 387–392.

Yan, H., Zhu, D., Xu, D., Wu, J. and Bian, X et al. (2008). A study on *Cordyceps militaris* polysaccharide purification, composition and activity analysis. Afr. J. Biotechnol., 7: 4004–4009.

Yang, F.Q., Feng, K., Zhao, J. and Li, S.P. (2009). Analysis of sterols and fatty acids in natural and cultured *Cordyceps* by one-step derivatization followed with gas chromatography-mass spectrometry. J. Pharm. Biomed. Anal., 49:1172–1188.

Yu, H.M., Wang, B.S., Huang, S.C. and Duh, P.-D. (2006). Comparison of protective effects between cultured *Cordyceps militaris* and natural *Cordyceps sinensis* against oxidative damage. J. Agric. Food Chem., 54: 3132–3138

Yu, R., Song, L., Zhao, Y., Bin, W., Wang, L. et al. (2004). Isolation and biological properties of polysaccharide CPS-1 from cultured *Cordyceps militaris*. Fitoterapia, 75: 465–472.

Zhang, D.Y., Tang, Q.J., He, X.Z., Wang, Y.P., Zhu, G.Y. et al. (2023). Antimicrobial, antioxidant, anti-inflammatory, and cytotoxic activities of *Cordyceps militaris* spent substrate. PLOS One, 18: e0291363. 10.1371/journal.pone.0291363.

Zhang, H.C., Deng, L.A., Yang, J., Yang, G.W., Fan, H.T. et al. (2022). Preparation and evaluation of a nanoemulsion containing cordycepin and its protective effect on skin. J. Dispers. Sci. Technol., 44: 1094–1102.

Nutraceutical Prospects

10

The Significance of *Cordyceps* in Medicine and Nutrition

Karol Jędrejko, Katarzyna Kała, Katarzyna Sułkowska-Ziaja* and *Bożena Muszyńska**

1. Introduction

Selected mushrooms are currently classified in some countries as functional foods, with documented scientific research showing a beneficial effect on human health. Based on the analysis of the chemical composition of the fruit bodies of edible mushrooms such as *Cordyceps* spp., it can be concluded that they are a complete food product, containing all the basic ingredients that are necessary for the growth and development of the human body and supporting health conditions. *Cordyceps militaris*, apart from its nutritional value, demonstrates potential medicinal properties. *C. militaris* was first described by Carl Linnaeus in 1753 as *Clavaria militaris*. Some synonyms of the scientific name: *Clavaria granulosa, Clavaria militaris, Corynesphaera militaris, Hypoxylon militare, Sphaeria militaris,* and *Torrubia militaris, Xylaria militaris* (Shrestha et al., 2005a). This chapter briefly highlights the nutritional and medicinal significance of *Cordyceps militaris* and *Ophiocordyceps sinensis*.

2. Characteristics of *Cordyceps militaris*

The mycelium of *C. militaris* develops 1–8 cm high, club-shaped, orange to red fruit bodies, which emerge out erect from the dead buried pupae. At

Jagiellonian University, Medical College, Faculty of Pharamacy, Department of Medicinal Plant and Mushroom Biotechnology; 9 Medyczna Street, 30-688 Kraków, Poland.
* Corresponding authors: karol.jedrejko@gmail.com; bozena.muszynska@uj.edu.pl

maturity, the perithecia surface appears rough and punctured. The inner tissue is whitish to pale-orange (Fig. 1). The asci are long, cylindrical, and the spores are hyaline, smooth, filiform, and septate. On some occasions, its anamorphic state, *Isaria*, is also found. A white mycelial mass encompasses the parasitized insect; however, these need not be in an anamorphic state of *C. militaris* (Shrestha et al., 2005a; Ng and Wang, 2010).

Fig. 1. Perithecia of *Cordyceps militaris* erected from a buried larva.

The genus *Cordyceps* is classified under the class Ascomycota and has a rich tradition of use in Asian ethnomedicine. It is recgnized as a natural adaptogenic rawmaterial with anti-fatigue and immune-stimulating properties.

Currently, all *Cordyceps* spp. that control the nervous systems of insects, turning them into "zombies", are classified as *Ophiocordyceps* spp. Among the numerous *Cordyceps* species (about 500 spp.), the most scientific research focuses on *Ophiocordyceps sinensis* (Davis et al., 2012).

The first to notice the health properties of the mushroom were Himalayan shepherds. They observed that their herds became strong and active after grazing in meadows rich in fungus. Mention of it dates back to around the 5th century AD. Since then, *Cordyceps* has been used in traditional Chinese medicine (TCM), and its price has been several times higher than the price of silver and gold. The inhabitants of China, India and Nepal were known to consume *O. sinensis* for centuries to adapt the body to difficult high-mountain climatic conditions, especially low temperatures, high atmospheric pressure, and reduced oxygen levels (Shweta et al., 2023). TCM recommends the use of *O. sinensis* in numerous human diseases such as cancer, blood circulation, diabetes, infectious or parasitic diseases, kidney disorders, liver disorders, respiratory diseases, and sexual dysfunctions (Das et al., 2010).

The uniqueness of *O. sinensis* results from its form of existence in the ambientnaturalenvironment.Itisendemictoareasabove3000mabovesealevel. *O. sinensis* is a parasite that attacks the larvae of the *Thitarodes* moth. These larvae reside about 15cm below the ground where they are "infected" by

O. sinensis. In late autumn, a chemical interaction occurs between the spores of the fungus and the cuticle of the moth larvae. Healthy larvae hatch normally, but infected larvae remain underground and migrate towards the surface, or rather, alight, due to the fungal attack. Native inhabitants collect *O. sinensis* during May to June, dry and grind it into powder, and mix it with water to be consumed as a warm therapeutic drink (de Menezes et al., 2023).

The scarcity of *O. sinensis* in its natural habitats, coupled with the lengthy process involved in its formation (from autumn to spring) has led to challenges in meeting the growing demand, for dietary supplements, functional foods, and nutraceuticals, particularly in regions such as Asia, Europe, and the United States (de Menezes et al., 2023). *C. militaris* is not only valued for its potential medicinal properties but is also used in culinary preparations. It is considered a gourmet ingredient in some cultures, especially in Asian cuisine.

3. *Cordyceps militaris* and *Ophiocordyceps sinensis*

Understanding the morphological differences and similarities between *C. militaris* and *O. sinensis* is crucial for taxonomic identification and potential applications, particularly in the context of traditional medicine. Both *C. militaris* and *O. sinensis* show a stromal structure, which is a visible fruiting body growing from the host insect. Perithecia, flask-shaped structures housing sexual spores (ascospores) during the reproductive phase, are evident in both species, serving as a common feature central to their sexual reproduction. Distinguishing features encompass the chromatic attributes of the stroma, with *C. militaris* exhibiting hues ranging from orange to red, while *O. sinensis* typically presents a dark stroma, variably colored in shades of brown to black. In *C. militaris*, perithecia are generally cylindrical or club-shaped, whereas *O. sinensis* showcases elongated and twisted perithecia (Shrestha et al., 2005a; Ng and Wantg, 2010; Olatunji et al., 2018). *C. militaris* demonstrates a broader host range, infecting diverse insect species, resulting in variable morphologies contingent upon the host. Conversely, O. sinensis primarily parasitizes ghost moth larvae (*Hepialidae*), and its morphological characteristics are intricately adapted to this specific host. The stromal supporting structure, or stem, of *C. militaris* exhibits variable lengths, while *O. sinensis* typically features a long and slender stem (Shrestha et al., 2005b). Geographically, *C. militaris* exhibits a global distribution, thriving in diverse environments spanning North America, Europe, and Asia. In contrast, *O. sinensis* is confined to high-altitude regions of the Tibetan Plateau, the Himalayas, and adjacent areas (Dong et al., 2016).

The differences between these species also concern their cultivation. Growing *O. sinensis* is demanding and has not been achieved under controlled conditions, leading to a reliance on wild harvesting. *C. militaris* can be grown relatively easily on artificial substrates, which increases its commercial availability. Additionally, the mycelium of this species can be easily obtained *in vitro*. Importantly, in the case of *O. sinensis* mycelium obtained *in vitro*, it was found that it does not synthesize cordycepin under these conditions and has a lower quantity of amino acids and mannitol compared to fruit bodies of this species originating from natural sites (Cui, 2015). The concentration of polysaccharides obtained from *in vitro* cultures of *O. sinensis* was higher than in the raw material of natural origin. *C. militaris* is an alternative to *O. sinensis* because the composition of active substances in *C. militaris* grown *in vitro* do not vary from each other. Based on the results of the analysis of the composition of cultivated *C. militaris* , it was shown that the quantity of bioactive ingredients cordycepin and polysaccharides is greater than their concentration in *O. sinensis* from the natural environment. The profile of 17 amino acids determined in the study in both species is similar, with a predominance of L-arginine and L-proline in *C. militaris* cultivation (Zeng et al., 2014).

The presence of nucleosides, cordycepin, and adenosine was confirmed in the fruit bodies of *C. militaris*. It was also shown that these ingredients in *C. militaris* are greater than in *O. sinensis*. The analysis revealed the presence of biologically active substances in the fruit bodies of *C. militaris*. These include amino acids, carotenoids (lutein and zeaxanthin), elements (magnesium, potassium, selenium, and sulfur), ergothioneine, glycolipids, glycoproteins, γ-aminobutyric acid, mannitol (also called cordycepic acid), phenolic compounds (phenolic acids and flavonoids), statins (lovastatin), sterols (ergosterol), vitamins, and xanthophylls (Hur, 2008; Yang et al., 2009; Das et al., 2010; Dong et al., 2013; Ashraf et al., 2020; Jędrejko et al., 2021).

4. Bioactive Compounds in Mycelium and Fruit Bodies

Investigations have shown that the distribution and concentration of bioactive compounds are not consistent in the fruit bodies *of C. militaris*. The external parts of fruit bodies contain the highest quantities of carotenoids, nucleosides, polysaccharides, organic compounds, and selenium (Jędrejko et al., 2021).

4.1 Amino Acids

The most important substances are amino acids. For this group of compounds, its noteworthy to discuss non-protein amino acids including ergothioneine and GABA. Ergothioneine and GABA are potent bioactive substances with important roles in the body (Ashraf et al., 2020; Jędrejko et al., 2021). GABA is a non-protein amino acid. It is synthesized in the human body through the conversion of glutamic acid (glutamate), facilitated by the enzyme glutamate decarboxylase (GAD). This is aided by vitamin B_6 as a cofactor. Functioning as an inhibitory neurotransmitter in the central nervous system, GABA influences various neural components, including the cerebellum, hippocampus, hypothalamus, striatum, and spinal cord. It plays a regulatory role in learning processes, memory, and sleep, including emotional aspects like stress and anxiety. Additionally, GABA exhibits anticonvulsant and myorelaxant activities (Boonstra et al., 2015; Li et al., 2015; Yamatsu et al., 2016; Hepsomali et al., 2020). In examinations of the mycelium and fruit bodies of *C. militaris*, the quantity of GABA was 68.6–180.1 mg/kg and 756.30 µg/g dry mass, respectively (Chen et al., 2012; Cohen et al., 2014).

Ergothioneine is a naturally occurring thiol derivative of the amino acid histidine, characterized by a unique sulfur-atom bridge between the imidazole ring and the thiol group (Fig. 2). Abundantly found in various biological systems, particularly in certain fungi and bacteria, ergothioneine exhibits potent antioxidant properties. It is known to act as a scavenger of ROS, thereby mitigating oxidative stress and defending cells from damage by free radicals. The unique chemical structure of ergothioneine allows it to efficiently transport and accumulate in tissues, especially those exposed to oxidative challenges. Research suggests potential health benefits associated with ergothioneine consumption, including neuroprotective effects and

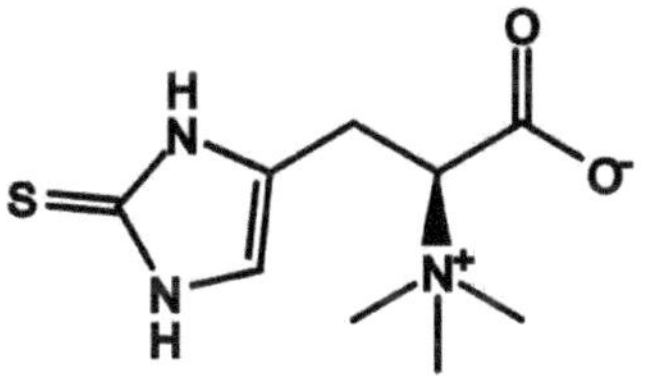

Ergothioneine

Fig. 2. Chemical structure of ergothioneine.

modulation of inflammatory responses (Chen et al., 2012; Jędrejko et al., 2021). The quantity of ergothioneine in the mycelium of *C. militaris* ranges from 130.6 mg/kg to 782.3 mg/kg dry mass, as determined in various studies (Chan et al., 2015). Additionally, a separate investigation reported an estimated concentration of 409.8 µg/g dry mass in the fruit bodies of *C. militaris* (Chen et al., 2012).

4.2 Bioelements

Bioelements such as Calcium (Ca), Magnesium (Mg), Iron (Fe),Nickel (Ni), Strontium (Sr), Sodium (Na), Sulfur (S), Titanium (Ti) , Phosphorus (P), Manganese (Mn), Zinc (Zn), Selenium (Se), Aluminium (Al), Silicon

(Si), Chromium (Cr), Gallium (Ga) ,Potassium (K), Vanadium (V) and Zirconium (Zr) are present in the fruit bodies of *C. militaris* .The presence of these elements was determined using various analytical methods. The high concentration of the above-mentioned elements, especially magnesium, potassium, selenium, and sulfur, makes *C. militaris* an alternative source of these bio elements in the human diet (Ashraf et al., 2020; Jędrejko et al., 2021). Thus, a proper diet is very important for physical growth and strength, mental development, performance, and productivity. This species can be considered one of the most important nutraceuticals, with multifaceted effects. With the enhanced interest in *C. militaris* in recent years for nutraceutical, biotechnological, or pharmaceutical purposes, further scientific validation is required to ascertain the benefits of dietary intake (Yang et al., 2009; Hur, 2008; Tuli et al., 2014; Kuo et al., 2015; Mehra et al., 2017; Soltani et al., 2018; Ashraf et al., 2020; Jędrejko et al., 2021).

4.3 *Carbohydrates*

D-mannitol is one of the important products of *C. militaris* metabolism. It is categorized as polyhydric alcohols, or polyols for short. D-mannitol in *C. militaris* is also known as cordycepic acid (Shawkat et al., 2012; Cohen et al., 2014). D–mannitol is used by *C. militaris* as a reserve of carbohydrates in the human body, a transporter of other compounds, and in osmoregulation and regulation of metabolic pathways. In the human body, it has diuretic, antitussive, and antioxidant properties (Meena et al., 2015). Depending on the location of polysaccharide synthesis in hyphae, they may present themselves as intracellular or extracellular secondary metabolites (Jędrejko et al., 2021). Different species of medicinal mushrooms produce different types of polysaccharides, which can be both soluble and insoluble in water (Zhu et al., 2015; Singdevsachan et al., 2016). The species *C. militaris* is also a source of such substances. Molecules of monosaccharides (simple sugars) are assembled into polysaccharides through glycosidic bonds. Some of them are made up of only residues of one type of monosaccharide, homopolysaccharides, while others belong to heteropolysaccharides and are made up of residues of different monosaccharides (Zhu et al., 2015; Singdevsachan et al., 2016; Ashraf et al., 2020). Polysaccharides that have been obtained from the *C. militaris* mycelium have different chemical structures. This structure is determined by the kind of monosaccharides that make up a given polysaccharide— their linear sequence, location of glycosidic bonds, spatial configuration, and the extent of cross-linking of the chain (Zhang et al., 2019). The differences in the qualitative and quantitative composition of simple sugars and the chemical structure observed in the polysaccharide fraction of *C. militaris* between normal environmental conditions and laboratory cultivation highlight the impact of growth conditions on the chemical composition of the fungus (Gu and Liang, 1987). The most important monosaccharides forming sugar polymers are glucose, mannose, and galactose. In scientific studies on *C. militaris* polysaccharides, additional simple sugars such as arabinose, rhamnose and xylose have been identified as reported in studies by Zhang et al., 2019, Ashraf et al., 2020 and Jędrejko et al., 2021. Research studies have demonstrated that polysaccharides including those found in *C. militaris*, exhibit a range of beneficial properties in both *in vitro* and *in vivo* experiments. These properties include anticancer, antioxidant, immunostimulatory, anti-inflammatory, hypoglycemic, hypolipidemic, and hepatoprotective activities as reported by Wang et al. (2015). The research revealed the immunostimulatory potential of the polysaccharide fraction extracted from *C. militaris*. The examined polysaccharides were found to elicit an immune response *in vitro* by enhancing the activity of macrophages, leading to the production of NO, ROS (reactive oxygen species), IL-1β, interferon (IFN-γ), TNF-α (tumor necrosis factor), T and B lymphocytes, NK (natural killer cells), and an augmentation in phagocytosis by macrophages. Other studies have shown the anticancer potential of polysaccharides, which resulted from the inhibition of growth as well as induced apoptosis in cancer cells (Shashidhar et al., 2013; Elkhateeb et al., 2019).

4.4 *Cordycepin*

Amino acid analysis indicated that fruit bodies contain high amounts of glutamic acid, lysine, proline, and threonine. It is also significant that the fatty acid profile indicates up to 70% unsaturated fatty acids (Hur, 2008; Yang et al., 2009; Ashraf et al., 2020). Importantly, the amounts of adenosine and cordycepin in fruit bodies ranged between 0.18–0.06% and 0.97–0.36%, respectively. The most common and specific bioactive substance found in *C. militaris* species is cordycepin, which has a potential impact on the human body (Tuli et al., 2014; Yang et al., 2014; Kuo et al., 2015; Mehra et al., 2017). Cordycepin (3′-deoxyadenosine) is a water-insoluble organic compound that is a structural analogue of the nucleoside, adenosine, except for the lack of a 3′-hydroxyl group (Fig. 3) (Ashraf et al., 2020; Jędrejko et al., 2021). The chemical formula of cordycepin is $C_{10}H_{13}N_5O_3$. Its molecular weight is 251.24 Da with a melting point between 228 and 231°C (Soltani et al., 2018). On an average, the quantity of cordycepin in *C. militaris* is about 2–3 mg/kg. This concentration is low compared to the commercial market requirements for health-promoting purposes; thus, different synthetic and semi-synthetic techniques of obtaining cordycepin have been described (Huang et al., 2018). Cordycepin can be synthesized chemically by substituting the OH group with a H group at the 3′ position

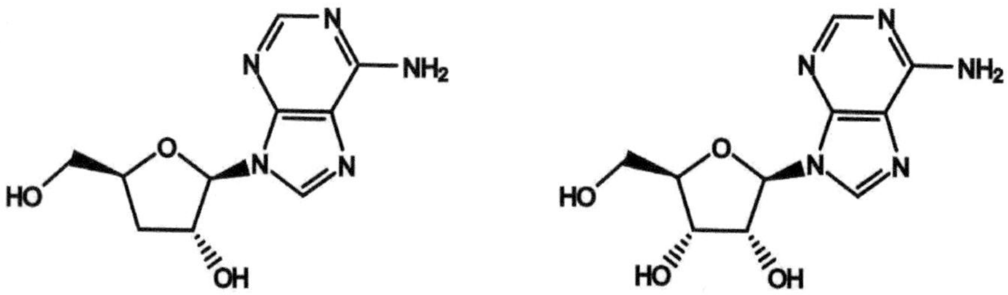

Cordycepin

Adenosine

Fig. 3. Chemical structure of cordycepin and adenosine.

of the ribofuranosyl group to produce the described adenosine analogue (Huang et al., 2018; Jin et al., 2018; Ashraf et al., 2020).

Cordycepin has an extremely broad range of pharmacological activities, such as anticancer, anti-inflammatory, antimicrobial, antioxidant, antiviral, ergogenic, hypoglycemic, hypolipidemic, immunomodulatory, and properties with regulation of spermatogenesis and steroidogenesis (Soltani et al., 2018). The best-known effect of *Cordyceps* spp. is its influence on the body's vital forces and increasing physical endurance. Many works were also published devoted to increasing physical endurance as a result of consuming *Cordyceps* spp. (Chen et al., 2010; Thongsawang et al., 2021; Ler et al., 2022). Cordycepin was isolated in 1950 from the species *C. militaris* (Liu et al., 2015). There are several published reports suggesting that cordycepin therefore has a multidirectional effect in that it competitively inhibits the course of DNA and RNA synthesis and metabolism, which affect the activity of adenosine deaminase as well as the mTOR signaling pathway, which are key to the proper functioning of the human body (Wong et al., 2010; Ashraf et al., 2020). In the case of cordycepin, it is particularly interesting to explore the mechanism of immune-related activity. Experiments conducted on mice proved that the mechanism of the immunomodulating or immunostimulatory effect of cordycepin results from the activation of cellular and humoral immunity (Tuli et al., 2014; Jędrejko et al., 2021).

4.5 *Phenolic Compounds*

Phenolic compounds are an important group of metabolites present in the fruit bodies of *C. militaris*. Mushrooms generally contain phenolic compounds, with varying concentrations in different parts such as the mycelium and fruiting bodies. They serve as potential antioxidant agents and defend against oxidative damage to enzymes, lipids, nucleic acids, and proteins. Vanillic and caffeic acids have the strongest antioxidant properties. The *p*–hydroxybenzoic, gallic, and protocatechuic acids present in this species and other edible mushrooms have not only antioxidant but also antibacterial, antifungal, anti-inflammatory, and antiviral properties, as documented *in vitro* as well as *in vivo* experiments (Muszyńska et al., 2013).

Another group of phenolic compounds determined in *C. militaris* and other *Cordyceps* spp. is flavonoids. They occur mainly in the form of glycosides. The antioxidant activity of flavonoids depends mainly on conjugated double bonds in the C-2 and C-3 positions and hydroxyl and carboxyl groups in the C-4 positions. The direct mechanism of the antioxidant potential of these compounds includes limiting their production in cells by impeding the activity of oxidative enzymes, releasing hydrogen from the carboxyl group, reducing peroxides as well as hydroxides, and scavenging reactive oxygen species. The indirect functions include the chelation of metal ions, which stops the formation of a reactive hydroxyl radical in cells. They are also known to increase the absorption of vitamin C in the digestive tract, thanks to which they achieve even stronger control over oxidative stress in the body, which translates into a protective effect on cells (Muszyńska et al., 2013; Huang et al., 2015; Joshi et al., 2019). The fruit body extract of *C. militaris* possesses flavonoids up to 1.56 mg/g. Furthermore, the quantity of flavonoids in the fruit bodies and mycelium rutin equivalent (RE) was 5.54 mg RE/g and 2.26 mg RE/g, respectively (Huang et al., 2015; Jędrejko et al., 2021).

4.6 Pigments

The presence of carotenoids, including xanthophylls, was confirmed in the fruit bodies of *C. militaris*. Carotenoids of the fruit bodies of *C. militaris* have an intense yellow-orange color. The main compounds from this group present in the fruit bodies of *C. militaris* are cordyxanthin, zeaxanthin, and lutein. These pigments support color (daytime) vision. Additionally, cordyxanthin, unlike other compounds in this group, dissolves in water, which facilitates its penetration into the retina, which means that there is great hope for the treatment of eye diseases, including the prevention of processes related to macular degeneration. Besides the protective effect of lutein and zeaxanthin on the visual organs, there are scientific reports indicating that supplementation of carotenoid pigments enhances cognitive functions and decreases cortisol levels and stress symptoms in young individuals as well as adults (Bovier et al., 2015; Renzi-Hammond et al., 2017; Stringhan et al., 2018).

4.7 Vitamins

The presence of water-soluble vitamins (vitamin B_1, vitamin B_2, vitamin B_3, vitamin B_{12}, vitamin C) and lipid-soluble vitamins (vitamin A and vitamin E) was found in the fruit bodies of *C. militaris*. Vitamin B_2 (riboflavin) as well as vitamin B_3 (niacin) help alleviate fatigue and tiredness while playing a key role in regulating energy metabolism. Vitamin C has an antioxidant function, helps in the proper production of collagen, increases the absorption of iron from the digestive tract, and likewise supports the proper functioning of the immune system. Vitamin A contributes to cell specialization and supports optimal vision, while vitamin E safeguards cells from oxidative stress (Ashraf et al., 2020; Jędrejko et al., 2021).

4.8 Other Bioactive Compounds

The remaining groups of bioactive compounds found in the fruit bodies of *C. militaris* are lectins. Structurally they are protein molecules conjugated with sugar fragments,forming glycoproteins. It has been proven that lectins have mitogenic activity. By binding to sugar residues on the cell surface, lectins initiate the process of cell sticking and agglutination. An additional group of bioactive compounds found in *C. militaris* are beauveriolides, which are assessed by their complex structure as cyclodepsipeptides. Beauveriolides in tests show anti-atherosclerotic activity and the capability to reduce the concentration of β-amyloid. In turn, militarinones are a little-known group of compounds grouped as alkaloids. Structurally, they are derivatives of pyridine or pyrrolidine-2,4-dione. So far, the antibacterial and cytotoxic activity of these compounds has been confirmed (Singh et al., 2010; Hassan et al., 2015; Chen et al., 2020). Pentostatin, found in the fruit bodies of *C. militaris*, is an analogue of the purine base hypoxanthine. Pentostatin has inhibitory activity against the enzyme adenine deaminase. This compound's anticancer and immunosuppressive activity have been proven . Cordymin, discerned within the fruit bodies of *C. militaris*, is characterized as a peptide exhibiting antibacterial and antifungal efficacy. Research suggests that cordymin may exhibit inhibitory effects against a range of microorganisms, making it a potential candidate for antimicrobial applications (Chen et al., 2020; Jędrejko et al., 2021).

5. Clinical Trials

5.1 COVID-19

O. sinensis and *C. militaris* affect the functions of the reproductive system. Long-term intake (12 weeks) of *O. sinensis* with Jujing pills effectively improves sperm quality in males (Yang et al., 2021). Intervention with *C. militaris* in patients with prostatic hyperplasia contributes to improving urinary flow, lessening micturition symptoms, and reducing the size of the prostatic gland (Hsieh et al., 2022). *O. sinensis* and *C. militaris* have been investigated for the treatment of COVID-19. *O. sinensis* (Jinshuibao) was administered three times a day over 2-weeks improving cardiopulmonary symptoms (such as chest tightness, dry cough, palpitation, shortness of breath, and sweating) in convalescent COVID-19 patients (Zhang et al., 2023a). Supplementation of *C. militaris* at a dose of 1500 mg/day (three capsules at 500 mg per day; 1 capsule contains 0.96 mg adenosine and 4.5 mg cordycepin) for 15 days in combination with standard treatment contributed to a shorter time period for improvement in clinical symptoms and likewise, and a higher and quicker number of recoveries in patients with mild to moderate COVID-19 infection (Dubhashi et al., 2023).

5.2 Cytotoxicity

Cordycepin demonstrates cytotoxic activity related to intracellular production of the triphosphate metabolite, 3′-dATP, inhibiting DNA and RNA synthesis. Although cordycepin has shown potent anti-tumor activity in pre-clinical studies, it has not been successful in clinical trials due to cellular uptake dependent on nucleoside transporters (hENT1), rapid enzymatic degradation by adenosine deaminase (ADA), and reliance on adenosine kinase (AK) for activation. However, the phosphoramidate derivative of cordycepin, signed as NUC-7738, enters cancer cells independent of the human equilibrative hENT1, and does not require phosphorylation by AK. Additionally, it is resistant to breakdown by ADA. NUC-7738 is a potential proapoptotic agent, investigated in a clinical trial on patients with advanced solid tumors (Schwenzer et al., 2019; Symeonides et al., 2020; Blagden et al., 2021).

5.3 Immunity

Fermented and powdered mycelium of *O. sinensis* (in the form of capsules) under the registered name Jinshuibao is a Chinese patent medicine and is often used as adjuvant therapy to support the effectiveness of standard drugs. Jinshuibao, combined with standard therapy, is used in the clinical treatment of asthma and renal diseases (Zhang et al., 2013; Li and Gaosi, 2020; Deng et al., 2021; Zhang et al., 2023b). *O. sinensis* and *C. militaris* are commonly used in traditional Chinese medicine as immune-enhancing agents. An increase in NK cell activity was demonstrated among healthy adults after an 8-week intervention with *O. sinensis* (1.68 g per day) (Jung et al., 2019). In another study, immunostimulatory effects associated with increased levels of IL-2, TNF-α, and IFN-γ, likewise enhanced lymphocyte proliferation. As a result NK cell activity, was observed in healthy males after a 4-week intake of *C. militaris* in the form of capsules at a daily dose of 1.5 g (Kang et al., 2015). Also, longer supplementation (12 weeks) of *C. militaris* among healthy adults with a history of at least two upper respiratory tract infections per year contributed to the immune-enhancing effect, which correlated with an increase in NK cell activity and an increase in IgA levels (Jung et al., 2019).

5.4 Other Benefits

C. militaris has been shown to demonstrate hepatoprotective effects. In a study where patients with mild liver dysfunction were supplemented with *C. militaris* at a dose of 1.5 g per day (containing cordycepin at a concentration of 1.9 mg/g) for 8-weeks it was observed to support liver function and protect against lipid accumulation (Heo et al., 2015). *O. sinensis* and *C. militaris* have also been studied as ergogenic aids. The intake of a multi-ingredient dietary supplement, including *O. sinensis* and other bioactive ingredients,did not result in improvements in body composition or exercise performance in young individuals over a 14 week period. The supplement also contained vitamin B12, chromium, and extracts from Astragalus root extract, Green Tea leaf, Ashwagandha root, and *R. rosea* root (Kreipke, 2021). A recent randomized, double-blind, placebo-controlled trial showed that 12-weeks consumption with 2 g daily of *O. sinensis* (as CordyMax Cs-4) was associated with improved aerobic performance in amateur marathoners (Savioli et al., 2022).

Hirsch et al., demonstrated that a 3-week dietary intervention of a mushroom blend (trade name, PeakO2®),containing the primary ingredient *C. militaris* and concomitant ingredients *Ganoderma lucidum, Hericium erinaceus, Lentinula edodes, Pleurotus eryngii,* and *Trametes versicolor*) at a dose of 4 g per day improved exercise tolerance. However, a 1-week intake of the mushroom blend had no significant effect on exercise performance in participants (Hirsch et al., 2017). Other *Cordyceps* spp., such as *C. cicadae,* have also been studied recently in humans. Oral supplementation of powdered mycelium of *C. cicadae* over 90 days contributed to the alleviation of dry eye symptoms through a decrease in tear osmolarity and tear electrolytes, as well as an increase in tear film breakup time (Chang et al., 2022).

6. Dietary Supplements and Functional Foods

Several dietary components or functional foods are prepared from the mushrooms *Cordyceps* spp., especially from *O. sinensis* and likewise from *C. militaris*. Commercially available products on the market are sold in various formulations, including capsules, powder, liquid (drops), or whole dried fruiting bodies. Dietary supplements may include standardized or non-standardized extracts of *Cordyceps* spp.

It's essential to note that in the European Union (EU), only *O. sinensis* (both the mycelium and the fruiting body, also known as the sporophorum) is classified as an authorized ingredient, and not considered novel, in food supplements. However, this classification applies exclusively to their use in food supplements. Any other food uses of this product, apart from food supplements, might be considered novel foods and consequently, may need to be authorized (European Commission 1).

C. militaris (Scarlet Caterpillar Club) in the EU, has been recognized as a medicinal mushroom and categorized as a novel food. an unauthorized ingredient in food supplements in accordance with the European Food Safety Authority's (EFSA) novel food catalogue and the European Commission's (EC)regulations (European Commission 1).

Commercially available dietary supplements or functional foods based on *Cordyceps* spp. are marketed as antiaging or longevity agents' energy and endurance support; nootropic and brain booster supplements; enhancement of the immune system; hepatic or renal protectors; and improvement of lung function. In the EU, all health claims for *O. sinensis*, such as antioxidant properties (in doses of 400–800 mg/day), helping to strengthen the body, an increase in endurance performance (in doses of 3 g of dried powder or equivalent extracts), and supporting the immune system, are not approved (European Commission 2; Yu et al., 2022).

7. Conclusion

Cordyceps species are recognized as valuable nutraceuticals owing to their functional food properties and ther potential ability to combat lifestyle diseases. They contain important bioactive compounds including cordycepin, carbohydrates, amino acids, pigments, phenolics, vitamins, and minerals. These fungi have shown the ability to enhance immunity, exhibit cytotoxic effects against cancer cells, and may have the potential to combat diseases such as COVID-19. Traditional Chinese medicine has long utilized *Cordyceps* and related species to address various human ailments , such as cancer, diabetes, kidney/liver/respiratory disorders, and sexual dysfunctions. *Cordyceps* spp. play a significant role in dietary supplements and are useful in the production of functional foods.

References

Ashraf, S.A., Elkhalifa, A.E.O., Siddiqui, A.J., Patel, M., Awadelkareem, A.M. et al. (2020). Cordycepin for health and wellbeing: A potent bioactive metabolite of an entomopathogenic *Cordyceps* medicinal fungus and its nutraceutical and therapeutic potential. Molecules, 25(12): 2735. 10.3390/molecules25122735.

Blagden, S., Haris, N.M., Boh, Z., Kazmi, F., Aroldi, F. et al. (2021). A first-in-human study of NUC-7738, a ProTide transformation of 3'-deoxyadenosine, in patients with advanced solid tumours (NuTide: 701). Ann. Oncol., 32: S619–S620. 10.1016/j.annonc.2021.08.1088

Boonstra E., de Kleijn R., Colzato L.S., Alkemade A., Forstmann B.U et al. (2015). Neurotransmitters as food supplements: The effects of GABA on brain and behavior. Front. Psychol., 6: 1520. 10.3389/fpsyg.2015.01520

Bovier E.R., Hammond B.R. (2015). A Randomized placebo-controlled study on the effects of lutein and zeaxanthin on visual processing speed in young healthy subjects. Arch. Biochem. Biophys., 572: 54–57.

Chan J.S.L., Barseghyan G.S., Asatiani M.D., Wasser S.P. (2015). Chemical composition and medicinal value of fruiting bodies and submerged cultured mycelia of caterpillar medicinal fungus *Cordyceps militaris* CBS-132098 (Ascomycetes) Int. J. Med. Mushrooms, 17: 649–659.

Chang, H.-H., Chang, W.-J., Jhou, B.-Y., Kuo, S.-Y., Hsu, J.-H. et al. (2022). Efficacy of *Cordyceps cicadae* (Ascomycota) Mycelium Supplementation for Amelioration of dry eye symptoms: A randomized, double-blind clinical pilot study. Int. J. Med. Mushrooms, 24(12): 57–67. 10.1615/IntJMedMushrooms.2022045307

Chen, B., Sun, Y., Luo, F. and Wang C. (2020). Bioactive metabolites and potential mycotoxins produced by *Cordyceps* fungi: A review of safety. Toxins, 12(6): 410. 10.3390/toxins12060410.

Chen, S., Li, Z., Krochmal, R., Abrazado, M., Kim, W. et al. (2010). Effect of Cs-4 (*Cordyceps sinensis*) on exercise performance in healthy older subjects: a double-blind, placebo-controlled trial. J. Altern. Compl. Med., 16(5): 585–590.

Chen, S.Y., Ho, K.J., Hsieh, Y.J., Wang, L.T. and Mau, J.L et al. (2012). Contents of lovastatin, γ-aminobutyric acid and ergothioneine in mushroom fruiting bodies and mycelia. LWT, 47: 274–278. 10.1016/j.lwt.2012.01.019.

Cohen, N., Cohen, J., Asatiani, M.D., Varshney, V.K., Yu, H.-T. et al. (2014). Chemical Composition and Nutritional and Medicinal Value of Fruit Bodies and Submerged Cultured Mycelia of Culinary-Medicinal Higher Basidiomycetes Mushrooms. Int. J. Med. Mushrooms, 16: 273–291.

Cui, J.D. (2015). Biotechnological production and applications of *Cordyceps militaris*, a valued traditional Chinese medicine. Crit. Rev. Biotechnol., 5: 475–484.

Das, S.K., Masuda, M., Sakurai, A. and Sakakibara, M. (2010). Medicinal uses of the mushroom *Cordyceps militaris*: Current state and prospects. Fitoterapia, 81(8): 961–968.

Davis, M., Sommer, R. and Menge, J. (2012). Field Guide to Mushrooms of Western North America. Berkeley: University of California Press, p. 472.

de Menezes, T.A., Aburjaile, F.F., Quintanilha–Peixoto, G., Tomé, L.M.R., Fonseca, P.L.C. et al. (2023). Unraveling the secrets of a double–life fungus by genomics: *Ophiocordyceps australis* ccmb661 displays molecular machinery for both parasitic and endophytic lifestyles. J. Fungi, 9(1): 110. 10.3390/jof9010110.

Deng, F., Zhong, S., Yu, C., Lin, C. and Cai, S. et al. (2021). Efficacy and safety of Jinshuibao capsule combined with beclomethasone propionate in the treatment of bronchial asthma. Minerva Surg., 77(3): 299–301. 10.23736/S2724-5691.21.09035-3.

Dong, C., Li, W.-J., Li, Z.-Z., Yan, W.-J., Li, R.-H. et al. (2016). *Cordyceps* industry in China: Current status, challenges and perspectives–Jinhu declaration for *Cordyceps* industry development. Mycosystema, 35: 1–15. 10.13346/j.mycosystema.150207.

Dong, J.Z., Ding, J., Yu, P.Z., Lei, C., Zheng, X.J. et al. (2013). Composition and distribution of the main active components in selenium-enriched fruit bodies of *Cordyceps militaris* link. Food Chem., 137(1–4): 164–167.

Dubhashi, S., Sinha, S., Dwivedi, S., Ghanekar, J., Kadam, S. et al. (2023). Early trends to show the efficacy of *Cordyceps militaris* in mild to moderate COVID inflammation. Cureus, 15(8): e43731. 10.7759/cureus.4373.

Elkhateeb W.A., Daba G.M., Thomas P.W. and Wen T.C. (2019). Medicinal mushrooms as a new source of natural therapeutic bioactive compounds. Egypt. Pharm. J., 18: 88–101. 10.4103/epj.epj_17_19.

European Commission 1. Novel Food Catalogue. https://ec.europa.eu/food/food-feed-portal/screen/novel-food-catalogue/search (accessed December 16, 2023).

European Commission 2. Register of Nutrition and Health Claims Made on Foods. https://food.ec.europa.eu/safety/labelling-and-nutrition/nutrition-and-health-claims_en (accessed December 16, 2023).

Gu, H.S. and Liang, M.Y. (1987). Study on the manual cultivation of *Cordyceps militaris*. Med. Inf. Lett., 5: 51–52.

Hassan M., Rouf R., Tiralongo E., May T. and Tiralongo J. et al. (2015). Mushroom lectins: Specificity, structure and bioactivity relevant to human disease. Int. J. Mol. Sci., 16: 7802–7838.

Heo, J.Y., Baik, H.W., Kim, H.J., Lee, J.M., Kim, H.W. et al. (2015). The efficacy and safety of *Cordyceps militaris* in Korean adults who have mild liver dysfunction. J. Clin. Nutr., 7(3): 81–86. 10.15747/jcn.2015.7.3.81

Hepsomali, P., Groeger, J.A., Nishihira J. and Scholey, A. (2020). Effects of oral gamma-aminobutyric acid (GABA) administration on stress and sleep in humans: A systematic review. Front. Neurosci., 14: 923. 10.3389/fnins.2020.00923.

Hirsch, K.R., Smith-Ryan, A.E., Roelofs, E.J., Trexler, E.T. and Mock, M.G. et al. (2017). *Cordyceps militaris* improves tolerance to high-intensity exercise after acute and chronic supplementation. J. Diet. Suppl., 14(1): 42–53. 10.1080/19390211.2016.1203386.

Hsieh, S.-A., Lin, T.-H., Wang, J.-S., Chen, J.-J. Hsu, W.-K. et al. (2022). The effects of *Cordyceps militaris* fruiting bodies in micturition and prostate size in benign prostatic hyperplasia patients: A pilot study. Pharmacol. Res. Mod. Chinese Med., 4: 100143. 10.1016/j.prmcm.2022.100143

Huang, S., Liu, H., Sun, Y., Chen, J., Li, X. et al. (2018). An effective and convenient synthesis of cordycepin from adenosine. Chem. Pap., 72: 149–160.

Huang, S.J., Lin, C.P., Mau, J.L., Li, Y.S. and Tsai, S.Y. et al. (2015). Effect of UV-B irradiation on physiologically active substance content and antioxidant properties of the medicinal caterpillar fungus *Cordyceps militaris* (Ascomycetes) Int. J. Med. Mushrooms, 17: 241–253.

Hur, H. (2008). Chemical Ingredients of *Cordyceps militaris*. Mycobiology, 36: 233–235.

Jędrejko, K.J., Lazur, J. and Muszyńska, B. (2021). *Cordyceps militaris*: an overview of its chemical constituents in relation to biological activity. Foods, 10(11): 2634. 10.3390/foods10112634

Jin, Y., Meng, X., Qiu, Z., Su, Y., Yu, P. et al. (2018). Anti-tumor and anti-metastatic roles of cordycepin, one bioactive compound of *Cordyceps militaris*. Saudi J. Biol. Sci., 25: 991–995.

Joshi, M., Sagar, A., Kanwar, S.S. and Singh, S. (2019). Anticancer, Antibacterial and Antioxidant Activities of *Cordyceps militaris*. Indian J. Exp. Biol., 57: 15–20.

Jung, S.J., Hwang, J.-H., Oh, M.-R. and Chae, S.-W. (2019). Effects of *Cordyceps militaris* supplementation on the immune response and upper respiratory infection in healthy adults: A randomized, double-blind, placebo-controlled study. J. Nutr. Health, 52(3): 258–267.

Jung, S.J., Jung, E.-S., Choi, E.-K., Sin, H.-S., Ha, K.-C. et al. (2019). Immunomodulatory effects of a mycelium extract of *Cordyceps* (*Paecilomyces hepiali*; CBG-CS-2): a randomized and double-blind clinical trial. BMC Cpmpl. Alter. Med., 19 (1): 77. 10.1186/s12906-019-2483-y.

Kang, H.J., Baik, H.W., Kim, S.J., Lee, S.G., Ahn, H.Y. et al. (2015). *Cordyceps militaris* enhances cell-mediated immunity in healthy Korean men. J. Med. Food. 18(10): 1164–1172.

Kreipke, V.C., Mofft, R.J, Ma, C.J.T. andOrmsbee, M.J. (2021). Effects of Concurrent Training and a Multi-Ingredient Performance Supplement Containing Rhodiola rosea and *Cordyceps sinensis* on Body Composition, Performance, and Health in Active Men." J. Diet. Suppl., 18 (6): 597–613.

Kuo, H.C., Huang, I.C. and Chen, T.Y. (2015). *Cordyceps* s.l. (Ascomycetes) species used as medicinal mushrooms are closely related with higher ability to produce cordycepin. Int. J. Med. Mushrooms, 17(11): 1077–1085.

Ler, H.Y., Tan, J.Y. and Chan, K.Q. (2022). Effect of *Cordyceps militaris*, arginine and citrulline supplementation on long distance runners in hot conditions. Proceedings of the 8th International Conference on Movement, Health and Exercise. Singapore, pp. 407–416.

Li, J., Zhang, Z., Liu, X., Wang, Y., Mao, F. et al. (2015). Study of GABA in Healthy Volunteers: Pharmacokinetics and Pharmacodynamics. Front. Pharmacol., 6: 260. 10.3389/fphar.2015.00260.

Li, Y. and Gaosi, X. (2020). Clinical efficacy and safety of Jinshuibao combined with ACEI/ARB in the treatment of diabetic kidney disease: a meta-analysis of randomized controlled trials. J. Ren. Nutr., 30(2): 92–100. 10.1053/j.jrn.2019.03.083.

Liu, Y., Wang, J., Wang, W., Zhang, H., Zhang, X. et al. (2015). The Chemical Constituents and Pharmacological Actions of *Cordyceps sinensis*. Evid. Based Compl. Alt. Med., 575063. 10.1155/2015/575063.

Meena, M., Prasad, V., Zehra, A., Gupta, V.K. and Upadhyay, R.S. et al. (2015). Mannitol metabolism during pathogenic fungal-host interactions under stressed conditions. Front. Microbiol., 6: 1019. 10.3389/fmicb.2015.01019.

Mehra, A., Zaidi, K.U., Mani, A. and Thawani, V. (2017). The health benefits of *Cordyceps militaris*—A review. KAVAKA, 48: 27–32.

Muszyńska, B., Sułkowska-Ziaja, K. and Ekiert, H. (2013). Phenolic Acids in Selected Edible Basidiomycota Species: *Armillaria mellea, Boletus badius, Boletus edulis, Cantharellus cibarius, Lactarius deliciosus and Pleurotus ostreatus*. Acta Sci. Pol. Hortorum Cultus., 12: 107–116.

Ng, T.B. and Wang, H.X. (2010). Pharmacological actions of *Cordyceps*, a prized folk medicine. J. Pharm. Pharmacol., 57(12): 1509–1519. 10.1211/jpp.57.12.0001.

Olatunji, O.J., Tang, J., Tola, A., Auberon, F., Oluwaniyi, O. et al. and Ouyang, Z. (2018). The genus *Cordyceps*: An extensive review of its traditional uses, phytochemistry and pharmacology. Fitoterapia, 129: 293–316.

Renzi-Hammond, L., Bovier, E., Fletcher, L., Miller, L., Mewborn, C. et al. (2017). Effects of a lutein and zeaxanthin intervention on cognitive function: A randomized, double-masked, placebo-controlled trial of younger healthy adults. Nutrients, 9: 1246. 10.3390/nu9111246.

Savioli, F.P., Zogaib, P., Franco, E., de Salles F.C.A., Giorelli, G.V. et al. (2022). Effects of *Cordyceps sinensis* supplementation during 12 weeks in amateur marathoners: A randomized, double-blind placebo-controlled trial. J. Herb. Med., 34: 100570. 10.1016/j.hermed.2022.100570.

Schwenzer, H.P., Symeonides, S.N., Plummer, E.R., Bond, G.P. and Patricia. S. et al. (2019). A first-in-human study of, NUC-7738, a 3'dA phosphoramidate, in patients with advanced solid tumors or lymphoma (NuTide 701). Develp. Ther. Tumor Biol., 37(15): 10.1200/JCO.2019.37.15_suppl.TPS314.

Shashidhar, M.G., Giridhar, P., Sankar K.U. and Manohar, B. (2013). Bioactive principles from *Cordyceps sinensis*: A potent food supplement—A review. J. Funct. Foods, 5(3): 1013–1030. 10.1016/j.jff.2013.04.018

Shawkat, H., Westwood, M.-M. and Mortimer, A. (2012). Mannitol: A review of Its clinical uses. *Continuing Education in Anaesthesia Critical Care & Pain*, 12(2): 82–85. 10.1093/bjaceaccp/mkr063.

Shrestha, B., Han, S.K., Lee, W.H., Choi, S.K., Lee, J.O. et al. (2005b). Distribution and in vitro Fruiting of *Cordyceps militaris* in Korea. Mycobiology, 33(4): 178–181.

Shrestha, B., Han, S.K., Yoon, K.S. and Sung, J.M. (2005a). Morphological Characteristics of Conidiogenesis in *Cordyceps militaris*. Mycobiology, 33(2): 69–76.

Shweta, Abdullah, S., Komal and Kumar, A. (2023), A brief review on the medicinal uses of *Cordyceps militaris*. Pharmacol. Res.—Mod. Chinese Med., 7: 100228. 10.1016/j.prmcm.2023.100228

Singdevsachan, S.K., Auroshree, P., Mishra, J., Baliyarsingh, B., Tayung, K. et al, (2016) Mushroom polysaccharides as potential prebiotics with their antitumor and immunomodulating properties: A review. Bioact. Carbohydr. Diet. Fibre, 7: 1–14.

Singh, R.S., Bhari, R. and Kaur, H.P. (2010). Mushroom lectins: Current status and future perspectives. Crit. Rev. Biotechnol., 30: 99–126.

Soltani, M., Malek, R.A., Elmarzugi, N.A., Mahoodally, M.F., Uy, D. et al. (2018). Cordycepin: A biotherapeutic molecule from medicinal mushroom. pp. 319–349. *In*: Singh, B.P., Lallawmsanga and Passari, A.K. (eds.). Biology of Macrofungi, Fungal Biology. Springer Nature, Basel, Switzerland.

Stringham, N.T., Holmes, P.V. and Stringham, J.M. (2018). Supplementation with macular carotenoids reduces psychological stress, serum cortisol, and sub-optimal symptoms of physical and emotional health in young adults. Nutr. Neurosci., 21: 286–296.

Symeonides, S., Aroldi, F., Harris, N.Md., Kestenbaumum, S., Plumer, R. et al. (2020). 600TiP—A first-in-human study of, NUC-7738, a 3'-dA phosphoramidate, in patients with advanced solid tumours (NuTide: 701). Ann. Oncol., 31(4): S462–S504. S501. 10.1016/annonc/annonc271

Thongsawang, S., Krataithong, T., ChorCharoenying, S., Norchai, P. and Nokkaew, N. et al. (2021). Applying *Cordyceps sinensis* to Boost Endurance Performance in Long-Distance Runners. J. Exer. Physiol. Online 24(3): 1–13.

Tuli, H.S., Sandhu S.S. and Sharma, A.K. (2014). Pharmacological and therapeutic potential of *Cordyceps* with special reference to Cordycepin. 3 Biotech., 4: 1–12.

Wang, L., Xu, N., Zhang, J., Zhao, H., Lin, L. et al. (2015). Antihyperlipidemic and Hepatoprotective Activities of Residue Polysaccharide from *Cordyceps militaris* SU-12. Carbohydr. Polym., 131: 355–362.

Wong, Y.Y., Moon, A., Duffin, R., Barthet-Barateig, A., Meijer, H.A. et al. (2010). Cordycepin inhibits protein synthesis and cell adhesion through effects on signal transduction. J. Biol. Chem., 285: 2610–2621.

Yamatsu, A., Yamashita, Y., Pandharipande, T., Maru, I. and Kim, M. et al. (2016). Effect of oral γ-aminobutyric acid (GABA) administration on sleep and its absorption in humans. Food Sci. Biotechnol., 25: 547–551.

Yang, F.Q., Feng, K., Zhao, J. and Li, S.P. (2009). Analysis of sterols and fatty acids in natural and cultured *Cordyceps* by one-step derivatization followed with gas chromatography-mass spectrometry. J. Pharm. Biomed. Anal., 49: 1172–1178.

Yang, S., Jin, L., Ren, X., Lu, J. and Meng, Q. (2014). Optimization of fermentation process of *Cordyceps militaris* and antitumor activities of polysaccharides *in vitro*. J. Food Drug Anal., 22: 468–476.

Yang, Z.-X., Chen, Y., Xue, J.-G., Huang, X.F. and Chang, F.-J. (2021). *Cordyceps sinensis* combined with Jujing Pills for male infertility. Zhonghua Nan Ke Xue (National Journal of Andrology) 27(1): 50–55.

Yu, X., Yan, D., Lan, Q., Fang, J., Ding, Z. et al. (2022). Efficacy and Safety of Jinshuibao Capsule in Diabetic Nephropathy: A Systematic Review and Meta-Analysis of Randomized Controlled Trials. Comput. Math. Methods Med., 9671768. 10.1155/2022/9671768.

Zeng, W.-B., Yu, H., Gem F., Yangm J.-Y., Chen, Z.H. et al. (2014). Distribution of nucleosides in populations of *Cordyceps cicadae*. Molecules, 19(5): 6123–6141. 10.3390/molecules19056123.

Zhang, J., Wen, C., Duan, Y., Zhang, H. and Ma, H. et al. (2019). Advance in *Cordyceps militaris* (Linn) Link polysaccharides: Isolation, structure, and bioactivities: A review. Int. J. Biol. Macromol., 132: 906–914. 10.1016/j.ijbiomac.2019.04.020.

Zhang, C.-Q., Yin, J.-Q., Xin, Q., Wangt, Y.Q. and Ge, Z.-M. et al. (2013). Jinshuibao capsule combined losartan potassium intervened early renal damage of hypertension patients of yin and yang deficiency: A clinical research. Zhongguo Zhong Xi Yi Jie Za Zhi (Chinese Journal of Integrated Traditional and Western Medicine) 33 (6): 731–735.

Zhang, Y., Dong, D., Youqin, Y., Zhang, H., Guangli, W. et al. (2023a). Effectiveness and safety of Jinshuibao capsules in treatment of residual cardiopulmonary symptoms in convalescent patients of coronavirus disease 2019: A pilot randomized, double-blind, placebo-controlled clinical trial. J. Trad. Chinese Med., 43(1): 134–139. 10.19852/j.cnki. jtcm.2023.01.012.

Zhang, H., Yuan, C., Sun, C. and Zhang, Q. (2023b). Efficacy of Jinshuibao as an adjuvant treatment for chronic renal failure in China: A meta-analysis. Medicine, 102(32): e34575. 10.1097/MD.0000000000034575.

Zhu, F., Du, B., Bian, Z. and Xu, B. (2015). Beta-glucans from edible and medicinal mushrooms: characteristics, physicochemical and biological activities. J. Food Compos. Anal., 41: 165–173.s

11

Cordycepin—From Diet to Medicine

Shallu Saini,[1] *Sonal Datta,*[1] *Hardeep Singh Tuli,*[1,*]
Chanderdeep Tandon,[2] *Simran Tandon*[3] and *Anil Kumar Sharma*[2,*]

1. Introduction

Although there are several types or modes of therapy, natural bioactive components are still preferred because they do not have any significant adverse effects. According to research, nearly 50% of prescription medications supplied in the US come from either naturally occurring or structurally altered or synthesized natural components. As a result, studying the chemistry of natural compounds is becoming increasingly paramount . Studies investigating the phytochemistry and pharmacological effects of medicinally important herbs, and the isolation and characterization of bioactive components, have been an important area of research. Nowadays, researchers can isolate, characterize and produce bioactive compounds from the medicinal mushroom, *Cordyceps*. It is currently a topic of intensive research to pinpoint the specific ingredients responsible for pharmacological activity in *Cordyceps militaris*, which has been utilized as crude medication for the well-being of humanity since the dawn of civilization. Since ancient times, it is well known that medicinal mushrooms produce biometabolites that are utilized as potential drugs to treat several ailments. Our diet might be changed to include more mushrooms since they contain antioxidants, which could prevent or reduce more than two-thirds of cancer-related

[1] Department of Bio-Sciences and Technology, Maharishi Markandeshwar (Deemed to be University), Mullana (Ambala), Haryana 133207, India.
[2] Department of Biotechnology, [3]Department of Health Sciences, Amity University, Sector 82 A, IT City Road, Block D, Sahibzada Ajit Singh Nagar, Mohali, Punjab 140306, India.
* Corresponding authors: hardeep.biotech@gmail.com; anibiotech18@gmail.com

fatalities. The exceptional pharmacological activity of *C. militaris*, one of the medicinally important mushrooms, necessitates further research before it can be used to treat humans. In some regions of Asia, the genus *Cordyceps* has been used medicinally for thousands of years (Gu et al., 2007). The Latin words 'cord' and 'ceps', which imply club and head, respectively, are the source of the name *Cordyceps*. Due to its internal spore production in an ascus-like sac, *C militaris* is a member of the Phylum Ascomycota, which is classified under the Order Hypocreales. It is an annual entomopathogenic fungus that frequently parasitizes lepidopteran insects and spiders. The biologically active substances in *Cordyceps*, particularly its extract, include cordycepin, exo-polysaccharides, and vitamins. The primary active ingredient that has received the most attention for its potential medical benefits is cordycepin, an adenosine analogue. This chapter highlights updated information about cordycepin and its emerging therapeutic potential with future prospects.

2. Chemistry and Extraction of Cordycepin

Cordycepin has a structure that is remarkably like cellular nucleoside adenosine and functions similarly to a nucleoside analogue. A purine (adenine) nucleoside molecule called cordycepin is joined to a ribofuranose molecule through a glycosidic bond. The crucial chemical step in the production of cordycepin involves substituting a hydrogen (H) for the hydroxyl (OH) group at position 3′ in the ribofuranosyl moiety, leading to the formation of a deoxy analogue of adenosine.

Studies have demonstrated that cordycepin is among the most significant and reliable naturally occurring components, possessing several therapeutic properties (Yue et al., 2013; Baoyan and Haibo, 2012; Patel and Ingalhalli, 2013). Due to its several biological and therapeutic effects, a significant amount of purified material is urgently required for additional research. Various extraction techniques have been designed to extract cordycepin from *C. militaris* fruit bodies, such as pressure extraction, ultrasound or microwave-assisted extraction, and reflux and soxhlet extraction. The following is a discussion of a few of them. Investigations by Li et al. (2012) were focused on the isolation of cordycepin from the cultivated mycelium of *C. militaris.* The extracted material was further purified by using cation exchange resin (CER) from LSD001. They discovered optimum desorption conditions: 0.2M NH_3 in 80% ethanol (v/v), 2 hours of desorption, 25°C, and a pH of 14. In another study, Zeng and Chen (2013) looked into the ideal conditions for cordycepin extraction from *C. militaris* waste media produced by column chromatography. *Cordyceps* wastewater was maintained to leach at 70°C for eight hours with a dry feed-to-water ratio of one gram to 20 ml. The cordycepin was further separated using polyamide column chromatography and macroporus resin XAD16. Similarly, Song et al. (2007) examined how to enhance the extraction of cordycepin from cultured *Cordyceps* by utilizing high-performance liquid chromatography with a photodiode-array detection (HPLC-DAD) coupled approach. The author reported the extraction of cordycepin with an ethanol concentration of 20.21% and a volume ratio of solvent to extracted sample at 33.13 mL g−1.

3. Current Advances in *Cordyceps* Research

As discussed above, cordycepin, the bioactive compound of *Cordyceps*, is a nucleoside analogue of adenosine, differing by only a single hydroxyl group, and has been shown to induce apoptosis, reduce inflammation, inhibit RNA transcription in cell cultures, and have anti-metastatic potential. For these reasons, it is extensively studied for its anti-cancer and anti-metastatic properties. There are several emerging applications of cordycepin, such as antioxidant potential, gut microbiome modulation, cosmeceutical applications, and neuroprotective effects.

Upatcha et al. demonstrated the antioxidant activity of cordycepin using human dermal fibroblasts (HDFs). The authors used medium-loaded nanoparticles (CMP) of *Cordyceps*, which could increase cell proliferation and reduce H_2O_2-induced ROS. It was noteworthy that this CMP inhibited H_2O_2-induced oxidative stress and induced autophagy to revive HDFs. Thus, nano-encapsulated *Cordyceps* extract enhances collagen synthesis and regeneration of skin through antioxidation as well as autophagy (Upatcha et al., 2023). In another exciting study, several bioactive compounds were identified in the ethanol extract of *Cordyceps*. Liquid chromatography-mass spectrometry was employed to analyze and identify 21 bioactive compounds in the ethanol extract (1−21) of *C. militaris*. Through a multi-platform computational approach, inhibitors against Alzheimer-related protein, acetylcholinesterase (PDB-4EY7), and β-amyloid protein (PDB-2LMN) were predicted. This study demonstrated the potential lead inhibitors against Alzheimer's from *Cordyceps* extract (Thai et al., 2023).

Cordycepin ameliorates type 2 diabetes mellitus (T2DM) by modulating the gut microbiota and metabolites. This study proves that *cordycpes* can be utilized as a food complement to relieve T2DM, providing an auspicious view for new functional foods (Liu et al., 2023). Additionally, studies on the BT549 and 4T1 cancer cell lines have indicated that cordycepin inhibits their proliferation, migration, and invasion. Western blotting was employed to assess the protein levels of N-Cadherin, E-Cadherin, TWIST1, SLUG, SNAIL1, and ZEB1 in samples that had been exposed to cordycepin. It's interesting to note that by suppressing the production of TWIST1 and SLUG, cordycepin can deactivate the EMT signalling pathway. The findings of this study suggest that cordycepin, by specifically targeting transcription factors that promote EMT, would be a promising therapeutic molecule for the treatment of triple-negative breast cancer (Wei et al., 2022).

4. Therapeutic Applications

4.1 Anti-Angiogenic Potential

The ability of *Cordyceps* to inhibit the formation of new blood vessels from existing ones is known as angiogenesis. Angiogenesis plays a role in the growth and multiplication of physiologically normal and malignant tissues, including placental development, embryogenesis, inflammation-associated diseases, and tumor growth (Folkman, 1995; Risau, 1997). Angiogenic proteins, including VEGF, FGF, and EGF, can be released by activated endothelial cells to cause angiogenesis in tumor cells (Jain, 2002; Kaya et al., 2000; Weidner et al. 1991). According to Ferrara et al. (2003) and Breier and Risau (1996), these variables are strongly expressed and linked to the development of several forms of different human malignancies. In order to cure and prevent cancer, substances having inhibitory effects on angiogenesis have been recognized as a promising approach. The efforts of the research fraternity have been to identify natural novel formulations with angiogenesis-inhibiting properties. The angiogenic potential of intact cells and purified components has been examined using chick embryo models. A variety of animal model systems have been created to understand the principles underlying angiogenesis. The specialized, highly vascularized chorioallantoic membrane (CAM) found in the avian embryo is a perfect indicator of whether a substance is anti- or pro-angiogenic. The chorion, allantois, and inner shell membrane tend to combine to produce the chorio-allantoic membrane, which participates in metabolic functions like respiration, digestion, and excretion. The pharmacological functions of *Cordyceps* were further enhanced by Won and Park in 2005. They made ethanolic extracts of fruit bodies (FBE) and cultured mycelia (CME) and reported the dose-dependent inhibitory effect through chorioallantoic membrane assays (CAM).

4.2 Anti-Cancer Potential

It has been reported that toxic nucleoside analogues were the first series of anti-cancer drugs for the treatment of malignancy. Cordycepin has been known to have therapeutic properties and characteristics that could exhibit anti-cancer potential. Cordycepin is structurally similar to adenosine, but it lacks a 3' hydroxyl group, which is known to increase its potency slightly. Cordycepin is well known to interact with lots of cellular machinery, including purine synthesis (Overgaard, 1964; Rottman and Guarino, 1964),

DNA replication (Holbein et al., 2009) and mTOR signalling pathways (Wong et al., 2010). Furthermore, the functions of apoptosis, cell cycle arrest, anti-metastatic activity, suppression of platelet aggregation, and pathways mediated by inflammatory mediators are other alternative molecular targets of cordycepin. Table 1 represents the mechanistic insight into the anticancer properties of cordycepin in various preclinical cancer studies.

4.3 Anti-Diabetic Effect

One of the primary disorders that has recently drawn attention is diabetes mellitus (DM). One of the main bioactivse components of *C. militaris*, cordycepin (molecular formula: C10H13N5O3), has been shown to lower blood sugar levels (Dong et al., 2014). The effects and mechanisms of cordycepin were examined in both healthy and oxidatively damaged INS-1 cells, as established by previous cell and molecular biology procedures. The results suggested that cordycepin might boost insulin production and secretion. The increased Ca2+ concentration and membrane depolarization brought on by the raised ATP level may be related to the process. The genetic expression of the genes that code for insulin, pancreatic duodenal homeobox factor 1, and glucose transporter 1 (GLUT1) was elevated by cordycepin. One of the essential bioactive components of *Cordyceps militaris*, cordycepin, was seen to lower blood sugar levels. INS-1 cells that were both healthy and oxidatively damaged were investigated in order to see the effect of cordycepin, which further paves the way to delineate the mechanism as far as the action of cordycepin is concerned. The results suggested that cordycepin might boost insulin production and secretion. The increased Ca2+ concentration and membrane depolarization brought on by the raised ATP level may be related to the process. Cordycepin is responsible for the increment of mRNA levels for the genes encoding glucose transporter 1 (GLUT1), insulin, and pancreatic duodenal homeobox factor 1 at the genetic level (Sun et al., 2022). Even though the precise mechanism involved in cordycepin's anti-diabetic effect is still unknown, a few studies have recognized a potential pathway. They found that cordycepin prevents the protein expression of inflammation-associated mediators, which stops NO and inflammation-associated cytokines, including TNF-, IL-1, and IL-6 from being produced in LPS-induced macrophages. Thus, the expression of the genes that regulate type 2 diabetes (PPAR and 11-HSD1) was reduced. As the cordycepin concentration increases, the expression of co-stimulatory molecules such as ICAM-1 and B7-1/-2 likewise dropped (Shin et al., 2009).

4.4 Anti-Inflammation Activity

It is well known that inflammation contributes to the development of cancer and controls its growth, as well as treatment therapy. According to previous

Table 1. Overview of cordycepin action in preclinical cancer studies.

Type of cancer	Cell line	Physiological effect	Mechanism of action	Reference
Lymphoma	U937 cells	Apoptosis induction	↑ Bax and ↓ Bcl-2, Bid, ↑ caspase 3 and 8, ↓ MMP, ↑ROS, ↑ G2/M, ↓cyclin dependent kinase 1, ↓ cyclin B1 phosphorylated ↑ p38, ↑ JNK and ↓ ERK phosphorylation	Shi et al., 2022
Oral squamous cancer	human OEC-M1 oral cancer cell line	Apoptotic induction	↑phosphatidylserine flipping and G2M	Wu et al., 2007
Breast Cancer	MCF-7 cells	Apoptotic induction	↑ caspases 7, 8, 9, ↑ Bax/Bcl-2 ratio, ↓ XIAP	Lee et al., 2019
Bladder cancer	T24 cells	Apoptotic induction	↑ caspase -3, -8 and -9, ↑ Bax/Bcl-2, ↓ PI3K/Akt	Kim et al., 2019
Renal cell carcinoma	Caki-1 cells	Apoptotic induction	↓ microRNA-21 and Akt, ↑ PTEN	Yang et al., 2017
Tongue cancer	CAL-27	Apoptotic induction	↑ Bax, caspase-3, caspase-9, caspase-12, and ↓ Bcl-2, ↑ ROS	Zheng et al., 2020
Cervical Cancer	SiHa and HeLa cervical cancer cell	Apoptotic induction	↓ CDK-2, CYCLIN-A2 and CYCLIN-E1, ↑ ROS	Tania et al., 2019
Liver cancer	HepG2 cells	Apoptotic induction	↑ FADP, ↓ BCL-2, MMP changes and ↓ cytochrome c	Shao et al., 2016
Neuroblastoma	SK-N-SH and BE(2)-M17 cells	Apoptotic induction	↑ caspase-3, ↑PARPs, ↑ LC3	Li et al., 2015
Oesophageal cancer	ECA109 and TE-1	Apoptotic induction	↑caspase -3, -9, ↑ PARPs, ↑ Bax, ↓ Bcl-2, Bax/Bcl-2, ↑ G2/M	Xu et al., 2019
Breast cancer	4T1 breast cancer cells	metastasis Inhibition	↓ CCL17, ↓ MMP-9, ↓ OPN, ↓ IL-33	Cai et al., 2018
Melanoma	B16-F0 mouse melanoma cells	metastasis Inhibition	↓ HGF	Kubo et al., 2010
Lung cancer	NCI-H1299 human lung cancer cell line, LLC cells mouse lung cancer cells	metastasis Inhibition	↓ 3β (GSK-3β), ↓ Akt, ↓ MMP-2 and MMP-9	Zhou et al., 2018
Melanoma	MeWo (HTB-65)	Angiogenesis inhibition and suppression of tumor growth	↑ GSK-3β, ↑ p38α, ↓ Akt1, ↓ VEGF	Ruma et al., 2014
Hepatocellular carcinoma	SMMC-7721 cells, HUVECs	Angiogenesis inhibition	↓ VEGF/VEGFR2, ↓ Akt, and ↓ ERK, CD31	Li et al., 2020

research, a variety of cytokines, chemokines, and their receptors are released by cancer cells, and these chemicals are crucial in initiating inflammatory responses. (Arias et al., 2007; Nelson et al., 2004, and Farrow et al., 2004). According to Rayburn et al. (2009), inflammatory diseases can also be treated with a number of anti-cancer medicines. Previously, it was found that the expression of cytokines such as Interleukin-6,8 and Macrophage Inflammatory Proteins-1 is likely to be enhanced in cancer (Chavey et al., 2007; Wang et al., 2009). Consequently, limiting inflammatory mediators is crucial to creating a strong strategy to combat and prevent cancer.

Various cytokines, chemokines, and their receptors are known to be expressed by cells, and these molecules play (Arias et al., 2007) an essential function in mediating inflammatory reactions (Farrow et al., 2004). It has been estimated that cancer-preventive agents may also be utilized to treat inflammation-associated aspects (Rayburn et al., 2009). Consequently, inflammatory mediators must be blocked to create a powerful strategy to combat and prevent cancer. Rao et al. (2010) have shown that cordycepin is a strong suppressor of free radical NO and cytokines (TNF- and IL-12). The research suggests that cordycepin's anti-inflammatory properties may aid in the prevention of cancer. Kim and colleagues' study in a mouse model suggests that cordycepin inhibits acute lung inflammatory reactions and the levels of related genes, such as adhesion molecules (ICAM and VCAM), cytokines and chemokines, and the chemokine receptor CXCR2. In a study, researchers investigated the anti-inflammatory potential of cordycepin using a mouse model. The detailed results of the study revealed the downregulation of pro-inflammatory cytokines and chemokines in the cellular microenvironment of the mouse model.

4.5 *Anti-Metastatic Activity*

A significant portion of the associated fatalities and the dismal prognosis for patients with all forms of cancer are caused by metastasis. The process of tumour metastasis is extremely complicated, and current metastasis models show that metastasis occurs late in the course of carcinogenesis (Rosell and Karachaliou, 2015). Matrix metalloproteinases (MMPs) are proteolytic enzymes that are thought to promote numerous alterations in the microenvironment during tumour growth, according to mounting evidence (Fig. 1). The type IV collagenases MMP-2 and MMP-9 are among the MMP family members that are principally responsible for the breakdown of the extracellular matrix (Kessenbrock et al., 2010). Another investigation found that the *C. militaris* fraction (CMF) may encourage the breakdown of β-catenin by glycogen synthase kinase 3β (GSK-3β), which in turn inhibits Akt's activation and reduces the expression of matrix metalloproteinases (MMP)-2 and MMP-9. These findings suggested that

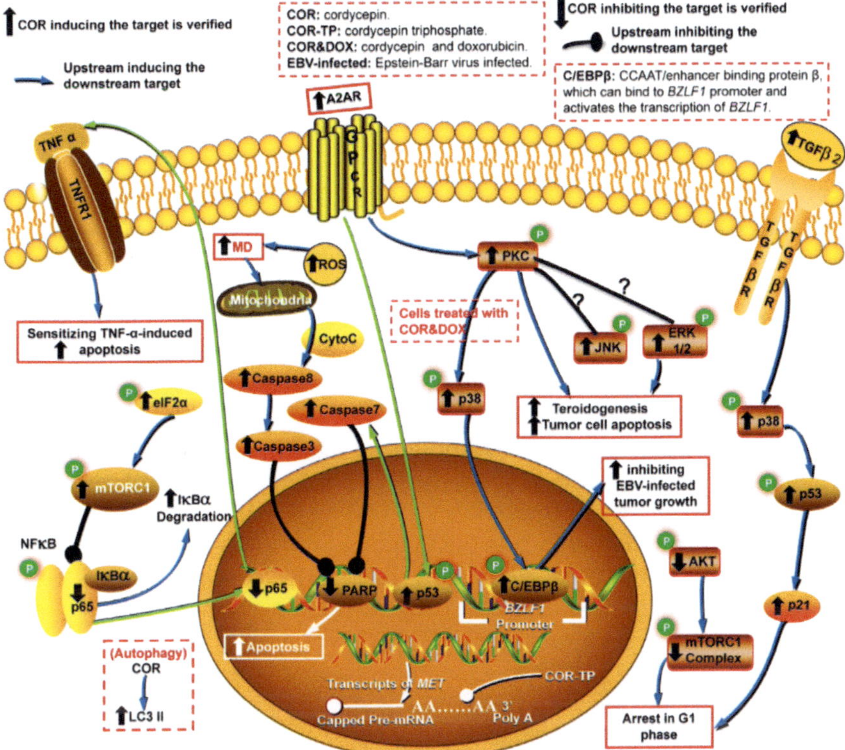

Fig. 1. Anti-cancer molecular targets of Cordycepin (Qin et al., 2019).

CMF might have applications in the Akt/GSK-3β/β-catenin pathway-mediated therapy of lung cancer metastasis (Zhou et al., 2018).

4.6 Anti-Platelet Aggregation

It is well recognized that platelets control the development, growth, and metastasis of tumors. According to earlier research (Tsuruo and Fujita, 2008; Goubran and Bournouf, 2012), cancer cells can activate platelets, which help them survive in the bloodstream during hematogenous migration by preventing the lysis of cancer cells via NK (Natural Killer Cells) and CTLs (Cytotoxic T-Lymphocytes). Tumour cell-induced platelet aggregation (TCIPA) has been defined through numerous molecular mechanisms, including GPIIb/IIIa, GP Ib-IX-V, P2Y receptors, and PAR receptors (Jackson, 2007; Bambace and Holmes, 2011). Cordycepin, according to Cho et al. (2006, 2007a), prevents platelet aggregation in the collagen-induced process by reducing [Ca2+] ions and thromboxane A2 [TXA2] and by increasing cellular levels of cAMP and cGMP. Also, the investigation exhibited that [Ca2+] ions and TXA2 up-regulation were each suppressed by 500-M cordycepin by 74% and 46%, respectively. Still, most of the molecular pathways that initiate and mediate platelet aggregation and secretions produced by tumor cells need to be uncovered.

4.7 Cardio-Protective Effect

Cordycepin has a cardioprotective effect on the inflammatory response that harms the healthy tissue surrounding the infarct. To create the myocardial I/R model, the left anterior descending artery of mice was momentarily ligated. The infarct size and apoptosis index were measured using the TTC/ Evans Blue staining and TUNEL assay. Echocardiography was used to assess the heart's health. Hypoxia and reoxygenation (H/R) were performed on newborn rat ventricular cardiomyocytes (NRVCs). Cell viability and necrosis were evaluated by MTS and LDH, respectively. According to the study's findings, cordycepin is known to significantly reduce apoptosis, decrease infarct size, and improve heart function in mice undergoing I/R injury. It also increased autophagy. Treatment with cordycepin clearly reduced the formation of ROS in NRVCs. Additionally, cordycepin partially stimulated autophagy by controlling the AMPK/mTOR pathway in the context of H/R damage (Xu et al., 2022).

4.8 Cell Cycle Regulation

Cordycepin is known to restrict the cell cycle at specific stages in addition to being involved in apoptotic pathways (Figure 1). Cyclin-dependent kinases (CDKs) are positively and negatively controlled by cyclins, and cyclin-dependent kinase inhibitors (CDKIs) are responsible for controlling the cell cycle. Studies on a human leukaemia cell have shown cordycepin decreased the expression of the proteins cyclin B1 and cyclin dependent kinase 1 (CDK1), causing arrest in the G2/M phase. (Shi et al., 2022). Furthermore, cordycepin causes an accumulation of cells in the S-phase due to inhibition of cyclin A2 and cyclin E expression. Additionally, cordycepin was reported to result in DNA damage along with Cdc25A degradation, pointing towards the Chk2-Cdc25A pathway, which was stimulated in cordycepin-induced S-phase arrest in NB-4 and leukaemia cells (Liao et al., 2015).

4.9 Induction of Apoptosis

The highly conserved programmed cell death known as apoptosis is driven in part by the caspase family of cysteine proteases, which cleave a particular subset of downstream substrates to produce the final apoptotic phenotype. Although a variety of upstream mediators activate a small number of upstream molecular pathways to start apoptosis, they mostly focus on a similar set of downstream pathways by way of effector caspase activation. Programmed cell death is another name for apoptosis. In previous studies, regulation of the expression of different proteins by cordycepin has been shown to play a functional role in triggering apoptosis (Figure 1). The Bcl-2 protein family, which includes both anti- and pro-apoptotic members, is one example of a protein involved in this planned cell death process. According to Kim et al. (2019), the activity of caspase-3 was not only boosted by cordycepin but also resulted in the cleavage of poly (ADP-ribose) polymerase. Caspases 8 and 9 are implicated in the start of the extrinsic and intrinsic apoptosis pathways, respectively. Additionally, cordycepin elevated the Bax/Bcl-2 ratio, truncated Bid, disrupted the mitochondria's integrity, and caused the release of cytochrome C into the cytoplasm. Additionally, cordycepin significantly inhibited the PI3K/Akt signalling pathway, whereas LY294002, a PI3K/Akt inhibitor, boosted cordycepin's ability to induce apoptosis. Similarly, cordycepin-facilitated cell death was reported in MCF-7 breast cancer cells as a result of lowered expression of the apoptosis X-linked inhibitor protein. This was followed by an increased frequency of caspase-7, -8, and -9 cleavage. Moreover, it resulted in an increased ratio of Bcl-2-associated X protein/B-cell lymphoma 2 (Bax/Bcl-2) protein expression. According to Yang et al. (2017), through the control of microRNA-21 and PTEN phosphatase, cordycepin promoted apoptotic cell death in renal cell cancer in the Caki-1 cell line.

4.10 *In vivo Anticancer Studies*

The first *in vivo* investigation to show cordycepin's anticancer characteristics was carried out in a mouse model, and it was established that cordycepin prolonged the survival of animals with Ehrlich mouse ascites tumors (Jeggar et al., 1961). In addition, from December 1997 to March 2001, a phase 1 clinical investigation utilizing cordycepin and pentostatin was conducted to cure patients with refractory acute or chronic myelogenous leukaemia. The most recent update for phase 2 clinical trials occurred in January 2009 (NCT00709215). It has been widely accepted that cordycepin from *Cordyceps* is known to possess anti-cancer properties. However, the molecular insights into cordycepin's anti-cancer properties in several cancers, including pancreatic cancer cell growth, are not completely understood. It has been established that cordycepin induces apoptosis by triggering cyto c, caspases 3, and 9. Additional research revealed that cordycepin hindered the MAPK pathway by preventing Ras expression and Erk phosphorylation. Additionally, cordycepin activated the Chk2 (checkpoint kinase 2) pathway, downregulated cyclin A2 and CDK2 phosphorylation, and resulted in DNA damage and S-phase arrest. Intriguingly, our group demonstrated that cordycepin could combine with FGFR2 with a very strong affinity

($K_D = 7.77 \times 10^{-9}$) and block the Ras/ErK pathway, inhibiting the development of pancreatic cancer cells (Xue-Ying et al., 2020).

4.11 Neuroprotective Potential

Traumatic brain injury (TBI) causes secondary injury, particularly white matter injury (WMI), which is extremely vulnerable to neuroinflammation and furthers unfavourable long-term consequences. Cordycepin is identified to play a vital role in various cell cycle phases; however, the cross-talk between immune cell invasion and the inflammatory cellular microenvironment is still needed to be explored vigorously (Kim et al., 2019). Although its long-term effects are unknown, an extract from the *Cordyceps militaris* plant has been shown to reduce TBI-induced neuroinflammation. Changing microglia/macrophage polarisation and preventing neutrophil infiltration after TBI are two ways that cordycepin has been shown to have long-term neuroprotective effects. These results offer great clinical potential for raising the quality of life in TBI patients (Wei et al., 2021).

5. Challenges in Applications of Cordycepin

Cordycepin has a wide range of pharmacological applications; therefore, a large quantity of a purified and well-characterized compound is required. However, the large-scale production of bioactive cordycepin is still a challenge. One of the challenges in producing biologically active cordycepin in large amounts is that cordycepin gets rapidly metabolized in the body into inactive form owing to its structural similarity to adenosine, thus limiting its potential as a drug for several indications. Adenosine deaminase (ADA) is a ubiquitous enzyme of the purine metabolism that catalyses the irretrievable deamination of adenosine and deoxyadenosine to inosine and deoxyinosine, respectively. The ADA catalyses the hydrolysis of adenosine to yield inosine as well as the hydrolysis of the derivatives of nucleoside adenosine, such as 2′-deoxyadenosine. Because of the high structural similarity of cordycepin to adenosine, it is also a substrate of ADA (Cristalli et al., 2001). Therefore, it is critical to design strategies to decrease deamination so that bioavailability and efficacy can be increased. To enhance solubility and efficacy, an adenosine deaminase inhibitor was administered alongside cordycepin. This approach aimed to inhibit the deamination of cordycepin and optimize its therapeutic effects (Li et al., 2015). In the current scenario, research investigations are in progress to explore medicinal aspects of cordycepin and its associated metabolites. In addition to these studies, we are also enhancing cordycepin production and the bioprocessing of its major derivatives so as to increase their bioavailability and efficacy. The nanoencapsulated *Cordyceps* extract has been shown to enhance collagen synthesis and regeneration of skin through antioxidation and autophagy (Upatcha et al., 2023). Similarly, researchers have improved the solubility and stability of cordycepin by encapsulating it with bovine serum albumin as β-cyclodextrin nanoparticles (Jaiswal et al., 2023). The cordycepin biosynthesis pathway in *Cordyceps* and its bioactivity are also well studied by Kunhorm et al. (2019). This group also proposed two bioprocesses: solid-state fermentation and liquid culture, for cosmeceutical applications and other uses (Kunhorm et al., 2019).

Novel modifications in the cordycepin structure and nanoformulations are essential not only to increase its clinical implications but also to strengthen its bioavailability. Heterologous expression of cordycepin in the *Escherichia coli* expression system and utilizing synthetic biology are other promising approaches to enhance cordycepin expression. These processes and modifications will undoubtedly help optimize the application of cordycepin and develop novel therapeutic strategies for several diseases.

Conclusion

The therapeutic potential of cordycepin to treat a variety of malignancies through several methods is well known. To be more precise, cordycepin can cause DNA damage, trigger apoptosis in cancer cells, and resist the cell cycle. These effects can either kill or restrict the proliferation of cancer cells. The immune system can be modulated, and autophagy can be induced by cordycepin. The metastasis of tumours is additionally prevented by cordycepin. There is a significant desire to identify and improve the bioavailability and structural activity of cordycepin using interdisciplinary biotechnological and chemical methods. There are several emerging applications of cordycepin; thus, a large quantity of biologically active compounds is required for preclinical and clinical studies. Therefore, improving the bioprocess methods for the production of cordycepin, developing efficient expression systems, and utilizing synthetic biology are promising approaches to enhance cordycepin expression and production. Furthermore, researchers are suggested to synthesize novel derivatives of cordycepin by adding active functional groups and preparing nanoformulations to enhance bioactivity as well as bioavailability.

References

Arias, J.I., Aller, M.A. and Arias, J. (2007). Cancer cell: using inflammation to invade the host. Molecular Cancer, 6: 1–10. 10.1186/1476-4598-6-29.

Bambace, N.M. and Holmes, C.E. (2011). The platelet contribution to cancer progression. J. Thromb. Haemost., 9: 237–249. 10.1111/j.1538-7836.2010.04131. x.

Baoyan, F. and Haibo, Z. (2012). Cordycepin: pharmacological properties and their relevant mechanisms. CellMed., 2: 14–1. 10.5667/tang.2012.0016.

Breier, G. and Risau, W. (1996). The role of vascular endothelial growth factor in blood vessel formation. Trends Cell Biol., 6: 454–456. 10.1016/0962-8924(96)84935-X.

Cai, H., Li, J., Gu, B., Xiao, Y., Chen, R. et al. (2018). Extracts of *Cordyceps sinensis* inhibit breast cancer cell metastasis via down-regulation of metastasis-related cytokines expression. J. Ethnopharmacol., 214: 106–112.10.1016/j.jep.2017.12.012.

Chavey, C., Bibeau, F., Gourgou-Bourgade, S., Burlinchon, S., Boissière, F. et al. (2007). Oestrogen receptor negative breast cancers exhibit high cytokine content. Breast Cancer Res., 9: 1–11. 10.1186/bcr1648.

Cho, H.J., Cho, J.Y., Rhee, M.H., Kim, H.S., Lee, H.S. et al. (2007). Inhibitory effects of cordycepin (3'-deoxyadenosine), a component of *Cordyceps militaris*, on human platelet aggregation induced by thapsigargin. J. Microbiol. Biotechnol., 17: 1134–1138.

Cho, H.J., Cho, J.Y., Rhee, M.H., Lim, C.R. and Park, H.J. et al. (2006). Cordycepin (3 '-deoxyadenosine) inhibits human platelet aggregation induced by U46619, a TXA2 analogue. J. Pharm. Pharmacol., 58: 1677–1682. 10.1211/jpp.58.12.0016.

Cristalli, G., Costanzi, S., Lambertucci, C., Lupidi, G., Vittori, S. et al. (2001). Adenosine deaminase: functional implications and different classes of inhibitors. Med. Res. Rev., 21: 105–128. 10.1002/1098-1128(200103)21:2<105::aid-med1002>3.0.co;2-u.

Dong, Y., Jing, T., Meng, Q., Liu, C., Hu, S. et al. (2014). Studies on the antidiabetic activities of *Cordyceps militaris* extract in diet-streptozotocin-induced diabetic Sprague-Dawley rats. Biomed Res. Int., 2014: 1–11. 10.1155/2014/160980.

Farrow, B., Sugiyama, Y., Chen, A., Uffort, E., Nealon, W. et al. (2004). Inflammatory mechanisms contributing to pancreatic cancer development. Ann. Surg., 239: 763. 10.1097%2F01.sla.0000128681.76786.07.

Ferrara, N., Gerber, H.P. and LeCouter, J. (2003). The biology of VEGF and its receptors. Nat. Med., 9: 669–676. 10.1038/nm0603-669.

Folkman, J. (1995). Angiogenesis in cancer, vascular, rheuma.toid and other disease. Nat. Med., 1: 27–30. 10.1038/nm0195-27.

Goubran, H.A. and Burnouf, T.P.R. (2012). Platelets, coagulation and cancer: Multifaceted interactions. Am. J. Med., 3: 130–140.

Gu, Y.X., Wang, Z.S., Li, S.X. and Yuan, Q.S. (2007). Effect of multiple factors on accumulation of nucleosides and bases in *Cordyceps militaris*. Food Chem., 102: 1304–1309. 10.1016/j.foodchem.2006.07.018.

Holbein, S., Wengi, A., Decourty, L., Freimoser, F.M., Jacquier, A. et al. (2009). Cordycepin interferes with 3' end formation in yeast independently of its potential to terminate RNA chain elongation. Rna, 15: 837–849. 10.1261/rna.1458909.

Jackson, S.P. (2007). The growing complexity of platelet aggregation. Blood, 109: 5087–5095. 10.1182/blood-2006-12-027698.

Jagger, D.V., Kredich, N.M. and Guarino, A.J. (1961). Inhibition of Ehrlich mouse ascites tumor growth by cordycepin. Cancer Res., 21: 216–220.

Jain, R.K. (2002). Tumor angiogenesis and accessibility: role of vascular endothelial growth factor. Semin. Oncol., 29: 3–9. 10.1016/S0093-7754(02)70063-8.

Jaiswal, J., Srivastav, A.K., Rajput, P.K., Yadav, U.C. and Kumar, U. et al. (2023). Integrating Synthesis, Physicochemical Characterization, and In Silico Studies of Cordycepin-Loaded Bovine Serum Albumin Nanoparticles. J. Agric. Food Chem., 71: 12225–12236. 10.1021/acs.jafc.3c03608.

Kaya, M., Wada, T., Akatsuka, T., Kawaguchi, S., Nagoya, S. et al. (2000). Vascular endothelial growth factor expression in untreated osteosarcoma is predictive of pulmonary metastasis and poor prognosis. Clin. Cancer Res., 6: 572–577.

Kessenbrock, K., Plaks, V. and Werb, Z. (2010). Matrix metalloproteinases: regulators of the tumor microenvironment. Cell, 141: 52–67. 10.1016/j.cell.2010.03.015.

Kim, S.O., Cha, H.J., Park, C., Lee, H., Hong, S.H. et al. (2019). Cordycepin induces apoptosis in human bladder cancer T24 cells through ROS-dependent inhibition of the PI3K/Akt signaling pathway. Biosci. Trends, 13: 324–333. 10.5582/bst.2019.01214.

Kim, Y.O., Kim, H.J., Abu-Taweel, G.M., Oh, J. and Sung, G.H. et al. (2019). Neuroprotective and therapeutic effect of *Cordyceps militaris* on ischemia-induced neuronal death and cognitive impairments. Saudi. J. Biol. Sci., 26: 1352–1357. 10.1016/j.sjbs.2018.08.011.

Kubo, E., Yoshikawa, N., Kunitomo, M., Kagota, S., Shinozuka, K. et al. (2010). Inhibitory effect of *Cordyceps sinensis* on experimental hepatic metastasis of melanoma by suppressing tumor cell invasion. Anticancer Res., 30: 3429–3433.

Kunhorm, P., Chaicharoenaudomrung, N. and Noisa, P. (2019). Enrichment of cordycepin for cosmeceutical applications: culture systems and strategies. Appl. Microbiol. Biotechnol., 103: 1681–1691. 10.1007/s00253-019-09623-3.

Lee, D., Lee, W. Y., Jung, K., Kwon, Y.S., Kim, D. et al. (2019). The inhibitory effect of cordycepin on the proliferation of MCF-7 breast cancer cells, and its mechanism: An investigation using network pharmacology-based analysis. Biomolecules, 9: 414. 10.3390/biom9090414.

Li, C., Yan, A., Cai, C. and Liu, Z. (2012). Fast determination of adenosine and cordycepin in *Cordyceps* and its deserted solid medium. Chin. J. Chromatogr., 30: 711–715. 10.3724/sp.j.1123.2012.02037.

Li, G., Nakagome, I., Hirono, S., Itoh, T. and Fujiwara, R. (2015). Inhibition of adenosine deaminase (ADA)-mediated metabolism of cordycepin by natural substances. Pharmacol. Res. Perspect., 3: 1–11. 10.1002/prp2.121.

Li, Y., Li, R., Zhu, S., Zhou, R., Wang, L. et al. (2015). Cordycepin induces apoptosis and autophagy in human neuroblastoma SK-N-SH and BE (2)-M17 cells. Oncol. Lett., 9: 2541–2547. 10.3892/ol.2015.3066.

Li, Z., Guo, Z., Zhu, J., Bi, S., Luo, Y. et al. (2020). *Cordyceps militaris* fraction inhibits angiogenesis of hepatocellular carcinoma in vitro and in vivo. Pharmacogn. Mag., 16: 169–176.

Liao, Y., Ling, J., Zhang, G., Liu, F., Tao, S. et al. (2015). Cordycepin induces cell cycle arrest and apoptosis by inducing DNA damage and up-regulation of p53 in Leukemia cells. Cell Cycle, 14: 761–771. 10.1080/15384101.2014.1000097.

Liu, X., Dun, M., Jian, T., Sun, Y., Wang, M. et al. (2023). *Cordyceps militaris* extracts and cordycepin ameliorate type 2 diabetes mellitus by modulating the gut microbiota and metabolites. Front. Pharmacol., 14: 1134429. 10.3389/fphar.2023.1134429.

Nelson, W.G., De Marzo, A.M., DeWEESE, T.L. and Isaacs, W.B. (2004). The role of inflammation in the pathogenesis of prostate cancer. J. Urol., Balt., 172: S6–S12. 10.1097/01.ju.0000142058. 99614.ff.

Overgaard-Hansen, K. (1964). The inhibition of 5-phosphoribosyl-1-pyrophosphate formation by cordycepin triphosphate in extracts of Ehrlich ascites tumor cells. Biochim. Biophys. Acta., 80: 504–507. 10.1016/0926-6550(64)90154-9.

Patel, K.J. and Ingalhalli, R.S. (2013). *Cordyceps militaris* (L.: Fr.) Link â€"An Important Medicinal Mushroom. J. Pharmacogn. Phytochem., 2: 315–319.

Qin, P., Li, X., Yang, H., Wang, Z.Y. and Lu, D. (2019). Therapeutic potential and biological applications of cordycepin and metabolic mechanisms in cordycepin-producing fungi. Molecules, 24: 2231. 10.3390/molecules24122231.

Rao, Y.K., Fang, S.H., Wu, W.S. and Tzeng, Y.M. et al. (2010). Constituents isolated from *Cordyceps militaris* suppress enhanced inflammatory mediator's production and human cancer cell proliferation. J. Ethnopharmacol., 131: 363–367. 10.1016/j.jep.2010.07.020.

Rayburn, E.R., Ezell, S.J. and Zhang, R. (2009). Anti-inflammatory agents for cancer therapy. Mol. Cell. Pharmacol., 1: 29. 10.4255%2Fmcpharmacol.09.05.

Risau, W. (1997). Mechanisms of angiogenesis. Nature, 386: 671–674. 10.1038/386671a0.

Rosell, R. and Karachaliou, N. (2015). Relationship between gene mutation and lung cancer metastasis. Cancer Metastasis Rev., 34: 243–248. 10.1007/s10555-015-9557-1.

Rottman, F. and Guarino, A.J. (1964) The inhibition of phosphoribosyl-pyrophosphate amidotransferase activity by cordycepin monophosphate, Biochim. Biophys. Acta., 89: 465–472. 10.1016/0926-6569(64)90072-0.

Ruma, I., Putranto, E.W., Kondo, E., Watanabe, R., Saito et al. (2014). Extract of *Cordyceps militaris* inhibits angiogenesis and suppresses tumor growth of human malignant melanoma cells. Int. J. Oncol., 45: 209–218. 10.3892/ijo.2014.2397.

Shao, L.W., Huang, L.H., Yan, S., Jin, J.D. and Ren, S.Y. et al. (2016). Cordycepin induces apoptosis in human liver cancer HepG2 cells through extrinsic and intrinsic signaling pathways. Oncol. Lett., 12: 995–1000. 10.3892/ol.2016.4706.

Shi, L., Cao, H., Fu, S., Jia, Z., Lu, X. et al. (2022). Cordycepin enhances hyperthermia-induced apoptosis and cell cycle arrest by modulating the MAPK pathway in human lymphoma U937 cells. Mol. Biol. Rep., 49: 8673–8683. 10.1007/s11033-022-07705-6.

Shin, S., Lee, S., Kwon, J., Moon, S., Lee, S. et al. (2009). Cordycepin suppresses expression of diabetes regulating genes by inhibition of lipopolysaccharide-induced inflammation in macrophages. Immune Netw., 9: 98–105. 10.4110/in.2009.9.3.98.

Song, J. F., Liu, C. Q., Li, D. J. and Jin, B. et al. Q. (2007). Optimization of cordycepin extraction from cultured *Cordyceps militaris* by HPLC-DAD coupled with uniform design. J. Chem. Technol. Biotechnol., 82: 1122–1126. 10.1002/jctb.1784.

Sun, H., Zhang, A., Gong, Y., Sun, W., Yan, B. et al. (2022). Improving effect of cordycepin on insulin synthesis and secretion in normal and oxidative-damaged INS-1 cells. Eur. J. Pharmacol., 920: 174843. 10.1016/j.ejphar.2022.174843.

Tania, M., Shawon, J., Saif, K., Kiefer, R., Khorram, M.S. et al. (2019). Cordycepin downregulates Cdk-2 to interfere with cell cycle and increases apoptosis by generating ROS in cervical cancer cells: In vitro and in silico study. Curr. Cancer Drug Targets, 19: 152–159. 10.2174/1568009618666180905095356.

Thai, N.M., Dat, T.T.H., Hai, N.T.T., Bui, T.Q., Phu, N.V. et al. (2023). Identification of potential inhibitors against Alzheimer-related proteins in *Cordyceps militaris* ethanol extract: experimental evidence and computational analyses. 3 Biotech, 13: 292. 10.1007/s13205-023-03714-9.

Tsuruo, T. and Fujita, N. (2008). Platelet aggregation in the formation of tumor metastasis. Proc. Jpn. Acad., Ser. B, 84: 189–198. 10.2183/pjab.84.189.

Upatcha, N., Kaokaen, P., Sorraksa, N., Phonchai, R., Kunhorm, P. et al. (2023). Nanoencapsulated *Cordyceps* extract enhances collagen synthesis and skin cell regeneration through antioxidation and autophagy. J. Microencapsul., 40: 1–15. 10.1080/02652048.2023.2198008.

Wang, D., DuBois, R.N. and Richmond, A. (2009). The role of chemokines in intestinal inflammation and cancer. Curr. Opin. Pharmacol., 9: 688–696. 10.1016/j.coph.2009.08.003.

Wei, C., Khan, M. A., Du, J., Cheng, J., Tania, M. et al. (2022). Cordycepin inhibits triple-negative breast cancer cell migration and invasion by regulating EMT-TFs SLUG, TWIST1, SNAIL1, and ZEB1. Front. Oncol., 12: 1–9. 10.3389/fonc.2022.898583.

Wei, P., Wang, K., Luo, C., Huang, Y., Misilimu, D. et al. (2021). Cordycepin confers long-term neuroprotection via inhibiting neutrophil infiltration and neuroinflammation after traumatic brain injury. J. Neuroinflammation, 18: 1–17. 10.1186/s12974-021-02188-x.

Weidner, N., Semple, J.P., Welch, W.R. and Folkman, J. (1991). Tumor angiogenesis and metastasis—correlation in invasive breast carcinoma. N. Engl. J. Med., 324: 1–8. 10.1056/NEJM199101033240101.

Won, S.Y. and Park, E.H. (2005). Anti-inflammatory and related pharmacological activities of cultured mycelia and fruiting bodies of *Cordyceps militaris*. J. Ethnopharmacol., 96: 555–561. 10.1016/j.jep.2004.10.009.

Wong, Y. Y., Moon, A., Duffin, R., Barthet-Barateig, A., Meijer, H.A. et al. (2010). Cordycepin inhibits protein synthesis and cell adhesion through effects on signal transduction. J. Biol. Chem., 285: 2610–2621. 10.1074/jbc.M109.071159.

Wu, W.C., Hsiao, J.R., Lian, Y.Y., Lin, C.Y. and Huang, B.M. et al. (2007). The apoptotic effect of cordycepin on human OEC-M1 oral cancer cell line. Cancer Chemother. Pharmacol., 60: 103–111. 10.1007/s00280-006-0354-y.

Xu, H., Cheng, J. and He, F. (2022). Cordycepin alleviates myocardial ischemia/reperfusion injury by enhancing autophagy via AMPK-mTOR pathway. J. Physiol. Biochem., 78: 401–413. 10.1007/s13105-021-00816-x.

Xu, J.C., Zhou, X.P., Wang, X.A., Xu, M.D., Chen, T. et al. (2019). Cordycepin induces apoptosis and G2/M phase arrest through the ERK pathways in esophageal cancer cells. J. Cancer, 10: 2415. 10.7150%2Fjca.32071.

Xue-Ying, L.I., Homng, T.A.O., Can, J.I.N., Zhen-Yun, D.U., Wen-Feng, L.I.A.O. et al. (2020). Cordycepin inhibits pancreatic cancer cell growth *in vitro* and *in vivo* via targeting FGFR2 and blocking ERK signaling. Chin. J. Nat. Med., 18: 345–355. 10.1016/S1875-5364(20)30041-8.

Yang, C., Zhao, L., Yuan, W. and Wen, J. (2017). Cordycepin induces apoptotic cell death and inhibits cell migration in renal cell carcinoma via regulation of microRNA-21 and PTEN phosphatase. Biomed. Res., 38: 313–320. 10.2220/biomedres.38.313.

Yue, K., Ye, M., Zhou, Z., Sun, W. and Lin, X. (2013). The genus *Cordyceps*: a chemical and pharmacological review. J. Pharm. Pharmacol., 65: 474–493. 10.1111/j.2042-7158.2012.01601.x.

Zeng, Y. and Chen, Z.H. (2013). Extraction of cordycepin from waste medium of *Cordyceps militaris* with macroporous resin column chromatography. J. Nat. Sci. Hunan Nor. Uni., 36: 66–69.

Zheng, Q., Sun, J., Li, W., Li, S. and Zhang, K. (2020). Cordycepin induces apoptosis in human tongue cancer cells *in vitro* and has antitumor effects *in vivo*. Arch. Oral Biol., 118: 104846. 10.1016/j.archoralbio.2020.104846.

Zhou, Q., Zhang, Z., Song, L., Huang, C., Cheng, Q., Bi, S. and Yu, R. (2018). *Cordyceps militaris* fraction inhibits the invasion and metastasis of lung cancer cells through the protein kinase B/glycogen synthase kinase 3β/βcatenin signaling pathway. Oncology Letters, 16(6): 6930–6939.

Plant Protection

12

Crop Protection Potential of Entomopathogenic *Cordyceps*

Hemraj Chhipa

1. Introduction

The interwining challenges of climate change and a burgeoning global population pose significant concerns. The rising demand for nutritious agricultural produce to sustain the growing populace is met with challenges due to the adverse impacts of climate change on crop production. Additionally, climate change supports the growth of crop pests and pathogens,exacerbating the complexities faced by the agricultural sector (Ozdemir, 2022). Beside this, insect pests alone are responsible for reducing the production of crops by 18–26% at the global level, which causes a loss of about $70 billion (Bhalla, 2021). The intensified quest for enhanced agricultural production has led to a surge in the application of chemical fertilizers and pesticides.However, overreliance on these chemicals not only amplifies crop yields but also contributes to the development of resistance in pests and pathogens against these chemicals (Kang et al., 2017; Wang et al., 2019). Sparks and Nauen (2015) reported that up to 586 species have been found to have insecticide resistance during the last century. Unjustifiable use of synthetic pesticides like organochlorines, organophosphates, pyrethroids and others to control insect pests adversely affects the environment. There is a need for sustainable intensification of agriculture to prevent environmental damage and increase food production

College of Horticulture and Forestry, Agriculture University, Kota, Jhalawar, Rajasthan 326023, India.
Email: hrchhipa8@gmail.com; hemraj@aukota.org

for a growing population (Crist et al., 2017). To mitigate the adverse effects of conventional chemicals, agricultural scientists are exploring alternative options,with the utilization of biofertilizers and biopesticides being one such avenue (Hussain et al., 2020). Microbial biopesticides have demonstrated promising results in managing chemical resistance in pests. However, the need for more options in biopesticides is emphasized, especially in light of changing environmental conditions, that can impact the maximum efficacy of bio-control agents (Praveen, 2023) (Table 1). Currently, 356 biopesticides (contain bacteria, fungi and viruses) including active biomolecules are registered in the US market and have shown active output against pathogenic fungi, insect pests and nematodes. Very few fungal biopesticides have been explored yet, and scientists are working to discover new strains throughout the globe (Arthurs and Dara, 2019).

Table 1. Biopesticides available on the market (Arthurs and Dara, 2019).

Biocontrol agent	Target pest/pathogen	Pathogen against
Bacteria		
Bacillus firmus	Plant parasitic nematode	Nematode
Bacillus sphaeicus	Mosquito larvae	Insect
Bacillus subtilis	*Botrytis* spp.	Fungi
Bacillus thuringiensis subsp. *kurstaki*	Caterpillars	Insect
Burkholderia renojensis	Broad spectrum Insects/ nematode	Insects/nematode
Chromobacterium subtsugae	Broad spectrum Insects	Insect
Paenibacillus popilliae	Various Insects	Insect
Pasteuria spp.	Plant parasitic nematodes	Nematode
Fungi		
Beauveria bassiana (Bals.-Criv.) Vuill.	Thrips, aphids and whiteflies	Insect
Coniothyrium minitans W.A. Campb.	*Sclerotinia* spp.	Fungi
Metarhizium brunneum Petch	Thrips, whiteflies and weevils	Insect
Purpureocillium lilacinum (Thom) Laungsa-ard, Houbraken, Hywel-Jones & Aamson	Plant parasitic nematodes	Nematode
Protozoa		
Paranosema locustae Canning	Grasshoppers	Insect
Vitrus		
Cydia pomonella granulosis virus	Codling moth	Insect
Plodia interpunctella granulovirus	Indian meal moth	Insect
Spodoptera exigua Nucleopolyhedrovirus	Beet armyworm	Insect

Entomopathogenic fungi represent a promising biopesticide option that does not harm non-targeted organisms and naturally eliminates target pathogens. Utilizing entomopathogenic fungi as insecticides could offer a sustainable solution for controlling insect pests (Venkatesh et al., 2022). Presently, very few fungi have been recognized such as *Beauveria bassiana, Cordyceps fumosorosea, Melipona quadrifasciata, Metarhizium anisopliae, Plebeia droryana* and *Scaptotrigona bipunctata* (Fatia et al., 2023). Entomopathogenic ascomycetes are also known to regulate populations of pests associated with plants (Quesada-Moraga et al., 2023). *Cordyceps* sp. has shown good potential as a bioinsecticide due to the presence of various bioactive compounds. Entomopathogenic fungi from the *Cordyceps* genus have shown good results in controlling various insect pests like Asian citrus psyllids, diamond-back moths, termites, red palm weevils, and whiteflies (Kepler et al., 2017; Gao et al., 2017; Keppanan et al., 2019). Entomopathogenic fungi penetrate into the target pest and grow rapidly to cover the insect pest and negatively affect their immune system preventing their growth (Hussain et al., 2020b). Various mycotoxins, like bassianolide, beauvericin, cyclosporins, destruxins, enniatins and isariotins have been reported from different entomopathogenic fungi, which play an important role in the virulence of entomopathogenic fungi against insect pests of agricultural crops (Ravindran et al., 2018; Wang et al., 2019). Along with this, *Cordyceps* has also shown pharmaceutical and medicinal functions in arrhythmias, asthenia, heart diseases, hyperglycemia, hyperlipidemia, hyposexuality, liver disease, night sweating, organ transplantation, renal dysfunction, renal failure and respiratory disease (Zhou et al., 2009). This chapter discusses the biopesticide potential of *Cordyceps* in plant protection.

2. *Cordyceps*

Cordyceps, the genus of club-fungus, plays a significant role in ecological balance. There are 680 species of *Cordyceps* reported from different climatic zones (Shashidhar et al., 2013). It contains a wide range of natural products and has shown promising results as a biocontrol agent against various plant diseases. Some *Cordyceps* species like *C. gansuensis, C. grasspara, C. guizhouensis, C. gunnii, C. kangdingensis, C. liangshanensis, C. militaris, C. nutans, C. shanxiensis, C. taishanensis* and *C. tricentric* have been reported for their medicinal potential by various authors (Zhou et al., 2009). Most of the *Cordyceps* are endo-parasitic in nature and grow inside the arthropod organisms (Badea and Vamanu, 2023).

Cordyceps produce various chemical compounds like adenosine, cordycepic acid, ergosterol, myriocin, polysaccharides and others, which are responsible for the pharmacological and biocontrol activity of this fungus (Fig. 1) (Abo Nouh et al., 2021). It is a well-known Chinese herb

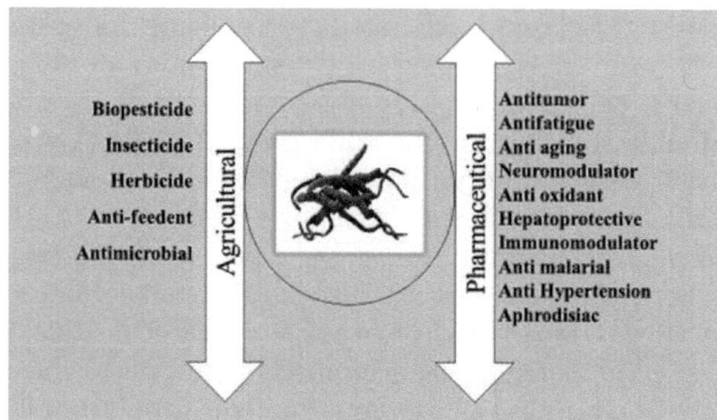

Fig. 1. Applications of *Cordyceps* in agriculture and pharmaceutical industries.

recognized for its anti-aging, anti-cancer, anti-diabetic, anti-fatigue, antihyperlipidemic, anti-inflammatory, antimicrobial, aphrodisiac, hypocholesterolemic, hypotensive, immune-boosting, kidney protection, liver-protecting agent, pro-sexual agent and vasorelaxation effects (Devi et al., 2020a and 2020b; Das et al., 2021). In addition to this, *Cordyceps* sinensis is used as a nutritive food in China due to the presence of essential amino acids and vitamins (vitamin B1, vitamin B2, vitamin B12, and vitamin K) (Wu et al., 2014).

3. Bioprospects of *Cordyceps* as a Biocontrol Agent

Cordyceps demonstrates significant activity against various agricultural pests, thanks to its rich composition of bioactive compounds such as adenosine, cordycepic acid, cordycepin, ergosterol, myriocin, polysaccharides and others. The presence of these compounds not only enhances *Cordyceps'* importance in agriculture but also extends its applications to the pharmaceutical industry (Table 2). Different species of *Cordyceps* have proven to be antifeedants, herbicides, and insecticides,contributing to crop protection and productivity improvements.

The application of *Cordyceps* has been summarized in following sections. The bioactive potential of different *Cordyceps* spp. are documented in Table 2.

3.1 Antifeedant

Cordyceps also produces some compounds that work as antifeedants against different insect pests. Li et al. (2022) isolated polyketides namely, Cordycicadins A–D from the liquid fermentation of *Cordyceps cicadae* (JXCH1) and found an antifeed activity of 57–65.4 µg/cm^2 of compounds Cordycicadins A and B against silkworm larvae (Bombyx mori).

Table 2. Bioactive potential of *Cordyceps* spp.

Species	Bioactive compound	Application	Reference
Cordyceps cicadae (Miq.) Massee (JXCH1)	Cordycicadins A-D	Antifeedant against silkworm larvae (*Bombyx mori*)	Li et al. (2022)
Cordyceps dipterigena Berk. & Broome	*Cordyceps*idone A and B	Antifungal activity	Varughese et al. (2012)
Cordyceps farinose (Holmsk.) Kepler, B. Shrestha & Spatafora	-	Against diamondback moth (*Plutella xylostella*)	Du et al. (2021a)
Cordyceps Fumosorosea (Wize) Kepler, B. Shrestha & Spatafora	Bassianolide	Against citrus pathogen -Asian citrus (*Diaphorina citri*)	Qasim et al. (2020), Luo et al. (2022)
	-	Against diamondback moth (*Plutella xylostella*)	Du et al. (2021a)
	-	Against diamondback moth (*Plutella xylostella*)	Lei et al. (2021)
	Beauvericir, Brevianamide F, Enniatin, 5-Methylmellein and Trichodermin	Against *Aphis craccivora* and *Bemisia tabaci*	Wu et al. (2021)
Cordyceps gunnii (Berk.) Berk.	Sesquiterpencids, 3,4-diacetoxy-12,13-epoxy-9-trichothecene-15-ol	Against nematodes	Qu et al. (2023)
Cordyceps javanica (Bally) Kepler, B. Shrestha & Spatafora	-	Epizootics among whiteflies	Wang et al., 2019
	-	Against Asian citrus (*Diaphorina citri*)	Avery et al. (2021)
Cordyceps militaris (L.) Fr.	Cordycepin	Herbicide	Quy et al. (2019)
	Cordycepin	Against *Plutella xylostella* larvae	Kim et al. (2002)
	Ergosterol	Antimicrobial activity	Abo Nouh et al. (2021)
	-	Antioxidant activity	Zhang et al. (2011)
Cordyceps tenuipes (Peck) Kepler, B. Shrestha & Spatafora	-	Against diamondback moth (*Plutella xylostella*)	Du et al. (2021a)

3.2 *Herbicide*

Chemical herbicides not only suppress the weed crop but also affect the soil properties, accumulation of xenobiotic compounds, increased loss of natural vegetation, groundwater contamination, and uncontrolled applications that generate resistance in weed (Berestetskiy, 2023). Thus, there is a need to explore bio-input to reduce the burden of chemical herbicides and maintain soil properties. A bioactive compound, cordycepin from *Cordyceps militaris* was shown to be an herbicide in a study conducted by Quy et al. (2019). They observed that at 40 ppm concentration, cordycepin inhibited the growth and germination of radish by 3.8 to 5.9-fold in comparison to benzoic acid and glyphosate by reducing chlorophyll and carotenoid content.

3.3 *Insecticide*

It is an entomopathogenic fungus that regulates the population of various insect pests causing disease in agricultural crops and damages the crop by feeding on leaves, reducing plant growth and stunting crop growth by lepidoptera potato tuber moth (PTM), Phthorimaea operculella (DBM), Plutella xylostella and others (Zheng et al., 2019). Similarly, Du et al. (2021a) found that three *Cordyceps* spp. (*C. farinosa*, *C. fumosorosea* and *C. tenuipes*) have shown pathogenicity against different larval stages of diamondback moth (DBM), *Plutella xylostella*, which is a devastating pest of agricultural crops, especially in the brassicaceous family, and that *Cordyceps* could be a potential biocontrol agent against DBM. *C. fumosorosea* has also shown biopesticide potential against the citrus pest Diaphorina citri nymphs and their adults and bioactive compounds beauverolides (B, C, F, I, and J) were shown to possess mycotoxin properties (Qasim et al., 2020). *Cordyceps javanica* was also found effective against citrus pest Asian citrus psyllid (AsCP) *Diaphorina citri* in combination with horticultural white oil, suppressed 90% of the population of AsCP after 14 days of treatment in comparison to only white oil treatment (Avery et al., 2021).

Further, Wu et al. (2021) also reported insecticidal activity of *C. fumosorosea* against *Aphis craccivora*, a major cowpea aphid and *Bemisia tabaci* (sucking pest and carrier of many viral diseases) in different regions of the world. They identified the major compounds trichodermin, 5-methylmellein, brevianamide F, and enniatin and beauvericin responsible for such activity in *C. fumosorosea*.

4. Diversity of Bioactive Compounds

Various bioactive compounds have been identified in *Cordyceps*, like alkaloids, cyclic peptides, flavonoids, phenolic compounds, polysaccharides, polyketides, terpenoid, sterol-like compounds, which make them important mushrooms for pharmacological and agricultural industries (Table 3) (Rastegari et al., 2019a and b; Das et al., 2021). It has been reported that cordycepin is one of the major bioactive compounds, along with adenosine and cordycepic acid, in *Cordyceps militaris* and *C. sinensis* (Deshmukh et al., 2020; Sornchaithawatwong et al., 2020). Quasim et al. (2020) identified bassianolide, beauvericin A, beauverolides, destruxin E, isaridins, isariotins A-C and tenuipyrone through ultra-performance liquid chromatography coupled with quadrupole time of flight mass spectrometer (UPLC-QTOF MS) technique in *C. fumosorosea*. Recently, Qasim et al. (2020) identified beauvericin A, beauverolide B, beauverolide C, beauverolide F, beauverolide I, beauverolide J, beauverolide L, destruxin E, isaridin A, isaridin B, isaridin G, isaridin E, isariotin A, isariotin B, isariotin C and tenuipyrone in *C. fumosorosea* through UPLC-QTOF MS (Qasim et. al., 2020). Similarly, compounds like beauvericin, beauverolides and 2,6-pyridindicarboxylic acid were also reported in insect *Aphis craccivora* and *Bemisia tabaci* control by *C. fumosorosea* (Wu et al., 2021).

5. Mechanism of Action as a Biocontrol Agent

The infections in insect hosts by entomopathogenic fungi like *Beauveria*, *Cordyceps* and *Metarhizium* have been reported by penetration, colonization and conidiogenesis (Moino et al., 2002). To prevent such infections, insects also activate their cellular and humoral immune systems against pathogenic hyphae, and cellular actions like phagocytosis, encapsulation, and nodulation occur in insect cells (Kwon et al., 2014; Lei et al., 2011). But to mislead the insect's humoral immunity, pathogens modulate the signal transduction pathway and generate changes in receptors, which are responsible for the identification of pathogens (Stokes et al., 2015). *M. brunneum* associated with coffee roots have a positive impact on plant protection against *Leucoptera coffeella* (Frazin et al., 2022). In potato plants, endophytic *Beauveria bassiana* is known to alleviate adverse impacts of salt stress (Tomilova et al., 2023).

It has been reported that *Cordyceps* action on insect pests started with the interaction of conidia or blastophore attachment with insect pests. Under favorable conditions, conidia grow and secrete digestive or cuticle degradative enzymes, which support the fungal mycelium to penetrate the insect cuticle portion. Further, fungus grows and penetrates through the insect's integument, blocks the various systems of the host, like the digestive and circulation systems and produces toxins that lead to the death of the insect pest (Wu et al., 2014). Further, Lei et al. (2021) expressed the process of *C. fumosorosea* infection in *Plutella xylostella* larvae using scanning and transmission election microscopy. They reported that *C. fumosorosea* adhered through the protrusions of the insect cuticle surface, and the conidia

Table 3. Bioactive compounds from *Cordyceps* used in the pharmaceutical industry.

Bioactive compound	Action	Reference
Adenosine	Signal transportation in nerves	Olatunji et al., 2016; Du et al., 2021b
Ascorbic acid, β-carotene, lycopene, phenolic compounds and polysaccharides	Anti-oxidant activity	Sharma et al., 2015
Bioxanthracenes	Antimalarial activity and cytotoxicity.	Jaturapat et. al., 2001; Olatunji et. al., 2018
Cicadapeptins and myriocin	Antifungal and antibiotics	Krasnoff et al., 2005; Ng et al., 2005
Cordycicadione (1), cordycicadin F (2) and 7-hydroxybassiatin (3)	Anti-cancer Activity	Fan et al., 2023
Cordycepic acid	Alleviation the inflammatory reaction that is induced by lipopolysaccharides, and fibrogenic response	Das et al., 2021; Yoon et al., 2017
Cordycepin	Anti-inflammatory properties in many diseases including asthma, hepatitis, atherosclerosis, acute lung injury (ALI), rheumatoid arthritis, and Parkinson's disease (PD)	Tan et al., 2021
Cordycepin and carotenoids	Bactericidal activity against all enteric pathogenic bacteria; *Escherichia coli*, *E. coli* O157:H7, Shigella dysenteriae, Salmonella Typhi, Vibrio cholerae and Bacillus cereus	Kaewkod et al., 2023
Cordymin	Antifungal and antiproliferative activity against breast cancer cells.	Wong et al., 2011; Wong et al., 2019
Cordyheptapeptide	Cytotoxicity against multiple cancer cell lines	Molnar and Carreira, 2010
Coumarins and bifusicoumarin A-D	Antitumor	Elshamy et al., 2023
Ergosterol and ergosteryl esters	Regulates permeability and fluidity	Sun et al., 2019
Fumosoroseain A and fumosoroseanoside A	Anti-microbial or antiaging activities	Buchter et al., 2020; Wei et al., 2022

Hypoxanthine	Naturally occurring purine derivative	Xiao and Xiong, 2013
Macrolides ($C_{10}H_{14}O_4$)	Bacterial infections.	Mago et al., 2023
Naphthoquinone	Possess cytotoxic, antibacterial, antiviral, antifungal, insecticidal, antipyretic and anti-inflammatory activities	Prathumpai et al., 2007
N6-(2-hydroxyethyl)-adenosine	Immunomodulatory, anti-inflammatory	Liu et al., 2023
Polysaccharide	Antitumor activity	Zhu et al., 2016; Zhu et al., 2012; Liu et al., 2019
Tenuipesone A, tenuipesone B and beauvericin	*Antimicrobial* activity against *Bacillus subtilis* and *Staphylococcus aureus*	Yoneyama et al., 2022
Uridine (1), adenosine (2), L-pyroglutamic acid (3) and lysinonorleucine (4), 1,3,5-trimethoxybenzene (5), D-mannitol (6), L-pyroglutamic acid methyl ester (7), tryptophan (8) and phenylalanine (9)	myelosuppression	Zhang et al., 2022
CGIP, CGIP I-II, CGIP-F1and F2 and polysaccharide	Immunosuppressive activity	Xiao et al., 2004

of blastospores started to germinate within four hours. Further germinating tubes and hyphae of *C. fumosorosea* reached the cuticular epidermis and penetrated into haemocoel. After 36 hours, hyphae covered the entire body cavity from inside to outside, and after 72 hours of penetration, more conidia developed and covered the entire pest cuticular surface. Lei et al. (2021) worked on unravelling the mechanism of action of *C. fumosorosea* against the lepidopterian insect *Plutella xylostella* and found that infection with *C. fumosorosea occurs* by melanotic encapsulation in DBM larvae. The encapsulation is initiated through invasion between the thoracic and abdominal segments, leg, and spiracles of the larva. The infection covered progressively the abdominal pleuron to whole thoracic segments, and hyphae penetrated from outside to inside and again outside to completely cover the larvae to make them immobile (Fig. 2). The invading fungus releases mucilage-like substances and toxic proteins to kill the insect stages and absorb all nutrients from the dead insects for the completion of their life cycle (Liu et al., 2015).

The secretion of bioactive compounds such as mycotoxin also inhibits or kills the growth of the target insect. It has been reported that bassianolide compounds from *C. fumosorosea* were found active against *D. citri* nymphs and their adults, a pest of the citrus family (Qasim et al., 2020). The mycotoxins have shown a harmful impact on animals and insects via muscle contraction through acetylcholine imbalance, phenoloxidase production, and a negative effect on insects' immune systems (Rosa et al., 2018).

Cordyceps also produces mycotoxin compounds, which have shown insecticidal and antifeedant activity against agricultural insect pests (Namasivayam et al., 2018). Penetration of entomopathogenic fungi releases various mycotoxins in the hemocoel, which is one of the main causes of insect death, but very few bioactive compounds in *Cordyceps* have been reported and their mode of action is unknown. The mycotoxin targets

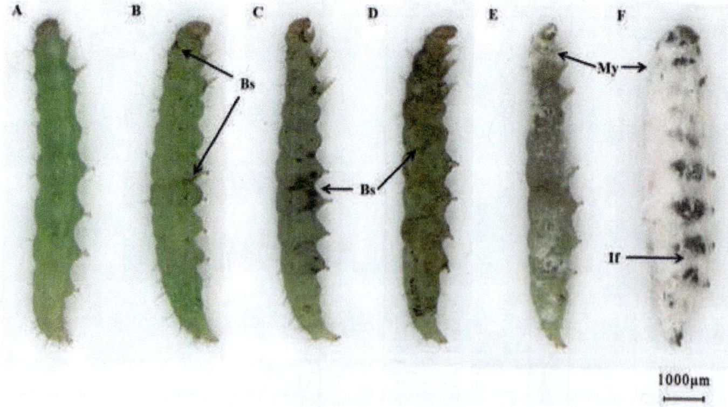

Fig. 2. Effect of *Cordyceps fumosorosea* strain IFCF01 on various larval stages of *Plutella xylostella* (Source: Lei et al., 2021; https://doi.org/10.3390/insects12020179).

the host immune-compromised cells and shows a toxic effect. Mycotoxin contaminated their surrounding environment and generated threats to animals, humans and insects. Mycotoxin shows lethal effects in agricultural insects through: (a) muscle contraction via acetylcholine imbalance; (b) phenoloxidase production; (c) the target immune system (Rosa et al., 2018; Niermans et al., 2019; Zhang et al., 2019).

Cordycepin is the most documented compound that inhibits signal communication in the immune system of pests and terminates nucleic acid synthesis in crop pests (Berestetskiy and Hu, 2021). While acting as a herbicide, cordycepin inhibits various cellular processes like ATP synthesis, gene expression, membrane permeability, nutrient absorption, stomatal opening and closure and various metabolic processes in the target plants (Quy et al., 2019). The activity of *Cordyceps* also depends on numerous environmental factors, including humidity, rain, solar radiation, temperature and the presence of other microbial communities (Hajek et al., 2017; Jaronski, 2009).

6. *Cordyceps* as a Potential Crop Bioprotectant

The control of insect pests and crop pathogenic organisms through chemical pesticides not only increases production but also increases the cost of crop production, has a negative impact on soil and farmers and reduces soil fertility, has a negative impact on untargeted organisms, and causes toxicity in water bodies (Rani et al., 2021; Tudi et al., 2021). The development of biopesticides provided a sustainable alternative to controlling crop pests and pathogens in an eco-friendly manner, alone or with chemicals or other biopesticides in an integrated pest management (IPM) system. The use of biopesticides not only reduces the burden of chemicals but also saves the lives of untargeted organisms, which are directly involved in soil fertility. There are many species of entomopathogenic fungi. *Cordyceps* sp. have been found effective against various nematodes, insects, and weeds due to the presence of different bioactive compounds (Lei et al., 2011; Woolley, 2018). Integration of *Cordyceps* with *B. bassiana* and *M. brunneum* in IPM may enhance the entomopathogenic activity and improve the rate of killing insect pests. The effective infection of pests through cordycepin-like compounds makes them a new bioprotectant against insect pests. But the activity at field level is also affected by environmental conditions. The efficacy of entomopathogenic fungi and their spores in controlling insect pests is also affected by environmental conditions like temperature, rainfall, and humidity. Optimum conditions favor the growth and antagonistic activities of entomopathogenic fungi. Unfavorable conditions like heavy rainfall prevent the blastospores from adhering to plant leaves due to the hydrophobic nature of fungal blastospores (Dunlap et al., 2005; Avery et al.,

2021). Such environmental factors affect the persistent output of biocontrol agents under field conditions. For maximum output of such biocontrol agents farmers need to be educated about their effective applications. Awareness and training of farmers about applications of bioprotectants may increase the output efficacy and farmers' belief in such products. The government should develop policies for effective biocontrol manufacturing and SOPs for their effective use by farmers. Increased output of bioprotectants will increase demand and further production at the commercial level. The growth of the biocontrol market will increase the area of organic farming, and the chemical toxicity of soil will reduce simultaneously.

Conclusion

Unjustified application of chemical pesticides negatively affected soil fertility and developed pesticide resistance in crop pests. Adaptation of ecofriendly biocontrol agents like *Cordyceps* can be a good alternative to chemical pesticides. Various species of *Cordyceps* (*C. farinosa, C. fumosorosea, C. javanica, C. militaris* and *C. tenuipes*) have shown active potential as insecticides, herbicides and antifeedant agents to prevent crop loss. The availability of bioactive compounds, especially cordycepic acid, cordycepins, ergosterol, myriocin and polysaccharides, increased their applications in agriculture and the pharmaceutical industry. *Cordyceps* has shown wide application in the pharmaceutical industry as an antimalarial, anti-cancer, anti-inflammatory, anti-microbial and immunosuppressive agent. The entomopathogenic nature of *Cordyceps* makes it effective against Asian citrus psyllid (ACP) *Diaphorina citri*; diamondback moth *P. xylostella*; potato tuber moth (PTM) *Phthorimaea operculella*; sucking pests *Aphis craccivora* and *Bemisia tabaci*; nematodes. The agricultural application of *Cordyceps* as not been studied much in comparison to its medicinal uses. The efficacy of *Cordyceps* against insect pests increased its potential as an effective biopesticide, but more research is needed to explore the hidden potentials of other species of *Cordyceps*.

References

Abo Nouh, F.A., Gezaf, S.A., Abo Nahas, H.H., Abo Nahas, Y.H., Vargas-De-La-Cruz, C. et al. (2021). Diversity of *Cordyceps* from Different Environmental Agroecosystems and Potential Applications. In Industrially Important Fungi for Sustainable Development: Volume 1: Biodiversity and Ecological Perspectives, pp. 207–236. Cham: Springer International Publishing.

Arthurs, S. and Dara, S.K. (2019). Microbial biopesticides for invertebrate pests and their markets in the United States. J. Invertebr. Pathol., 165: 13–21.

Avery, P.B., Duren, E.B., Qureshi, J.A., Adair, R.C. Jr, Adair, M.M. et al. (2021). Field efficacy of *Cordyceps javanica*, white oil and spinetoram for the management of the Asian Citrus Psyllid, *Diaphorina citri*. Insects, 12(9): 824.

Badea, T.I. and Vamanu, E. (2023). Medicinal Properties of Honey and *Cordyceps* Mushrooms. Nutraceuticals, 3(4): 499–512.

Berestetskiy, A. (2023). Modern approaches for the development of new herbicides based on natural compounds. Plants, 12(2): 234.

Berestetskiy, A. and Hu, Q. (2021). The chemical ecology approach to reveal fungal metabolites for arthropod pest management. Microorganisms, 9(7): 1379.

Bhalla (2021). https://www.weforum.org/agenda/2021/06/climate-change-insects-pests-crops agriculture/

Buchter, C., Koch, K., Freyer, M., Baier, S., Saier, C. et al. (2020). The mycotoxin beauvericin impairs development, fertility, and life span in the nematode *Caenorhabditis elegans* accompanied by increased germ cell apoptosis and lipofuscin accumulation. Toxicol. Lett., 334: 102–109.

Crist, E., Mora, C. and Engelman, R. (2017). The interaction of human population, food production, and biodiversity protection. Science, 356: 260–264.

Das, G., Shin, H.S., Leyva-Gómez, G., Prado-Audelo, M.L.D., Cortes, H. et al. (2021). *Cordyceps* spp.: A Review on Its Immune-Stimulatory and Other Biological Potentials. Front. Pharmacol., 11: 602364. doi:10.3389/fphar.2020.602364.

Deshmukh, L., Sharma, A.K. and Sandhu, S.S. (2020). Contrive Himalayan soft gold *Cordyceps* species: a lineage of Eumycota bestowing tremendous pharmacological and therapeutic potential. Curr Pharmacol Rep., 6: 155–166.

Devi, R., Kaur, T., Guleria, G., Rana, K.L., Kour et al (2020a). Fungal secondary metabolites and their biotechnological application for human health. pp. 147–161. *In*: Rastegari, A.A., Yadav, A.N., Yadav, N. (eds.). Trends of microbial biotechnology for sustainable agriculture and biomedicine systems: perspectives for human health. Elsevier, Amsterdam.

Devi, R., Kaur, T., Kour, D., Rana, K.L., Yadav, A. et al. (2020b). Beneficial fungal communities from different habitats and their roles in plant growth promotion and soil health. Microb. Biosyst., 5: 21–47.

Du, C., Wu, J., Cuthbertson, A.G., Hamid Bashir, M., Sun, T. et al. (2021a). Morphological, molecular and virulence characterisation of six *Cordyceps* spp. isolates infecting the diamondback moth, *Pluttela xylostella*. Biocontrol. Sci. Technol., 31(4): 373–386.

Du, J., Kan, W., Bao, H., Jia, Y., Yang, J. et al. (2021b). Interactions between adenosine receptors and cordycepin (3′-deoxyadenosine) from *Cordyceps militaris*: possible pharmacological mechanisms for protection of the brain and the amelioration of covid-19 pneumonia. Journal of Biotechnology and Biomedicine, 4(2): 26–32.

Dunlap, C.A., Biresaw, G. and Jackson, M.A. (2005). Hydrophobic and electrostatic cell surface properties of blastospores of the entomopathogenic fungus *Paecilomyces fumosoroseus*. Colloids. Surf. B. Biointerfaces., 46(4): 261–266.

Elshamy, A.I., Mohamed, T.A., Yoneyama, T., Noji, M., Ban, S. et al. (2023). Bifusicoumarins AD: Cytotoxic 3S dihydroisocoumarins from the entomopathogenic fungus *Cordyceps bifusispora* (NBRC108997). Phytochem, 212: 113743.

Faita, M.R., Pereira, E. and Poltronieri, A.S. (2023). Effect of entomopathogenic fungus-formulated bioinsecticides on stingless bees (Hymenoptera: Apidae: Meliponini) in the laboratory. J. Apic. Res., pp. 1–8

Fan, J., Liu, P., Zhao, K. and Chen. H.P. (2023). Three previously undescribed metabolites from *Cordyceps cicadae* JXCH-1, an entomopathogenic fungus. Nat. Prod. Bioprospect., 13(1): 46.

Franzin, M.L., Moreira, C.C., da Silva, L.N.P., Martins, E.F., Fadini, M.A.M. et al. (2022). *Metarhizium* associated with coffee seedling roots: Positive effects on plant growth and protection against *Leucoptera coffeella*. Agriculture, 12: 2030. https://doi.org/10.3390/agriculture12122030.

Gao, T., Wang, Z., Huang, Y., Keyhani, N.O. and Huang, Z. et al. (2017). Lack of resistance development in *Bemisia tabaci* to *Isaria fumosorosea* after multiple generations of selection. Sci. Rep., 7: 42727.

Hajek, A.E. and Meyling, N.V. (2017). Fungi. pp. 327–377. *In*: Hajek, A.E. and Shapiro Ilan, D.I. (eds.). Ecology of Invertebrate Diseases. Wiley: Oxford, UK.

Hussain, A. and AlJabr, A.M. (2020a). Potential synergy between spores of *Metarhizium anisopliae* and plant secondary metabolite, 1-chlorooctadecane for effective natural acaricide development. Molecules, 25: 1900.

Hussain, A., Ali, M.W., AlJabr, A.M. and Al-Kahtani, S.N. (2020b). Insights into the *Gryllus bimaculatus* immune-related transcriptomic profiling to combat naturally invading pathogens. J. Fungi, 6: 232.

Jaronski, S.T. (2009). Ecological factors in the inundative use of fungal entomopathogens. Bio. Control., 55: 159–185

Jaturapat, A., Isaka, M., Hywel-Jones, N.L., Lertwerawat, Y., Kamchonwongpaisan et al. (2001). Bioxanthracenes from the insect pathogenic fungus *Cordyceps pseudomilitaris* BCC 1620 I. Taxonomy, fermentation, isolation and antimalarial activity. J. Antibiot., 54(1): 29–35.

Kaewkod, T., Ngamsaoad, P., Mayer, K.O., Cheepchirasuk, N., Promputtha, I. et al. (2023). Evaluation of the biological activity of *Cordyceps militaris* as an antioxidant, antibacterial, and anti-inflammatory agent for natural healthcare. Preprint https://doi.org/10.21203/rs.3.rs-3111990/v1.

Kang, W.J., Koo, H.N., Jeong, D.H., Kim, H.K., Kim, J. et al. (2017). Functional and genetic characteristics of Chlorantraniliprole resistance in the diamondback moth, *Plutella xylostella* (Lepidoptera: Plutellidae). Entomol. Res., 47: 394–403.

Kepler, R.M., Luangsa-Ard, J.J., Hywel-Jones, N.L., Quandt, C.A., Sung, G.H. et al. (2017). A phylogenetically-based nomenclature for Cordycipitaceae (Hypocreales). IMA Fungus 8: 335–353.

Keppanan, R., Krutmuang, P., Sivaperumal, S., Hussain, M., Bamisile, B.S. et al. (2019). Synthesis of mycotoxin protein IF8 by the entomopathogenic fungus *Isaria fumosorosea* and its toxic effect against adult *Diaphorina citri*. Int. J. Biol. Macromol., 125: 1203–1211.

Kim, J. R., Yeon, S. H., Kim, H. S. and Ahn, Y.J. (2002). Larvicidal activity against Plutella xylostella of cordycepin from the fruiting body of *Cordyceps militaris*. Pest Manag. Sci., 58: 713–717. doi:10.1002/ps.508.

Krasnoff, S.B., Reátegui, R.F., Wagenaar, M.M., Gloer, J.B. and Gibson, D.M. et al. (2005). Cicadapeptins I and II: New Aib-Containing Peptides from the Entomopathogenic Fungus *Cordyceps h eteropoda*. J. Nat. Prod., 68(1): 50–55.

Kwon, H., Bang, K. and Cho, S. (2014). Characterization of the hemocytes in larvae of *Protaetia brevitarsis seulensis*: Involvement of Granulocyte-Mediated Phagocytosis. PLoS One, 9: e103620.

Lei, Y., Hussain, A., Guan, Z., Wang, D., Jaleel, W. et al. (2021). Unraveling the Mode of Action of *Cordyceps fumosorosea*: Potential Biocontrol Agent against *Plutella xylostella* (Lepidoptera: Plutellidae). Insects, 12: 179.

Lei, Y.Y., He, Y.R. and Lu, L.H. (2011). Physiological defense responses of *Plutella xylostella* (Lepidoptera: Plutellidae) larvae infected by entomopathogenic fungus *Isaria fumosorosea*. Acta Entomol. Sin., 54: 887–893.

Li, X., Chen, H.P., Zhou, L., Fan, J., Awakawa, T. et al. (2022). Cordycicadins A–D, Antifeedant Polyketides from the Entomopathogenic Fungus *Cordyceps cicadae* JXCH1. Org. Lett., 24(47): 8627–8632.

Liu, L., Zhang, J., Chen, C., Teng, J., Wang, C. et al. (2015). Structure and biosynthesis of fumosorinone, a new protein tyrosine phosphatase 1B inhibitor firstly isolated from the entomogenous fungus *Isaria fumosorosea*. Fungal Genet. Biol., 81: 191–200.

Liu, M., Li, X., Huai, M., Yang, Y. and Dong, C. et al. (2023). Cultivation, Bioactive Metabolites, and Application of Caterpillar Mushroom *Cordyceps militaris*: Current State, Issues, and Perspectives. In Fungi and Fungal Products in Human Welfare and Biotechnology (pp. 187–210). Singapore: Springer Nature Singapore.

Liu, X.C., Li, H., Kang, T., Zhu, Z.Y., Liu, Y.L. et al. (2019). The effect of fermentation conditions on the structure and anti-tumor activity of polysaccharides from *Cordyceps gunnii*. RSC Adv., 9(32): 18205–18216.

Luo, Y., Wu, S., He, X., Wang, D., He, Y. et al. (2022). Identification of a *Cordyceps fumosorosea* fungus isolate and its pathogenicity against Asian citrus psyllid, *Diaphorina citri* (Hemiptera: Liviidae). Insects, 13(4): 374.

Mago, P., Kaur, S., Srivastava, I., Mehrotra, R., Kaur, K. et al. (2023). A review on unravelling the medicinal properties of magical mushrooms: *Cordyceps militaris*. Biomed., 43(5): 1368–1375.

Moino Jr, A., Alves, S.B., Lopes, R.B., Neves, P.M.O.J., Pereira, R.M. et al. (2002). External development of the entomopathogenic fungi *Beauveria bassiana* and *Metarhizium anisopliae* in the subterranean termite *Heterotermes tenuis*. Sci. Agric., 59: 267–273.

Molnár, I. and Carreira, E.M. (2010). Natural Product Reports Current developments in natural products chemistry. Nat. Prod. Rep., 27: 1235–1240.

Namasivayam, S.K.R., Bharani, R.S.A. Karunamoorthy, K. (2018). Insecticidal fungal metabolites fabricated chitosan nanocomposite (IM-CNC) preparation for the enhanced larvicidal activity-An effective strategy for green pesticide against economic important insect pests. Int. J. Biol. Macromol., 120: 921–944.

Ng, T.B. and Wang, H.X. (2005). Pharmacological actions of *Cordyceps*, a prized folk medicine. J. Pharm. Pharmacol., 57(12): 1509–1519.

Niermans, K., Woyzichovski, J., Kröncke, N., Benning, R. and Maul, R. et al. (2019). Feeding study for the mycotoxin zearalenone in yellow mealworm (Tenebrio molitor) larvae—investigation of biological impact and metabolic conversion. Mycotoxin. Res., 35: 231–242.

Olatunji, O.J., Feng, Y., Olatunji, O.O., Tang, J., Ouyang, Z. et al. (2016). Neuroprotective effects of adenosine isolated from *Cordyceps cicadae* against oxidative and ER stress damages induced by glutamate in PC12 cells. Environ. Toxicol. Pharmacol., 44: 53–61.

Olatunji, O.J., Tang, J., Tola, A., Auberon, F., Oluwaniyi, O. et al. (2018). The genus *Cordyceps*: An extensive review of its traditional uses, phytochemistry, and pharmacology. *Fitoterapia*, 129: 293–316.

Ozdemir, D. (2022). The impact of climate change on agricultural productivity in Asian countries: a heterogeneous panel data approach. Environ Sci Pollut Res, pp.1-13.

Parveen, S.S. (2023). Entomopathogenic fungi as a sustainable biopesticide: Current status and future prospects. J. Entomol. Res., 47(4): 666–671.

Prathumpai, W., Kocharin, K., Phimmakong, K. and Wongsa, P. (2007). Effects of different carbon and nitrogen sources on naphthoquinone production of *Cordyceps* unilateralis BCC 1869. Appl. Biochem. Biotechnol., 136: 223–232.

Qasim, M., Islam, S.U., Islam, W., Noman, A., Khan, K.A. et al. (2020). Characterization of mycotoxins from entomopathogenic fungi (*Cordyceps fumosorosea*) and their toxic effects to the development of asian citrus psyllid reared on healthy and diseased citrus plants. Toxicon., 188: 39–47.

Qu, S.L., Xie, J., Wang, J.T., Li, G.H., Pan X.R. et al. (2023). Activities and metabolomics of *Cordyceps gunnii* under different culture conditions. Front. Microbiol., 13: 1076577.

Quesada-Moraga, E., Garrido-Jurado, I., González-Mas, N. and Yousef-Yousef, M. 2023). Ecosystem services of entomopathogenic ascomycetes. J. Inverte. Pathol., 201: 108015. https://doi.org/10.1016/j.jip.2023.108015.

Quy, T.N., Xuan, T.D., Andriana, Y., Tran, H.D., Khanh, T.D. et al. (2019). Cordycepin isolated from *Cordyceps militaris*: Its newly discovered herbicidal property and potential plant-based novel alternative to glyphosate. Molecules, 24(16): 2901.

Rani, L., Thapa, K., Kanojia, N., Sharma, N., Singh et al. (2021). An extensive review on the consequences of chemical pesticides on human health and environment. J. Clean. Prod., 283: 124657.

Rastegari A.A., Yadav, A.N. and Yadav, N. (2019a). Genetic manipulation of secondary metabolites producers. pp. 13–29. *In*: Gupta, V.K. and Pandey, A. (eds.). New and future developments in microbial biotechnology and bioengineering. Elsevier, Amsterdam.

Rastegari A.A., Yadav, A.N., Yadav, N. and Sarshari N.T. (2019b). Bioengineering of secondary metabolites. pp. 55–68. *In*: Gupta, V.K. and Pandey, A. (eds.). New and future developments in microbial biotechnology and bioengineering. Elsevier, Amsterdam.

Ravindran, K., Sivaramakrishnan, S., Hussain, M., Dash, C.K., Bamisile, B.S. et al. (2018). Investigation and molecular docking studies of Bassianolide from *Lecanicillium lecanii* against *Plutella xylostella* (Lepidoptera: plutellidae). Comp. Biochem. Physiol. C Toxicol. Pharmacol., 206: 65–72

Rosa, E., Woestmann, L., Biere, A. and Saastamoinen, M. (2018). A plant pathogen modulates the effects of secondary metabolites on the performance and immune function of an insect herbivore. Oikos, 127: 1539–1549.

Sharma, S. K., Gautam, N. and Atri. N.S. (2015). Optimization, composition, and antioxidant activities of exo-and intracellular polysaccharides in submerged culture of *Cordyceps gracilis* (Grev.) Durieu & Mont. Evid Based Complementary Altern. Med., 2015.

Shashidhar, M.G., Giridhar, P., Sankar, K.U. and Manohar, B. (2013). Bioactive principles from *Cordyceps sinensis*: a potent food supplement—a review. J. Funct. Foods, 5: 1013–1030

Sornchaithawatwong, C., Kunthakudee, N., Sunsandee, N. and Ramakul, P. (2020). Selective extraction of cordycepin from *Cordyceps militaris*—optimisation, kinetics and equilibrium studies. Ind. Chem. Eng., 2020: 1–13.

Sparks, T.C. and Nauen, R. (2015). IRAC: Mode of action classification and insecticide resistance management. Pestic. Biochem. Physiol., 121: 122–128.

Stokes, B.A., Yadav, S., Shokal, U., Smith, L.C. and Eleftherianos, I. et al. (2015). Bacterial and fungal pattern recognition receptors in homologous innate signaling pathways of insects and mammals. Front. Microbiol., 6: 19.

Sun, X., Feng, X., Zheng, D., Li, A., Li, C. et al. (2019). Ergosterol attenuates cigarette smoke extract-induced COPD by modulating inflammation, oxidative stress and apoptosis *in vitro* and *in vivo*. Clin. Sci., 133(13): 1523–1536.

Tan, L., Song, X., Ren, Y., Wang, M., Guo, C. et al. (2021). Anti-inflammatory effects of cordycepin: A review. Phytother. Res., 35(3): 1284–1297.

Tomilova, O.G., Kryukova, N.A., Efimova, M.V. and Kolomeichuk, L.V. (2023). The endophytic entomopathogenic fungus *Beauveria bassiana* alleviates adverse effects of salt stress in potato plants. Horticulturae 2023, 9,1140. https://doi.org/10.3390/horticulturae9101140.

Tudi, M., Daniel Ruan, H., Wang, L., Lyu, J., Sadler, R., Connell, D., Chu, C., Phung, D.T. et al. (2021). Agriculture development, pesticide application and its impact on the environment. Int. J. Environ. Res. Public Health, 18(3): 1112.

Varughese, T., Rios, N., Higginbotham, S., Arnold, A.E., Coley, P.D. et al. (2012). Antifungal depsidone metabolites from *Cordyceps dipterigena*, an endophytic fungus antagonistic to the phytopathogen *Gibberella fujikuroi*. Tetrahedron Lett., 53: 1624–1626.

Venkatesh, G., Priya, P.S., Anithaa, V., Dinesh, G.K., Velmurugan, S. et al. (2022). Role of entomopathogenic fungi in biocontrol of insect pests. pp. 539–548. *In*: Soni, R., Suyal, D.C. and Goel, R. (eds.). Plant Protection, De Gruyter, Berlin/Boston.

Wang, X., Xu, J., Wang, X., Qiu, B., Cuthbertson, A.G. et al. (2019). *Isaria fumosorosea*-based zero-valent iron nanoparticles affect the growth and survival of sweet potato whitefly, Bemisia tabaci (Gennadius). Pest Manag. Sci., 75 (8): 2174–2181.

Wang, X., Wang, J., Cao, X., Wang, F., Yang, Y. et al. (2019). Long-term monitoring and characterization of resistance to chlorfenapyr in *Plutella xylostella* (Lepidoptera: Plutellidae) from China. Pest Manag. Sci., 75: 591–597.

Wei, J., Zhou, X., Dong, M., Yang, L., Zhao, C. et al. (2022). Metabolites and novel compounds with anti-microbial or antiaging activities from *Cordyceps fumosorosea*. AMB Express, 12(1): 1–14.

Wong, J.H., Ng, T.B., Wang, H., Cheung, R.C., Ng, C.C.W. et al. (2019). Antifungal proteins with antiproliferative activity on cancer cells and HIV-1 enzyme inhibitory activity from medicinal plants and medicinal fungi. Curr. Protein Pept. Sci., 20(3): 265–276.

Wong, J.H., Ng, T.B., Wang, H., Sze, S.C.W., Zhang, K.Y. et al. (2011). Cordymin, an antifungal peptide from the medicinal fungus *Cordyceps militaris*. Phytomedicine, 18(5): 387–392.

Woolley, V.C. (2018). Elucidating the natural function of cordycepin, a secondary metabolite of the fungus *Cordyceps militaris*, and its potential as a novel biopesticide in Integrated Pest Management (Doctoral dissertation, University of Warwick).

Wu, J., Yang, B., Xu, J., Cuthbertson, A.G. S. and Ali, S. et al. (2021). Characterization and Toxicity of Crude Toxins Produced by *Cordyceps fumosorosea* against *Bemisia tabaci* (Gennadius) and *Aphis craccivora* (Koch). Toxins, 13: 220

Wu, S., Reddy, G.V. and Jaronski, S.T. (2014). Advances in microbial insect control in horticultural ecosystem. In Sustainable Development and Biodiversity 2; (Nandawani, D., Ed.), Springer: Basel, Switzerland, pp. 223–252.

Xiao, J.H. and Xiong, Q. (2013). Nucleosides, a valuable chemical marker for quality control in traditional Chinese medicine *Cordyceps*. Recent. Pat. Biotechnol., 7(2): 153–166.

Xiao, J.H., Liang, Z.Q., Liu, A.Y., Chen, D.X., Xiao, Y. et al. (2004). Immunosuppressive activity of polysaccharides from *Cordyceps gunnii* mycelia in mice *in vivo/vitro*. J. Food Agric. Environ., 2(3): 69–73.

Yoneyama, T., Elshamy, A.I., Yamada, J., El-Kashak, W.A., Kasai, Y. et al. (2022). Antimicrobial metabolite of *Cordyceps tenuipes* targeting MurE ligase and histidine kinase via in silico study. Appl. Microbiol. Biotechnol., 106(19–20): 6483–6491.

Yoon, D.H., Han, C., Fang, Y., Gundeti, S., Han Lee, I.S. et al. (2017). Inhibitory activity of *Cordyceps bassiana* extract on LPS induced inflammation in RAW 264.7 cells by suppressing NF-κB activation. Nat. Prod. Sci., 23(3): 162–168.

Zhang, J., Yu, Y., Zhang, Z., Ding, Y., Dai, X. et al. (2011). Effect of polysaccharide from cultured *Cordyceps sinensis* on immune function and anti-oxidation activity of mice exposed to 60Co. Int. Immunopharmacol., 11(12): 2251–2257

Zhang, H.H., Luo, M.J., Zhang, Q.W., Cai, P.M., Idrees, A. et al. (2019). Molecular characterization of prophenoloxidase-1 (PPO1) and the inhibitory effect of kojic acid on phenoloxidase (PO) activity and on the development of Zeugodacus tau (Walker) (Diptera: tephritidae). Bull. Entomol. Res., 109: 236–247.

Zhang, Y., Liu, J., Wang, Y., Sun, C., Li, W. et al. (2022). Nucleosides and amino acids, isolated from *Cordyceps sinensis*, protected against cyclophosphamide-induced myelosuppression in mice. Nat. Prod. Res., 36(23): 6056–6059.

Zheng, Y., He, S., Chen, B., Su, Z., Wang, W. et al. (2019). Time concentration-mortality response of potato tuber moth pupae to the biocontrol agent *Cordyceps tenuipes*. Biocontrol. Sci. Technol., 29(10): 965–978.

Zhou, X., Gong, Z., Su, Y., Lin, J. and Tang, K. et al. (2009). *Cordyceps* fungi: natural products, pharmacological functions and developmental products. J. Pharm. Pharmacol., 61(3): 279–291.

Zhu, Z.Y., Liu, N., Si, C.L., Liu, Y., Ding, L.N. et al. (2012). Structure and anti-tumor activity of a high-molecular-weight polysaccharide from cultured mycelium of *Cordyceps gunnii*. Carbohydrate polymers, 88(3): 1072–1076.

Zhu, Z.Y., Liu, X.C., Tang, Y.L., Dong, F.Y., Sun, H.Q. et al. (2016). Effects of cultural medium on the formation and antitumor activity of polysaccharides by *Cordyceps gunnii*. J. Biosci. Bioeng, 122(4): 494–498

Safety and Products

13

Heavy Metals of *Cordyceps* and Allied Species

*Purnendu Paul** and *Prakash Pradhan**

1. Introduction

Heavy metals are generally metallic elements with a high density compared with water (Fergusson, 1990). Heavy metals (HMs) are widespread contaminants found in the environment stemming from either natural courses (weathering of metal-containing rocks and volcanic outbreaks) or anthropogenic activities, including mining, coal combustion, waste disposal and pesticides application. A total of 54 heavy metals are found in nature, out of which heavy metals like cadmium (Cd), lead (Pb) and mercury (Hg) have detrimental effects on organisms (Osman et al., 2019). Metalloids, including arsenic (As) are also encompassed within the category of heavy metals (HMs), that can cause toxicity at very low levels of exposure (Duffus, 2002). Fungi, including edible species of the genera *Agaricus, Ganoderma, Cordyceps, Ophiocordyceps* and *Pleurotus* are excellent biosorbents that can absorb heavy metals from the environment. The cell wall material of these fungi contains a high level of functional groups that can interact and bind with heavy metals. (Dhankhar et al., 2011). Therefore, this could potentially present a significant human health risk when consuming these heavy metal-laden edible fungi. (Concha et al., 1998). The *Cordyceps* and allied species (CAs) constitute an entomopathogenic group of edible fungi, whose

West Bengal Biodiversity Board, Prani Sampad Bhawan, 5th Floor, LB-2, Sector-III, Salt Lake, Kolkata, West Bengal, India – 700 106.
* Corresponding authors: purno08psb@gmail.com; shresthambj@gmail.com

representatives include the caterpillar fungus *O. sinensis* (Dongchong xiacao), *C. militaris* (BeiChongCao) and others having a long history of safe use in Chinese medicine and having bioactive properties such as immunological modulation (Zhou et al., 2009, Shi et al., 2009), anti-cancer (Nakamura et al., 2015), anti-inflammatory activities (Won et al., 2005). The natural habitat for *Cordyceps* and allied species (CAs), where they parasitize the larvae of insects and eventually grow into the characteristic fruiting body, is indeed quite restricted. It primarily thrives in the high-altitude regions of the Tibetan Plateau and the Himalayas. Due to its limited natural habitat and the increasing demand for *Cordyceps* and allied species in the market, there has been a significant impact on local ecosystems and their population.

The increase in demand has led to growing enthusiasm for the artificial cultivation of *Cordyceps* and allied species in recent times. The cultivation of *Cordyceps* and allied species in controlled environments has the potential to alleviate the environmental pressures on their natural habitats and help preserve the wild populations of the fungus. Nevertheless, there is a concern regarding the potential absorption of heavy metals by *Cordyceps* and allied species during the growth process, as they may uptake these contaminants from the cultivation medium or substrate. With the surge in agricultural and industrial activities, the environmental pollution caused by heavy metals (HMs) in soil, water, and air has significantly accelerated (Callender, 2003). Heavy metals (HMs) are usually not biodegradable in nature and can persist in the environment for a very long time, even after the elimination of their point sources (Gall et al., 2015), causing severe impacts on living organisms (Zhao et al., 2019; Zhang et al., 2019; Hausladen et al., 2018; Itabashi et al., 2019). They can induce oxidative damage to cells (Thatoi et al., 2014; Bienert et al., 2018), cause structural harm to the cell (Cao et al., 2019), lead to the deactivation of enzymes (Balistrieri et al., 2018), affect the cellular uptake of nutrients (Murthy et al., 2014), and are also capable of destroying DNA structure (Nicolaus et al., 2016).

This review contemplates with the different parameters in terms of heavy metals in *Cordyceps* and allied species. Initially, it delves into the nature, distribution, market presence, and uses of common *Cordyceps* and allied species. Subsequently, it explores the diversity of heavy metals across different *Cordyceps* and allied species, elucidating their sources and the potential effects of these metals on them. The review further examines the bioaccessibility and speciation of heavy metals, contributing to a nuanced understanding of their presence in *Cordyceps* and allied species. A critical aspect is the health risk assessment, which evaluates the potential impact on consumers. The discussion extends to the existing limits for toxic heavy metals and the actual metal composition within *Cordyceps* and allied species. By doing so, the review aims to outline the current state of knowledge and identifies gaps in heavy metal studies within *Cordyceps* and allied species. It encourages technological advancements, sets standards, and fosters health risk assessments—both *in vitro* and *in vivo*. The overarching goal is to facilitate the safe utilization of *Cordyceps* and allied species for a healthy future.

2. *Cordyceps* - Distribution and Market

Cordyceps (Fr.) Link (family includes Cordycipitaceae, Ophiocordycipitaceae, and partial Clavicipitaceae) contains approximately 1300 species and has a range of diverse species, morphology, and variety of insect hosts (Sung et al., 2007). The species are mostly distributed in Asian countries like China, India, Japan, Korea and Nepal in the humid and tropical environments in other parts of the world (Olatunji et al., 2018).

Ophiocordyceps sinensis (Berk.) G.H. Sung, J.M. Sung, Hywel-Jones, and Spatafora (Ophiocordycipitaceae) (syn. *Cordyceps sinensis*) is primarily found in the Tibetan Plateau and its adjacent regions, which encompass areas such as Tibet, Gansu, Qinghai, Sichuan, and Yunnan Provinces in China (Yan et al., 2014; Li et al., 2019a). It also occurs in specific locations along the southern side of the Himalayan Mountain range, including regions in Bhutan, India, and Nepal. The fungus is typically distributed at altitudes not lower than 3000 to 5000 m above sea level (Shashidhar et al., 2013; Pradhan et al., 2024). A recent investigation has shown that the geographic range of the fungus will noticeably contract, with an upward shift in attitude towards the central regions of the Plateau (Chang and Wasser, 2017).

Cordyceps militaris (L.) Fr. (Cordycipitaceae) (the type species, *Cordyceps*), commonly known as the orange caterpillar mushroom, stands as the second most extensively studied species within its genus (Phull et al., 2022). This naturally occurring fungus thrives in mountainous regions at altitudes ranging from 150 to 2215 m asl (Pradhan et al., 2016), displaying a global distribution and exhibiting ease of artificial cultivation. Shrestha et al. (2012) have demonstrated the feasibility of cultivating both the stroma and submerged culture mycelia of *C. militaris*. Notably, this species shares medicinal properties comparable to those of its renowned allied species, *Ophiocordyceps sinensis*. Its adaptability to artificial cultivation methods further positions *C. militaris* as a promising subject for research and application (Pradhan et al., 2024). The fundamental distinction between the genera *Ophiocordyceps* and *Cordyceps* lies in the fact that *Ophiocordyceps* produces complete ascospores that do not separate into part spores. In contrast, *Cordyceps* either produces perithecia, whether immersed or superficial and perpendicular to the stromal surface, with the ascospores disassembling into part spores upon reaching maturity (Sung et al., 2007).

There are over 2,300 species of edible and medicinal fungi in the world (Beulah et al., 2013) and global mushroom production totals around 40 million tons, with China alone contributing approximately 31.7 million tons in 2017, constituting nearly 47% of the world's mushroom supply (Chang and Wasser, 2012; Chang and Wasser, 2017). The annual global production of CAs ranges from approximately 83.2 and 182.5 tons (Winkler, 2009). China stands as the leading producer, yielding between 80 to 175 tons of CAs annually. Following China, Nepal produces an estimated 1.0 to 3.2 tons, while India contributes approximately 1.7 to 2.8 tons, and Bhutan's production ranges from 0.5 to 1.5 tons (Shrestha et al., 2013). Currently, the price of natural *Ophiocordyceps* exceeds that of gold, with high-quality products priced as much as US$ 60,000 per kilogram (Lei et al., 2015). This high price has fostered a "*Cordyceps* industry" whose global market price is estimated at US$ 5 billion to $11 billion (Shrestha et al., 2012).

3. Common *Cordyceps* and Uses

According to MycoBank (https://www.mycobank.org), the *Cordyceps* and allied species are very diverse, mostly endoparasites of insects and other arthropods, while some of them are parasitic on other fungi as well (Nikoh et al., 2000). The most common *Cordyceps* and allied species that are economically important are *Ophiocordyceps sinensis* and *Cordyceps militaris*, which are highly valued for their medicinal benefits in Traditional Chinese medicinal and Chinese herbal medicine. Apart from these other *Cordyceps* and allied species having economic importance are *Cordyceps kyushuensis*, *C. pruinosa*, *C. bassiana*, *C. cicadae*, *C. gunnii*, *C. guangdongensis*, and *C. ophioglossoides* and others. (Zhu et al., 1998; Zhang et al., 2014). Common uses of *Cordyceps* and allied species are presented in Table 1.

4. Heavy metals in *Cordyceps*

CAs are known to accumulate heavy metals through various mechanisms, including absorption from the substrate, translocation from the host organism, and concentration in the fruiting body. Numerous studies have focused on analyzing the essential metal, non-essential metal and heavy metal content of CAs that are presented in Table 2 (Zuo et al., 2013; Zhang et al., 2013; Wei et al., 2017; Zhou et al., 2018; Fenglin et al., 2020; Xiao et al., 2021; Liu et al., 2021; Wu et al., 2015). The studies mainly utilize conventional techniques, including Atomic Absorption Spectroscopy (AAS) (Zuo et al., 2013), Inductively Coupled Plasma Mass Spectrometry (ICP-MS) (Zhang et al., 2013; Wei et al., 2017; Zhou et al., 2018; Fenglin et al., 2020; Xiao et al., 2021; Liu et al., 2021), and Inductively Coupled Plasma Atomic Emission Spectroscopy (ICP-AES) (Wu et al., 2015). ICP-MS is considered an advanced

Table 1. Uses of *Cordyceps* species.

Cordyceps	Use	Reference
Cordyceps bassiana Z.Z. Li, C.R. Li, B. Huang & M.Z. Fan	Used for skin conditions such as dermatitis and eczema. [k]	[k] Olatunji et al., 2018
Cordyceps cicadae (Miq.) Massee	Used to treat infantile convulsions, elevation of temperature, antitumor, immunoregulation, and reno-protection [l]	[l] Olatunji et al., 2018
Cordyceps guangdongensis T.H. Li, Q.Y. Lin & B. Song	Used against fatigue, avian influenza, inflammation, renal failure, and oxidation. [o]	[o] Yan et al., 2013
Cordyceps gunnii (Berk.) Berk.	Immunomodulatory and Anti-aging activities. [m] [n]	[m] Zhu et al., 2012b [n] Zhu et al., 2014
Cordyceps japonica Lloyd	Used against Cancer and boosting stamina. [t]	[t] Song et al., 2007
Cordyceps kyushuensis A. Kawam.	Used for treatment of fatigue, night sweating, hyperglycemia, hyperlipidemia, asthenia after severe illness, respiratory disease, renal dysfunction, arrhythmias, and other heart and liver diseases. [q]	[q] Zhang et al., 2013
Cordyceps militaris (L.) Fr.	Used as an energy enhancer, aphrodisiac source, and treatment for respiratory conditions. Additionally,it exhibits hypoglycemic, anti-inflammatory, antitumor, antibacterial, antifungal, antioxidant, and immuno-protective properties. [h] [i]	[h] Chou et al., 2014 [i] Das et al., 2010
Cordyceps ophioglossoides (J.F. Gmel.) Fr.	Exhibit antitumor, estrogenic, and anti-aging properties, for application in births to avoid excessive bleeding in women. [p] [q]	[p] Kawagishi et al., 2004
Cordyceps pruinosa Petch	Used for curing stomach diseases. [j]	[j] Wu et al., 2015
Ophiocordyceps sinensis (Berk.) G.H. Sung, J.M. Sung, Hywel-Jones, and Spatafora	Energy level and endurance enhancer, boost cellular immunity. [a], used in treatment of patients suffering from severe acute respiratory syndrome (SARS). [b], use as antitumor, cerebroprotective [c], gastroprotective[d], immunomodulatory [e,f], antiproliferative [g].	[a] Shashidhar et al., 2013 [b] Chiu et al., 2016a [c] Liu et al., 2011 [d] Chen et al., 2013 [e] Akaki et al., 2009 [f] Zhang et al., 2011 [g] Zhang et al., 2007
Cordyceps sobolifera (Hill ex Watson) Berk. & Broome	Used for immunopotentiation, fever reduction, relieving convulsion, and renoprotective activities [r]	[r] Zhong et al., 2009
Cordyceps taii Z.Q. Liang & A.Y. Liu	Exhibit Immunomodulatory, antiaging, antioxidant, antimutagenicity, cytotoxic and antimicrobial properties. [s]	[s] Xiao et al., 2012

Table 2. Metals and elements detected in *Cordyceps* species (ND = Not Detected; NS= Not Studied; [a] Zhou et al., 2018, [b] Fenglin et al., 2020, [c] Zuo et al., 2013, [d] Wei et al., 2017, [e] Wu et al., 2015, [f] Zhang et al., 2013, [g] Liu et al., 2021).

Element	*Cordyceps sinensis*	*Cordyceps militaris*	*Cordyceps kyushuensis*
Essential Elements			
Sodium (Na)	NS	NS	Detected [f]
Potassium (K)	NS	NS	Detected [f]
Magnesium (Mg)	Detected [d]	NS	Detected [f]
Calcium (Ca)	Detected [d]	NS	Detected [f]
Zinc (Zn)	Detected [d]	Detected [b]	Detected [f]
Iron (Fe)	Detected [d]	NS	Detected [f]
Copper (Cu)	Detected [a] [c][d]	Detected [b]	Detected [f]
Manganese (Mn)	Detected [d]	Detected [b]	Detected [f]
Vanadium (V)	Detected [d]	NS	Detected [f]
Chromium (Cr)	Detected [c][d]	Detected [e]	Detected [f]
Nickel (Ni)	Detected [d]	Detected [e]	Detected [f]
Cobalt (Co)	Detected [d]	NS	Detected [f]
Molybdenum (Mo)	Detected [d]	NS	Detected [f]
Selenium (Se)	ND [d]	NS	Detected [f]
Tin (Sn)	Detected [d]	NS	Detected [f]
Strontium (Sr)	Detected [d]	NS	NS
Non-essential Elements			
Aluminium (Al)	Detected [d]	NS	NS
Silver (Ag)	Detected [d]	Detected [b]	NS
Barium (Ba)	Detected [d]	NS	Detected [f]
Titanium (Ti)	ND [d]	NS	NS
Phosphorus (P)	Detected [d]	NS	NS
Heavy Metals			
Mercury (Hg)	Detected [a]		Detected [f]
Arsenic (As)	Detected [a] [c] [d][g]	Detected [e]	Detected [f]
Cadmium (Cd)	Detected [a] [c] [d]	Detected [e]	Detected [f]
Lead (Pb)	Detected [a] [c] [d]	Detected [e]	Detected [f]
Antimony (Sb)	Detected [d]	NS	NS

technology in heavy metal analysis due to its high detection sensitivity, capability for multiple element analysis, and isotope analysis. On the other hand, techniques like AAS and ICP-AES are also used simultaneously, but their sensitivity is lower compared to ICP-MS. Literature reveals that most studies have used ICP-MS to determine the presence of heavy metals in the sample. However, the running cost of ICP-MS is the highest compared to other techniques (Kagaya et al., 2009). Thus, it is evident that the detection of heavy metals in CAs is very expensive, possibly leading to a limited number of studies in this area.

In recent times, some new technologies have evolved and gained popularity for detecting heavy metals, including Laser-Induced Breakdown Spectrometry (LIBS) (De Lucia et al., 2011), Nanomaterial-Modified Aptamer-Based Electrochemical Sensors, Portable Atmospheric Pressure Discharge Plasma (APDP), Ratiometric Photoelectric Sensors, and Quantum Dots-Based Fluorescent Probes (Zuo et al., 2023). These techniques are considered advanced as they allow real-time detection of heavy metals. However, in the case of CAs for heavy metal detection, no study has been found that utilizes these newer technologies. Heavy metal presence has been detected of *Cordyceps* and allied species, including *O. sinensis*, *C. militaris*, and *C. kyushuensis*, according to the present literature review. Apart from these three species, no substantial research has been conducted to detect heavy metals in other *Cordyceps* and allied species. Therefore, there is a dearth in *Cordyceps* and allied species research regarding heavy metals, presenting a significant opportunity for exploring this aspect.

5. Standards for Heavy Metals in *Cordyceps*

The presence of heavy metals in food poses a significant health risk, leading to various health problems (Mitra et al., 2022). In response to this concern, many international and national regulatory bodies have established limits for heavy metal concentrations in food (FSSR, 2011), including CAs (Guo et al., 2017; NPC, 2019; Bhalla et al., 2022). These limits are crucial in ensuring that the levels of heavy metals in consumables remain within safe and permissible bounds for the well-being of public health. The literature suggests that the limits for arsenic (As) are exceeded in *Ophiocordyceps sinensis* according to various standards. Additionally, the lead (Pb) levels in *O. sinensis* are approaching the permissible limits set by different standards. However, there is a notable variance in these limit standards across different authorities, underscoring the need to establish a unified global standard for heavy metal limits, specifically in CAs. This standardization is essential to guaranteeing the quality of these remarkable gourmet nutraceutical commodities.

Table 3. Standards for toxic heavy metals in foodstuffs.

Heavy metal	Limit	Standard	Reference
Pb	10.0 mg/kg	ISO 18664: 2015 Traditional Chinese Medicine	Guo et al., 2017
As	4.0 mg/kg		
Cd	2.0 mg/kg		
Hg	3.0 mg/kg		
Pb	5.0 mg/kg	National Pharmacopoeia Commission, 2019 (NPC)	NPC, 2019
As	2.0 mg/kg		
Cd	1.0 mg/kg		
Hg	0.2 mg/kg		
Pb	10.0 mg/kg	Food and Agriculture Organization of the United Nations and World Health Organization (FAO/WHO)	Bhalla et al., 2022
As	10.0 mg/kg		
Cd	0.3 mg/kg		
Hg	1.0 mg/kg		
Pb	10 mg/kg	The Food Safety Standards Authority of India (FSSAI)	Food Safety and Standards (Contaminants, Toxins and Residues) Regulations, 2011; Version -V (2020)
As	5 mg/kg		
Cd	1.5 mg/kg		
Hg	1 mg/kg		

Heavy metal contamination in natural products, such as *Cordyceps* and allied species, is a critical issue as it directly affects human health. The toxicity of natural drugs depends on the mode and frequency of their usage. Different modes of consuming *Cordyceps* and allied species , such as in the form of supplements, extracts, or traditional preparations, may result in varying degrees of heavy metal exposure. Additionally, the frequency of consumption plays a crucial role in determining the cumulative exposure to these metals over time. Therefore, to standardize limits for heavy metals in CAs, it is imperative to take into consideration the mode and frequency of consumption for a more accurate determination of these standards.

Across various standards and regulations, it has been observed that the limits for arsenic in *O. sinensis* are consistently exceeded. Arsenic is a highly toxic metalloid, and its occurrence in food can lead to severe health issues. The proximity of lead levels in *O. sinensis* to permissible limits as defined by different standards is also a cause for concern. Lead toxicity is associated with a range of health problems, particularly impacting the nervous system and cognitive functions (Balali-Mood, 2021).

The significant divergence in limit standards from one authority to another emphasizes the need for a globally accepted standard for heavy metal concentrations in *Cordyceps* and allied species. A unified standard would provide clarity and consistency, ensuring that products of CAs meet the same rigorous criteria regardless of their origin. Establishing such a standard is crucial not only for consumer safety but also for international trade and the reputation of CAs products in the global market.

5.1 *Ophiocordyceps sinensis*

Table 2 illustrates the diversity of heavy metals detected in *O. sinensis*. However, studies on *O. sinensis* have revealed that the presence of heavy metals differs among samples sourced from different regions. The literature primarily focuses on various regions in China, including Tibet province, Sichuan province (Zuo et al., 2013), and Qinghai province (Zhou et al., 2018). A study by (Zuo et al., 2013) found the range of heavy metals to be As (0.49–13.88 mg/kg), Cd (0.05–0.26 mg/kg), Hg (0.00–0.21 mg/kg), Pb (0.04–5.12 mg/kg), Cr (0.02–1.75 mg/kg), and Cu (9.30–38.40 mg/kg). Wei (2017) conducted a more comprehensive study, determining 24 essential, non-essential, and toxic heavy metals from samples collected from five different places in China. This study provided a broader range of heavy metals and revealed their concentrations in both the caterpillar body and mycelia of *O. sinensis* (Table 4). Notably, five toxic heavy metals, As, Cd, Sb, Hg and Pb were found in caterpillar bodies and mycelia, with varying concentrations across samples. Arsenic(As) ranged from 0.37 to 32 mg/kg, Cadmium(Cd) from 0.014 to 0.163 mg/kg, Mercury(Hg) from 0.004 to 0.014 mg/kg, Lead(Pb) from 0.22 to 1.63 mg/kg, and Antimony(Sb) from 0.0004 to 0.034 mg/kg. Interestingly, the study observed significantly higher levels of Arsenic (As) in caterpillar bodies compared to mycelia. The quantities of Cadmium (Cd) and Lead (Pb) varied among samples, while Mercury (Hg) and Antimony (Sb) were only detected in a few samples, albeit in low quantities, with higher amounts in mycelia compared to caterpillar bodies.

Additionally, Zhou's (2018) study, conducted in Qinghai province, China, found variations in metal levels across different parts of the fungus. The average content of metals in *O. sinensis* was as follows: Copper (Cu) - 15.02 mg/kg in the caterpillar and 18.63 mg/kg in the stroma; Lead (Pb) - 1.39 mg/kg in the caterpillar and 2.06 mg/kg in the stroma; Arsenic (As) - 13.47 mg/kg in the caterpillar and 1.49 mg/kg in the stroma; Cadmium (Cd) - 0.06 mg/kg in the caterpillar and 0.07 mg/kg in the stroma; Mercury (Hg) - 0.08 mg/kg in the caterpillar and 0.22 mg/kg in the stroma. Stroma exhibited higher quantities of Copper (Cu) ,Lead (Pb), Cadmium (Cd), and Mercury (Hg) as compared to the caterpillar body. However, Arsenic (As) was present in higher proportions in the caterpillar, supporting the findings of Wei (2017).

5.2 *Cordyceps militaris*

The fruiting body and mycelia of *C. militaris* have been found to absorb heavy metals, as indicated in studies (Table 2). However, similar to *O. sinensis*, there are variations in the range of absorption. *C. militaris* is well-known as a substitute for *O. sinensis* and can be easily cultivated at

Table 4. Ranges of metals and elements in *Cordyceps* species ([a] Zuo et al., 2013 [b] Wei at al., 2017 [c] Zhou et al., 2018 [d] Fenglin et al., 2020 [e]Wu et al., 2015 [f] Zhang et al.,. 2013).

Species	Concentrations (range/average)	Compliance with standards
O. sinensis (Tibet province, Sichuan province) [a]	As: 0.49–13.88 mg/kg, Cd: 0.05–0.26 mg/kg, Hg: 0.00–0.21 mg/kg, Pb: 0.04–5.12 mg/kg, Cr: 0.02–1.75 mg/kg, Cu: 9.30–38.40 mg/kg	As exceeds limit by all standards
O. sinensis (Certified stores located in different region of China) [b]	Mg: 805.1–1749 mg/kg, Al: 73.0–705.8 mg/kg, P: 969.3–7185.0 mg/kg, Ca: 122.7–995.5 mg/kg, V: 0.4098–3.7599 mg/kg, Cr: 0.0877–2.3151 mg/kg, Mn: 18.4–80.7 mg/kg, Fe: 138.9–1747.3 mg/kg, Co: 0.0998–1.2792 mg/kg, Ni: 0.2766–3.7608 mg/kg, Cu: 5.7631–15.2888, Zn: 55.5–155.3 mg/kg, As: 0.37–32 mg/kg, Se: 0.034 mg/kg, Sr: 0.794–4.299 mg/kg, Mo: 0.032–2.326 mg/kg, Ag: 0.005–0.759 mg/kg, Cd: 0.014–0.163 mg/kg Sn: 0.001–0.048 mg/kg Sb: 0.004–0.034 mg/kg, Ba: 3.451–13.10 mg/kg, Hg: 0.004–0.014 mg/kg, Ti: 0.003–0.004 mg/kg, Pb: 0.22–1.63 mg/kg	As content in Caterpillar host exceeds limit by all standards
O. sinensis (Qinghai province) [c]	Cu: 15.02–18.63 mg/kg, Pb: 1.39–2.06 mg/kg, As: 1.49–13.47 mg/kg, Cd: 0.06–0.07 mg/kg, Hg: 0.08 mg/kg–0.22 mg/kg	As content in Caterpillar host exceeds limit by all standards
C. militaris (Jilin province) [d]	As: 0.0047 mg/kg, Pb: 0.22 mg/kg, Cd: 0.0041 mg/kg Cu: 9.1mg/kg, Ag: 0.024, Zn: 5.2, Mn: 1.6	All within the limit
C. militaris (Shenyang province) [e]	Ni: 0.2394 mg/kg, Cr: 0.0703 mg/kg, Pb: 0.8369 mg/kg, Cd: 0.0751 mg/kg, As: 0.100 mg/kg, Hg: ND	All within the limit
C. kyushuensis (Shandong province) [f]	Na: 876–1420 mg/kg, K: 530–946 mg/kg, Ca: 238–547 mg/kg, Mg: 117–986 mg/kg, Zn: 27.43–84.81 mg/kg, Fe: 39.20–114.72 mg/kg, Cu: 3.56–14.50 mg/kg, Mn: 3.12–34.21 mg/kg, V: 1.94–20.97 mg/kg, Cr: 3.21–18.62 mg/kg, Ni: 1.35–2.86 mg/kg, Co: 0.46–2.77 mg/kg, Mo: 0.61–0.74 mg/kg, Se: 0.42–2.73 mg/kg, Ba: 1.89–8.12 mg/kg, Sn: 0.15–0.76 mg/kg, As: 0.38–0.64 mg/kg, Pb: 0.51–1.09 mg/kg, Cd: 0.30–1.70 mg/kg, Hg: 0.10–0.40 mg/kg	All within the limit

lower altitudes. A study of Fenglin (2020), conducted on cultivated fruiting bodies from Jilin province, revealed the presence of arsenic (As), cadmium (Cd), copper (Cu), lead (Pb), manganese (Mn), silver (Ag) and zinc (Zn) in the fruit bodies of *Cordyceps militaris*.

The content of arsenic (As) in the cultivated fruit bodies of *Cordyceps militaris* was 0.0047 mg/kg, the content of lead (Pb) was 0.22 mg/kg, and the content of cadmium (Cd) was determined to be 0.0041 mg/kg. This raises concerns about heavy metal contamination, even in cultivated *Cordyceps* and allied species. However, the study did not discuss the sources of these heavy metals, assuming they may come from the growth medium. A study on rice grain-based cultivation of *C. militaris* indicated that the fungus can absorb heavy metals from the rice grain medium (Chen et al., 2011).

Another study conducted in Shenyang province on the finding of heavy metals in *Cordyceps militaris* during different seasons showed average values for Nickel(Ni) (0.2394 mg/kg), Chromium (Cr) (0.0703 mg/kg), Lead (Pb) (0.8369 mg/kg), Cadmium (Cd) (0.0751 mg/kg), and Arsenic (As) (0.100 mg/kg). The sequence of heavy metals in *Cordyceps militaris* was found to be Pb(Lead) > Ni(Nickel) > As(Arsenic) > Cd(Cadmium) > Cr(Chromium), with Hg(Mercury) not detected in most samples. Overall, these two studies revealed that the quantity of heavy metals in *C. militaris* is much less compared to *O. sinensis*, both in natural and cultivated fruiting bodies (Wu et al., 2015).

5.3 *Cordyceps kyushuensis*

A study conducted by Zhang et al. (2013) analyzed 20 different elements in both natural and cultured *Cordyceps kyushuensis*. These elements included macro elements such as Na, K, Ca, Mg, and micro elements Zn, Fe, Cu, Mn, V, Cr, Ni, Co, Mo, Ba, Sn and Se, as well as toxic elements such as Cd, Pb, Hg, and As. The study revealed that Ba was present in substantial quantities in all three samples, namely natural stroma, cultured stroma, and natural worm, ranging from 1.89 to 8.12 mg/kg. On the other hand, Sn, Pb, Cd, and Hg were found in very low quantities. Interestingly, Cd and Hg were not noticed in the cultured stroma, indicating that maintaining quality control of the culture medium could help mitigate the chances of heavy metal contamination.

One unique aspect of the study by Zhang et al. (2013) was that the heavy metal content was reported to be higher in the stroma compared to the caterpillar host for all 20 metals. This finding contrasts with previous studies conducted on *Cordyceps* and allied species. The significance of this observation lies in supporting the theory that the metal profile in Cordyceps and allied species not only varies from place to place but also differs among species.

6. Source of Heavy Metals in *Cordyceps*

In the environment, the sources of heavy metals can be both natural and anthropogenic. Anthropogenic sources pose a significant threat. Heavy metal contamination stems from various human-related activities, such as polluted water, the use of pesticides or herbicides, the application of phosphate-based fertilizers, sewage discharge, coal power plants, and the emission of particulate matter from vehicles, among others (Rai et al., 2019). In China, approximately 53% of provinces experience soil pollution ranging from moderate to high levels, primarily due to anthropogenic influences (Shifaw, 2018). This highlights the substantial impact of human activities on the environment, particularly concerning the presence of heavy metals in the soil.

Recent research has confirmed that *Cordyceps* and allied species possess the capability to absorb various minerals, particularly heavy metals, from their growth environment, including atmospheric surroundings, the culture medium and water (Fenglin et al., 2020). In some instances, this absorption is heightened due to the prolonged life of larvae, where the fungus's conidia infect the larvae, leading to the inadvertent accumulation of metals and metalloids during the larvae's feeding process. This phenomenon may result in elevated levels of these elements in *Cordyceps* and allied species (He et al., 2015).

Several studies have indicated that heavy metal contamination in *Cordyceps* and allied species could be influenced by distinct growth environments, encompassing factors such as host organism, humidity, pH, illumination climate, natural soil, organic matter content of artificial medium, temperature, and the wild worm (Zhang et al., 2013; Chen et al., 2018b; Liu et al., 2021). Additionally, research has confirmed the existence of robust correlations between heavy metal quantities in the soil and those found in *Cordyceps* and allied species (Chen et al., 2018a). This suggests that the environmental conditions in which *Cordyceps* and allied species grow play a crucial role in determining their heavy metal content.

Furthermore, the selectivity of *Cordyceps* and allied species in absorbing heavy metals has been documented in the literature. Some heavy metals, such as mercury (Hg) and chromium (Cr), exhibit moderate absorption, while others, notably arsenic (As) and copper (Cu), demonstrate a high degree of absorption by *Cordyceps* and allied species (Zuo et al., 2013). Also, some strains of CAs have a greater capability to absorb a particular heavy metal than other *Cordyceps* and allied species strains (Chen et al., 2018a).

7. Effect of Heavy Metals on the *Cordyceps*

Fungi can survive in adverse conditions due to their cellular structure and they are more tolerant against heavy metals because of their altered cellular mechanisms (Gajewska et al., 2022). The heavy metals are broadly divided into two groups: (a) essential heavy metals that are crucial for physiological growth but toxic when they exceed their quotas [e.g., iron, zinc, chromium and others]; (b) non-essential heavy metals with unknown biological functions that adversely affect the organisms [e.g., cadmium, lead, mercury and others] (Gajewska et al., 2022). So, at the same time, heavy metals are essential for growth and are toxic as well. The fungi can tolerate different concentrations of heavy metal exposure up to 800 ppm and some strains of *Penicilium simplicissimum* even show higher growth at high concentrations of heavy metals (Anahid et al., 2011).

A study conducted by Chen (2011) on *C. militaris* found that the growth of fruit bodies and the production of bioactive metabolites were sensitive to the source of the medium and its heavy metal content. Various sources of rice, serving as a substrate, exhibited varying quantities of heavy metals, and the presence of these metals significantly inhibited fruit body growth and metabolite production. Specifically, the heavy metals Lead (Pb), Cadmium(Cd), and Mercury(Hg) had a notable dose-dependent inhibitory effect on fruit body weight and height. Interestingly, there were no synergistic effects observed when all heavy metals were blended together. The heavy metals exerted a significant impact on the production of bioactive components, including cordycepin, cordycepic acid, polysaccharides, and free amino acids. Under the influence of 10 ppm of various metals, almost 99% of these components were inhibited. Cadmium (Cd) and mercury (Hg) were identified as the most toxic heavy metals for *C. militaris*, with Mercury(Hg) exhibiting a relatively lower tolerance for fruit body height. The inhibition was attributed to the disruption of chitin formation, a crucial factor for fungal cell wall integrity and growth (Chen et al., 2011).

8. Bioavailability and Bioaccessibility of Heavy Metals

Bioavailability usually refers to "the dose fraction of a pollutant that reaches the systemic circulation and can be taken up by an organism" (Rostami and Juhasz, 2011). According to the Encyclopedia of Environmental Health (2011), bioaccessibility is defined as "the fraction of total metal that is accessible (or soluble) in the gastrointestinal tract and is generally determined *in vitro* by incubating a sample in solutions that mimic, successively, the chemical conditions encountered in the human stomach and intestine". The two terms are very similar and in the case of heavy metal estimation, they are very useful as well. The heavy metal content in *Cordyceps* and allied species is usually not absorbed 100% by the digestive system (Zhou et al., 2018). Therefore, knowledge about bioavailability and bioaccessibility helps prevent the overestimation of health risks caused by heavy metals.

There are numerous studies on bioavailability and bioaccessibility of heavy metals in rice, wheat and vegetables from both *in vivo* and *in vitro* systems (Özcan et al., 2023; Zulkafflee et al., 2019; Zuo et al., 2023). However, there are limited studies for the bioaccessibility and bioavailability of *Cordyceps* and allied species and all the studies are based on *in vitro* methods using simulation studies (Zhou et al., 2018; Li et al., 2019b). The study of Zhou et al. (2018) investigated the bioaccessibility of heavy metals in various parts of *O. sinensis* using a simulated gastric juice method. The results revealed distinct differences in the bioaccessibility of different heavy metals in different parts of the organism. The average bioaccessibility of copper (Cu), lead (Pb), arsenic (As), mercury (Hg), and cadmium (Cd) varied across different parts of *O. sinensis*. In the whole organism, Copper(Cu) exhibited a bioaccessibility level of 41.29%, 40.11% for Lead (Pb), 64.46% for Arsenic (As), 18.91% for Mercury (Hg), and 81.14% for Cadmium (Cd). These values changed significantly in the caterpillar body as well as stroma. In the caterpillar body, the bioaccessibilities of Arsenic(As), Cadmium(Cd), Copper(Cu), Mercury(Hg) and Lead (Pb) were 48.26%, 42.92%, 66.15%, 12.86%, and 87.07%, respectively, while in the stroma, they were 38.30%, 30.53%, 30.18%, 7.46%, and 82.30%, respectively.

Notably, Copper (Cu), Arsenic (As),Mercury (Hg), and Cadmium (Cd) demonstrated higher bioaccessibility in gastric extracts compared to intestinal extracts, indicating their solubility in gastric juices. However, the bioaccessibility of Lead (Pb) was lower in gastric juice. The study also observed a direct proportionality between the quantity of dissolved heavy metals and their total content in the sample.

9. Arsenic Speciation Study in *Cordyceps*

Literature studies have revealed that the presence of arsenic (As) in *Cordyceps* and allied species sometimes exceeds permissible limits (Zuo et al., 2013; Wei et al., 2017; Zhou et al., 2018; Liu et al., 2021). Moreover, research has emphasized that the toxicity of heavy metals depends significantly on their speciation (Zuo et al., 2023). The attention given to the presence of Aesenic(As) in *Cordyceps* and allied species is due to its potential adverse health effects. Chemical speciation of Arsenic(As) includes: arsenate [As(V)], arsenite [As(III)], arsenobetaine (AsB), arsenocholine (AsC), dimethylarsinic acid (DMA), methylarsonic acid (MMA) and others. Studies have shown that inorganic arsenic, including As(III) and As(V), is more toxic than MMA and DMA. However, organic arsenic, along with AsB and AsC, is generally non-toxic. Therefore, determining the total Arsenic(As) content may not accurately reflect the sample's toxicity (Chen et al., 2018).

The reported range of arsenic concentrations in *Cordyceps* and allied species, viz. 9.00–10.18 mg/kg (Zhou et al., 2018), 5.77–13.20 mg/kg (Xiao et al., 2019), 4.00–5.25 mg/kg (Guo et al., 2018a), 0.82–5.68 mg/kg (Liu et al., 2021), 4.7–15.5 mg/kg (Li et al., 2019b), and 2.12–15.51 mg/kg (Zuo et al., 2018) suggest that its quantity in *Cordyceps* and allied species surpasses the standards. However, the total concentration of Arsenic(As) does not necessarily correlate with human toxicity. Inorganic arsenic is typically responsible for toxicity. According to literature, inorganic arsenic constitutes 8.69% of total arsenic (tAs) (Zhou et al., 2018b), 4.47–11.42% of tAs (Xiao et al., 2019), 6.0–8.3% of tAs (Guo at al., 2018), 7.21–23.17% of tAs (Liu et al., 2021), and 4% of tAs (Li et al., 2019b). Among inorganic arsenic species, As(III) and As(V) contribute the most. Concentrations of As(III) range from 0.41 to 0.73 mg/kg in whole *O. sinensis*, 0.07 to 0.08 mg/kg in stroma, and 0.48 to 0.64 mg/kg in caterpillar body (Zhou et al., 2018), 0.126 to 0.473 mg/kg (Xiao et al., 2019), 0.20 to 0.27 mg/kg (Guo et al., 2018b), 0.07 to 0.31 mg/kg (Liu et al., 2021), and 0.05 to 0.2 mg/kg (Li et al., 2019b).

Similarly, concentrations of As(V) range from 0.25 to 0.35 mg/kg in whole *O. sinensis*, 0.39 to 0.50 mg/kg in stroma, and 0.28 to 0.32 mg/kg in caterpillar body (Zhou et al., 2018), 0.180 to 0.0006 mg/kg (Xiao et al., 2019), 0.07 to 0.13 mg/kg (Guo, 2018b), 0.09 to 0.19 mg/kg (Liu et al., 2021), and 0.07 to 0.3 mg/kg (Li et al., 2019b). Unknown arsenic species were found to be predominantly present in the sample, constituting 91.31% of tAs (Zhou et al., 2018) and 89.0–92.3% (Guo et al., 2018b). These unknown arsenic species can easily convert to free inorganic arsenic (iAs) (Li et al., 2019b), potentially contributing to overall toxicity. It's worth noting that for the detection of inorganic arsenic, different extraction methods yield different results. The authors have suggested that a solution of HNO_3 is superior for recovering inorganic arsenic compared to simulated gastrointestinal juice (Xiao et al., 2019). Notably, the two potentially toxic organic allied species, MMA and DMA, were found to be insignificant in all the studies.

Literature indicates that, aside from arsenic, the bioaccessibility of heavy metal species has not been thoroughly investigated. Elements like lead (Pb), cadmium (Cd), mercury (Hg), and antimony (Sb) exist in various chemical forms, each with the potential to impact human health. Additionally, some metals, such as chromium (Cr), can have beneficial effects on health, but their hexavalent form (Cr^{6+}) is highly toxic to humans. Mercury, in particular, exhibits different forms, including organic mercury, inorganic mercury, monomethyl mercury, and dimethyl mercury (Rao et al., 2007; Zhang et al., 2013). The water-soluble nature and approximately 15% bioavailability of inorganic mercury make it a significant candidate for bioavailability studies in *Cordyceps* and allied species. Inorganic mercury compounds primarily affect the kidneys and can traverse the blood-brain barrier (Park et al., 2012). Therefore, it becomes evident that studying the species of heavy metals is crucial to obtaining a comprehensive understanding of the potential toxicity associated with *Cordyceps* and allied species. It is not only arsenic but also other heavy metals, each existing in diverse chemical forms, that may impact human health.

Moreover, it is imperative to explore different *in vitro* bioaccessibility methods and compare them with *in vivo* digestion techniques. Various methods can be employed to measure bioaccessibility include: *In Vitro* Gastrointestinal (IVG), Physiologically-Based Extraction Test (PBET), Solubility Bioaccessibility Research Consortium (SBRC) and the Unified Bioaccessibility Method (UBM) (Xu et al., 2023). Evaluating *Cordyceps* and allied species heavy metal bioaccessibility through these methods can provide a more precise understanding of the diverse heavy metal content. However, the absence of standardized methods for bioaccessibility or speciation studies poses a challenge.

10. Health Risk of Heavy Metals in *Cordyceps*

The exposure of heavy metals in humans mainly occurs by consuming food with contaminated heavy metals through the food chain (Zuo et al., 2023). According to Mitra et al. (2022), approximately 90% of heavy metal exposure is estimated to originate from food sources in humans (Mitra et al., 2022). The actual exposure level gives more accurate results about the effects of consuming heavy metals on health in comparison to the heavy metal content in the fungus itself. Many literatures revealed that the health implications were caused by heavy metals in consumed *Cordyceps* and allied species. According to Zhou et al. (2018), *O. sinensis* is usually consumed for long periods, i.e., 2–3 months per year. Their study showcases the health risk caused by five heavy metals (Cu, Pb, As, Cd and Hg) by conducting experiments of DI (Dietary Intake), DE (Dietary Exposure), THQ (The Target Hazard Quotient) and BTHQ (THQ×Bioaccessibility) values for the duration of 90 days with the highest daily dose recommended by Chinese Pharmacopoeia, 2015, 9 g/day, for an individual with 62 kg body weight and an exposure duration of 50 years. The results suggested that there was a very nominal chance of risk caused by heavy metals in the human body.

However, the greater value of arsenic made it a significant contributor to toxicity among the other heavy metals. The authors also suggested not to take *O. sinensis* for more than 90 days to avoid damage. In another study, Guo et al. (2018b) focused only on total As (tAs) by considering the mean value of Estimated Daily Intake (EDI), Hazard Quotient (HQ) and cancer risk (CR). The study revealed that the values of EDI, HQ and CR are within the standard limit and according to the authors, long-time consumption is not recommended. Another study by Li et al. (2019b) on arsenic assessed EDI, HQ and CR also showed that long-time consumption has health risks. They also recommended reducing the frequency and usage to overcome the risk (Li et al., 2019b). However, the study by Zuo et al. (2018) found that the Hazard index (HI) suggested that there was a risk of consumption of *O. sinensis*, the CR value suggested that even one month of exposure has a risk of cancer and the Target hazard quotient (THQ) suggested that 2 months of exposure has no risk on human health, but more than that may create a hazard. Therefore, all the literature suggests that there are very few chances of health risk for consumption of *Cordyceps* and allied species for a limited time of consumption, which is a maximum of three months per year with recommended consumption value.

11. Conclusion and Future Perspectives

Cordyceps and allied species can be vulnerable to contamination by heavy metals from their growth environment, including soil, water, and culture medium, making them prone to absorbing contaminants. To ensure the safe use of *Cordyceps* and allied species and limit potential health implications arising from heavy metal contamination, it is essential to channelize marketing of heavy metal-screened products of *Cordyceps* and allied species , and to facilitate that, there is a need to develop newer technologies with simplicity and cost-effectiveness, specifically designed to detect heavy metal contamination in *Cordyceps* and allied species. Further, thorough investigations, into the bioaccessibility of heavy metals in *Cordyceps* and allied species are essential for conducting a comprehensive health risk assessment and categorize weight- and age-dependent maximum dosages based upon inputs from traditional knowledge of pharmacopoeias and modern scientific insights. The use of controlled and high-quality culture media can effectively prevent heavy metal contamination in *Cordyceps* and allied species, helping to monitor product safety and quality. Concerning the safe use of products based on *Cordyceps* and allied species , it is recommended to introduce a unified global standard of safe levels for heavy metal quantity to ensure food safety and control quality. Further, the promotion of artificial cultivation of *Cordyceps* and allied species as well as promoting safe usage of their bioactive compounds as a replacement of whole fungal/fungal-insect system may be a game-changer to mitigate heavy metal contamination and ensure the sustainable use of *Cordyceps* and allied species.

References

Akaki, J., Matsui, Y., Kojima, H., Nakajima, S., Kamei, K. et al. (2009). Structural analysis of monocyte activation constituents in cultured mycelia of *Cordyceps sinensis*, Fitoterapia, 80(3): 182–187. https://doi.org/10.1016/j.fitote.2009.01.007.

Anahid, S., Yaghmaei, S. and Ghobadinejad, Z. (2011). Heavy metal tolerance of fungi. Sci. Iranica, 18(3): 502–508, https://doi.org/10.1016/j.scient.2011.05.015.

Balali-Mood, M., Naseri, K., Tahergorabi, Z., Khazdair, M.R. and Sadeghi, M. et al. (2021). Toxic mechanisms of five heavy metals: mercury, lead, chromium, cadmium, and arsenic. Front. Pharmacol., 12: 643972. https://doi.org/10.3389/fphar.2021.643972.

Balistrieri, L.S., Mebane, C.A., Cox, S.E., Puglis, H.J., Calfee, R.D. et al. (2018). Potential toxicity of dissolved metal mixtures (Cd, Cu, Pb, Zn) to early life stage white sturgeon (*Acipenser transmontanus*) in the Upper Columbia River, Washington, United States. Environ. Sci. Technol., 52(17): 9793–9800. https://doi.org/10.1021/acs.est.8b02261.

Beulah, H., Margret, A.A. and Nelson, J. (2013). Marvelous medicinal mushrooms. Int. J. Pharma. Bio. Sci., 3(1): 611–615.

Bhalla, A. and Pannu, A.K. (2022). Are Ayurvedic medications store house of heavy metals? Toxicol. Res., 11(1): 179–183. https://doi.org/10.1093/toxres/tfab124.

Bienert, G.P. and Tamás, M.J. (2018). Editorial: Molecular mechanisms of metalloid transport, toxicity and tolerance. Front. Cell Dev. Biol., 6: 99. https://doi.org/10.3389/fcell.2018.00099

Callender, E. (2003). Heavy metals in the environment-historical trends. Treat. Geochem., 9: 67–105. https://doi.org/10.1016/B0-08-043751-6/09161-1.

Cao, X., Ma, C., Zhao, J., Musante, C., White, J.C. et al. (2019). Interaction of graphene oxide with co-existing arsenite and arsenate: adsorption, transformation and combined toxicity. Environ. Int., 131: 104992. https://doi.org/10.1016/j.envint.2019.104992.

Chang, S.T. and Wassar, S.P. (2012). The role of culinary medicinal mushroom on human welfare with a pyramid model for human health. Int. J. Med. Mushrooms, 14(2): 95–134. https://doi.org/10.1615/intjmedmushr.v14.i2.10.

Chang, S.T. and Wasser, S.P. (2017). The cultivation and environmental impact of mushrooms. The oxford Research Encyclopedias: Environmental Science, Oxford University Press, Oxford, pp. 43. https://doi.org/10.1093/acrefore/9780199389414.013.231.

Chen, P.X., Wang, S., Nie, S. and Marcone, M. (2013). Properties of *Cordyceps sinensis*: a review. J. Func. Foods, 5(2): 550–569. https://doi.org/10.1016/j.jff.2013.01.034.

Chen, S., Yuan, B., Xu J., Chen, G., Hu, Q et al. (2018). Simultaneous separation and determination of six Arsenic species in Shiitake (*Lentinus edodes*) mushrooms: Method development and applications. Food Chem., 262: 134–141. https://doi.org/10.1016/j.foodchem.2018.04.036.

Chen, Y., Jiang, Z., Li, J. and Sun, Z. (2018). Study on the Arsenic Content of *Ophiocordyceps sinensis* in Sichuan Province. Med. Pl., 9(1): 47–49. https://doi.org/10.19600/j. cnki. issn2152 - 3924.2018.01.012.

Chen, Y.S., Liu, B.L. and Chang, Y.N. (2011). Effects of light and heavy metals on *Cordyceps militaris* fruit body growth in rice grain-based cultivation. Korean J. Chem. Eng., 28: 875–879. https://doi.org/10.1007/s11814-010-0438-6.

Chiu, C.P., Hwang, T.L., Chan, Y., El-Shazly, M., Wu, T.Y. et al. (2016). Research and development of *Cordyceps* in Taiwan. Food Sci. Hum. Wellness, 5: 177–185. https://doi. org/10.1016/j.fshw.2016.08.001.

Chou, S.M., Lai, W.J., Hong, T.W., Lai, J.Y., Tsai, S.H. et al. (2014). Synergistic property of cordycepin in cultivated *Cordyceps militaris*-mediated apoptosis in human leukemia cells. Phytomedicine, 21(12): 1516–1524. https://doi.org/10.1016/j.phymed.2014.07.014.

Concha, G., Vogler, G., Lezcano, D., Nermell, B. and Vahter, M. et al. (1998). Exposure to inorganic arsenic metabolites during early human development. Toxicol. Sci., 44(2): 185–190. https://doi.org/10.1006/toxs.1998.2486.

Das, S.K., Masuda, M., Sakurai, A. and Sakakibara, M. (2010). Medicinal uses of the mushroom *Cordyceps militaris*: Current state and prospects. Fitoterapia, 81(8): 961–968. https://doi. org/10.1016/j.fitote.2010.07.010.

De, L.Jr. F.C. and Gottfried, J.L. (2011). Rapid analysis of energetic and geo-materials using LIBS. Mater. Today, 14(6): 274–281. https://doi.org/10.1016/S1369-7021(11)70142-0.

Dhankhar, R. and Hooda, A. (2011). Fungal biosorption—an alternative to meet the challenges of heavy metal pollution in aqueous solutions. Environ. Technol., 32: 467–491. https://doi. org/10.1080/09593330.2011.572922.

Duffus, J.H. (2002). Heavy metals-a meaningless term? Pure Appl. Chem., 74(5): 793–807. https://doi.org/10.1351/pac200274050793.

Fenglin, L., Li, L., Xiaokun, J., Yan, L. and Hengwei, L. et al. (2020). Analysis of seven mineral elements in cultivated fruiting bodies of *Cordyceps militaris*. E3S Web Conf 185: 04019. https://doi.org/10.1051/e3sconf/202018504019.

Fergusson, J.E. (1990). The Heavy Elements: Chemistry, Environmental Impact and Health Effects. Pergamon Press, Oxford, p. 614. https://doi.org/10.2134/jeq1991.00472425002000040028x.

FSSR, 2011, Food Safety and Standards (Contaminants, Toxins and Residues) Regulations, 2011; Version # V (2020). https://www.fssai.gov.in.

Gajewska, J., Floryszak-Wieczorek, J., Sobieszczuk-Nowicka, E., Mattoo, A. and Arasimowicz-Jelonek, M. et al. (2022). Fungal and oomycete pathogens and heavy metals: An inglorious couple in the environment. IMA Fungus. 13: 6. https://doi.org/10.1186/s43008-022-00092-4.

Gall, J.E., Boyd, R.S. and Rajakaruna, N. (2015). Transfer of heavy metals through terrestrial food webs: A review. Environ. Monit. Assess., 187(4): 201. https://doi.org/10.1007/s10661-015-4436-3.

Guo, L.P., Zhou, L., Wang, S., Kang, C.Z., Hao, Q.X. et al. (2017). Statistic analysis of heavy metal residues in Chinese crude drugs with the international standards of Chinese medicine -Chinese herbal medicine heavy metal limit. Sci. Technol. Rev., 35(11): 91–98. https://doi.org/10.3981/j.issn.1000-7857.2017.11.013

Guo, L.X., Zhang, G.W., Li, Q.Q., Xu, X.M. and Wang, J.H. et al. (2018a). Novel arsenic markers for discriminating wild and cultivated *Cordyceps*. Molecules, 23(11): 2804–2819. https://doi.org/10.3390/molecules23112804

Guo, L.X., Zhang, G.W., Wang, J.T., Zhong, Y.P. and Huang, Z.G. et al. (2018b). Determination of arsenic species in *Ophiocordyceps sinensis* from major habitats in China by HPLC-ICP-MS and the edible hazard assessment. Molecules, 23(5): 1012. http://dx.doi.org/10.3390/molecules23051012.

Hausladen, D.M., Alexander-Ozinskas, A., Mcclain, C. and Fendorf, S. (2018). Hexavalent chromium sources and distribution in California groundwater. Environ. Sci. Technol., 52(15): 8242–8251. https://doi.org/10.1021/acs.est.7b06627.

Itabashi, T., Li, J., Hashimoto, Y., Ueshima, M., Sakanakura, H. et al. (2019). Speciation and fractionation of soil arsenic from natural and anthropogenic sources: Chemical extraction, scanning electron microscopy, and Micro-XRF/XAFS investigation. Environ. Sci. Technol., 53 (24): 14186–14193. https://doi.org/10.1021/acs.est.9b03864.

Kagaya, S., Maeba, E., Inoue, Y., Kamichatani, W., Kajiwara, T. et al. (2009). A solid phase extraction using a chelate resin immobilizing carboxymethylated pentaethylenehexamine for separation and preconcentration of trace elements in water samples. Talanta, 79(2): 146–152. https://doi.org/10.1016/j.talanta.2009.03.016.

Kawagishi, H., Okamura, K., Kobayashi, F. and Kinjo, N. (2004). Estrogenic substances from the mycelia of medicinal fungus *Cordyceps ophioglossoides* (ehrh.) Fr. (Ascomycetes). Int. J. Med. Mushrooms, 6: 249–252. https://doi.org/10.1615/intjmedmushr.v6.i3.40

Li, X., Liu, Q., Li, W., Li, Q., Qian, Z. et al. (2019a). A breakthrough in the artificial cultivation of Chinese cordyceps on a large-scale and its impact on science, the economy, and industry. Crit. Rev. Biotechnol., 39(2): 181–191. https://doi.org/10.1080/07388551.2018.1531820.

Li, Y., Liu, Y., Han, X., Jin, H. and Ma, S. (2019b). Arsenic Species in *Cordyceps sinensis* and Its Potential Health Risks. Front. Pharmacol., 10: 1471. https://doi.org/10.3389/fphar.2019.01471.

Liu, Y., Shi, M., Liu, X., Xie, J., Yang, R. et al. (2021) Arsenic transfer along the soil-sclerotium-stroma chain in Chinese cordyceps and the related health risk assessment. PeerJ, 9: 11023. https://doi.org/10.7717/peerj.11023.

Liu, Z., Li, P., Zhao, D., Tang, H. and Guo, J. (2011). Anti-inflammation effects of *Cordyceps sinensis* mycelium in focal cerebral ischemic injury rats. Inflammation, 34(6): 639—644. https://doi.org/10.1007/s10753-010-9273-5.

Mitra, S., Chakraborty, A.J., Tareq, A.M., Emran, T.B., Nainu, F. et al. (2022). Impact of heavy metals on the environment and human health: Novel therapeutic insights to counter the toxicity. J. King Saud Univ. Sc., 34(3): 101865. https://doi.org/10.1016/j.jksus.2022.101865.

Murthy, S., Bali, G. and Sarangi, S.K. (2014). Effect of Lead on Growth, Protein and Biosorption Capacity of *Bacillus Cereus* Isolated from Industrial Effluent. J. Environ. Biol., 35(2): 407–411.

Nakamura, K., Shinozuka, K. and Yoshikawa, N. (2015). Anticancer and antimetastatic effects of cordycepin, an active component of *Cordyceps sinensis*. J. Pharmacol. Sci., 127(1): 53–56. https://doi.org/10.1016/j.jphs.2014.09.001.

National Pharmacopoeia Commission (2019). 0212 general Principles for the verification of medicinal materials and decoction Pieces. In: Pharmacopoeia of the People's Republic of China (Part IV) (Revised Draft). http://english.nmpa.gov.cn.

Nicolaus, B., Poli, A., Di Donato, P., Romano, I., Laezza, G. et al. (2016). Pb2+ Effects on growth, lipids, and protein and dna profiles of the thermophilic bacterium *Thermus thermophilus*. Microorganisms, 4(4): 45. https://doi.org/10.3390/microorganisms4040045.

Nikoh, N. and Fukatsu, T. (2000). Interkingdom host jumping underground: phylogenetic analysis of entomoparasitic fungi of the genus *Cordyceps*. Mol. Biol. Evol., 17(4): 629–638. https://doi.org/10.1093/oxfordjournals.molbev.a026341.

Olatunji, O.J., Tang, J., Tola, A., Auberon, F., Oluwaniyi, O. et al. (2018). The genus *Cordyceps*: An extensive review of its traditional uses, phytochemistry and pharmacology. Fitoterapia, 129: 293–316. https://doi.org/10.1016/j.fitote.2018.05.010.

Osman, G.E.H., Abulreesh, H.H., Elbanna, K., Shaaban, M.R., Samreen, S. et al. (2019). Recent progress in metal-microbe interactions: Prospects in bioremediation. J. Pure Appl. Microbiol., 13(1): 13–26. https://doi.org/10.22207/jpam.13.1.02.

Özcan, M.M., Yılmaz, F.G. and Kulluk, D.A. (2023). The accumulation of element and heavy metal concentrations in different parts of some carrot and radish types. Environ. Monit. Assess., 195(6): 754. https://doi.org/10.1007/s10661-023-11364-w.

Park, J.D. and Zheng, W. (2012). Human exposure and health effects of inorganic and elemental mercury. J. Prev. Med. Public Health, 45(6): 344–352. https://doi.org/10.3961/jpmph.2012.45.6.344.

Phull, A.R., Ahmed, M. and Park, H.J. (2022). *Cordyceps militaris* as a bio functional food source: pharmacological potential, anti-inflammatory actions and related molecular mechanisms. Microorganisms, 10(2): 405. https://doi.org/10.3390/microorganisms10020405.

Pradhan, P., De, J. and Acharya, K. (2024). Strategies in artificial cultivation of two entomopathogenic fungi *Cordyceps militaris* and *Ophiocordyceps sinensis*. pp. 1–31. *In*: Singh, S.K., Kumar, D., Shamim, Md. and Sharma, R. (eds.). Applied Mycology for Agriculture and Foods: Industrial Applications. Apple Academic Press, Canada and USA.

Pradhan, P., Dutta, A., Paloi, S., Roy, A. and Acharya, K. et al. (2016). Diversity and distribution of macrofungi in the Eastern Himalayan ecosystem. EurAsian J. Biosci., 10: 1–12. https://doi.org/10.5053/ejobios.2016.10.0.1.

Rai, P.K., Lee, S.S., Zhang, M., Tsang, Y.F. and Kim, K. et al. (2019) Heavy metals in food crops: Health risks, fate, mechanisms, and management. Environ. Int., 125: 365-385, https://doi.org/10.1016/j.envint.2019.01.067.

Rao, R.N. and Talluri, M.V. (2007). An overview of recent applications of inductively coupled plasma-mass spectrometry (ICP-MS) in determination of inorganic impurities in drugs and pharmaceuticals. J. Pharm. Biomed. Anal., 43(1): 1–13. https://doi.org/10.1016/j.jpba.2006.07.004.

Rostami, I. and Juhasz, A.L. (2011). Assessment of Persistent Organic Pollutant (POP) Bioavailability and Bioaccessibility for Human Health Exposure Assessment: A Critical Review. Crit. Rev. Environ. Sci. Technol., 41(7): 623–656. https://doi.org/10.1080/10643380903044178.

Shashidhar, M.G., Giridhar, P., Sankar, K.U. and Manohar, B. (2013). Bioactive principles from *Cordyceps sinensis*: A potent food supplement—A review. J. Func. Foods, 5(3): 1013–1030. https://doi.org/10.1016/j.jff.2013.04.018.

Shi, B., Wang, Z., Jin, H., Chen, Y.W., Wang, Q. et al. (2009). Immunoregulatory *Cordyceps sinensis* increases regulatory T cells to TH17 cell ratio and delays diabetes in NOD mice. Int. Immunopharmacol., 9(5): 582–586. https://doi.org/10.1016/j.intimp.2009.01.030

Shifaw, E. (2018). Review of heavy metals pollution in China in agricultural and urban soils. J. Health Pollut., 8(18): 180607. https://doi.org/10.5696/2156-9614-8.18.180607.

Shrestha, B., Zhang, W., Zhang, Y. and Liu, X. (2012). The medicinal fungus *Cordyceps militaris*: research and development. Mycol. Prog., 11: 599–614. https://doi.org/10.1007/s11557-012-0825-y.

Shrestha, U.B. and Bawa, K.S. (2013). Trade, harvest, and conservation of caterpillar fungus (*Ophiocordyceps sinensis*) in the Himalayas. Biol. Conserv., 159: 514–520. https://doi.org/0.1016/j.biocon.2012.10.032.

Song, W.S., Jun, D.Y., Kim, J.S., Park, H.S., Kim, J.G. et al. (2007). Suppressive effect of ethyl acetate extract of *Paecilomyces japonica* on cell cycle progression of human acute leukemia Jurkat T cell clone overexpressing Bcl-2. Food Chem., 100: 99–107. https://doi.org/10.1016/j.foodchem.2005.09.026.

Sung, G.H., Hywel-Jones, N.L., Sung, J.M., Luangsa-ard, J.J., Shrestha, B. et al. (2007). Phylogenetic classification of *Cordyceps* and the clavicipitaceous fungi. Stud. Mycol., 57: 5–59. https://doi.org/10.3114/sim.2007.7.01.

Thatoi, H., Das, S., Mishra, J., Rath, B.P. and Das, N. et al. (2014). Bacterial chromate reductase, a potential enzyme for bioremediation of hexavalent chromium: A review. J. Environ. Manag., 146: 383–399. https://doi.org/10.1016/j.jenvman.2014.07.014.

Wei, X., Hu, H., Zheng, B., Arslan, Z., Huang, H.C. et al. (2017). Profiling metals in *Cordyceps sinensis* by using inductively coupled plasma mass spectrometry. Anal. Methods, 9(4): 724–728. https://doi.org/10.1039/C6AY02524B.

Winkler, D.C. (2009). Fungus (*Ophiocordyceps sinensis*) production and sustainability on the Tibetan Plateau and in the Himalayas. Asian Med., 5: 291–316. https://doi.org/10.1163/157342109X568829.

Won, S.Y. and Park, E.H. (2005). Anti-inflammatory and related pharmacological activities of cultured mycelia and fruiting bodies of *Cordyceps mililaris*. J. Ethnopharmacol., 96(3): 555–561. https://doi.org/10.1016/j.jep.2004.10.009.

Wu, P., Tao, Z., Liu, H., Jiang, G., Ma, C. et al. (2015). Effects of heat on the biological activity of wild *Cordyceps sinensis*. J. Trad. Chin. Med. Sci., 2: 32–38. https://doi.org/10.1016/j.jtcms.2014.12.005.

Xiao, H.J., Xiao, D.M., Chen, D.X., Xiao, Y., Liang, Z.Q. et al. (2012). Polysaccharides from the medicinal mushroom *Cordyceps taii* show antioxidant and immunoenhancing activities in a D-galactose-induced aging mouse model. Evid. Based Compl. Alt. Med., 273435. https://doi.org/10.1155/2012/273435.

Xiao, Y., Li, C., Xu, W. et al. (2021). Arsenic content, speciation, and distribution in wild *Cordyceps sinensis*. Evid. Based Compl. Alt. Med., 6651498. https://doi.org/10.1155/2021/6651498.

Xu, X., Wang, J., Wu, H.H., Lu, R. and Cui, J. et al. (2023). Bioaccessibility and bioavailability evaluation of heavy metal(loid)s in ginger *in vitro*: Relevance to human health risk assessment. Sci. Tot. Environ., 857(2): 159582.https://doi.org/10.1016/j.scitotenv.2022.159582.

Yan, J.K., Wang, W.Q. and Wu, J.Y. (2014). Recent advances in *Cordyceps sinensis* polysaccharides: mycelial fermentation, isolation, structure, and bioactivities: A review. J. Func. Foods, 6: 33–47. https://doi.org/10.1016/j.jff.2013.11.024.

Yan, W., Li, T., Lao, J., Song, B. and Shen, Y. et al. (2013). Anti-fatigue property of *Cordyceps guangdongensis* and the underlying mechanisms. Pharmaceut. Biol., 51: 614–620. https://doi.org/10.3109/13880209.2012.760103.

Zhang, G., Zhao, Y., Liu, F., Ling, J., Lin, J. et al. (2013). Determination of essential and toxic elements in *Cordyceps kyushuensis* Kawam by inductively coupled plasma mass spectrometry. J. Pharm. Biomed. Anal., 72: 172–176. https://doi.org/10.1016/j. jpba.2012.08.007.

Zhang, H.W., Lin, Z.X., Tung, Y.S., Kwan, T.H., Mok, C.K. et al. (2014). *Cordyceps sinensis* (a traditional Chinese medicine) for treating chronic kidney disease. Cochrane Database Syst. Rev., 2014(12): CD008353. https://doi.org/10.1002/14651858.CD008353.pub2.

Zhang, J., Yu, Y., Zhang, Z., Ding, Y., Dai, X. et al. (2011). Effect of polysaccharide from cultured *Cordyceps sinensis* on immune function and anti-oxidation activity of mice exposed to 60Co. Int. Immuno. Pharmacol., 11: 2251–2257. https://doi.org/10.1016/j. intimp.2011.09.019.

Zhang, Q. and Wu, J. (2007). *Cordyceps sinensis* mycelium extract induces human premyelocytic leukemia cell apoptosis through mitochondrion pathway. Exp. Biol. Med., 232: 52–57.

Zhang, Y., Hou, D., O'Connor, D., Shen, Z., Shi, P. et al. (2019). Lead Contamination in Chinese Surface Soils: Source Identification, Spatial-Temporal Distribution and Associated Health Risks. Crit. Rev. Environ. Sci. Technol., 49 (15): 1386–1423. https://doi.org/10.1080/10643 389.2019.1571354.

Zhao, L., Hu, G., Yan, Y., Yu, R., Cui, J. et al. (2019). Source apportionment of heavy metals in urban road dust in a continental city of Eastern China: Using Pb and Sr isotopes combined with multivariate statistical analysis. Atmos. Environ., 201: 201–211. https:// doi.org/10.1016/j.atmosenv.2018.12.050.

Zhong, S., Huijuan, P., Fan, L., Lu, G., Wu, Y. et al. (2009). Advances in research of polysaccharides in *Cordyceps* species. Food Technol. Biotechnol., 47: 304–312.

Zhou, L., Wang, S., Hao, Q., Kang, L., Kang, C. et al. (2018). Bioaccessibility and risk assessment of heavy metals, and analysis of arsenic speciation in *Cordyceps sinensis*. Chin. Med., 13: 40. https://doi.org/10.1186/s13020-018-0196-7.

Zhou, X., Gong, Z., Su, Y., Lin, J. and Tang, K. et al. (2009). *Cordyceps* fungi: natural products, pharmacological functions and developmental products. J. Pharm. Pharmacol., 61(3): 279–291. https://doi.org/10.1211/jpp/61.03.0002.

Zhu, J.-S., Halpern, G.M. and Jones, K. (1998). The scientific rediscovery of an ancient Chinese herbal medicine: *Cordyceps sinensis* Part I. J. Altern. Compl. Med., 4: 289–303. https://doi. org/10.1089/acm.1998.4.3-289.

Zhu, Z.Y., Chen, J., Si, C.L., Liu, N., Lian, H.Y. et al. (2012). Immunomodulatory effect of polysaccharides from submerged cultured *Cordyceps gunnii*. Pharm. Biol., 50: 1103–1110. https://doi.org/10.3109/13880209.2012.658114.

Zhu, Z.Y., Pang, W., Li, Y.Y., Ge, X.R., Chen, L.J. et al. (2014). Effect of ultrasonic treatment on structure and antitumor activity of mycelial polysaccharides from *Cordyceps gunnii*. Carbohydr. Polym., 114: 12–20. https://doi.org/10.1016/j.carbpol.2014.07.068.

Zulkafflee, N.S., Redzuan, M.N.A., Hanafi, Z., Selamat, J., Ismail, M.R. et al. (2019). Heavy metal in paddy soil and its bioavailability in rice using in vitro digestion model for health risk assessment. Int. J. Environ. Res. Public Health, 16(23): 4769. https://doi.org/10.3390/ ijerph16234769.

Zuo, H.L., Chen, S.J., Zhang, D.L., Zhao, J., Yang, F. et al. (2013). Quality evaluation of natural *Cordyceps sinensis* from different collecting places in China by the contents of nucleosides and heavy metals. Anal. Meth., 5: 5450–5456. https://doi.org/10.1039/c3ay40622a.

Zuo, T.T., Li, Y.L., Jin, H.Y., Gao, F., Wang, Q. et al. (2018). HPLC–ICP-MS speciation analysis and risk assessment of arsenic in *Cordyceps sinensis*. Chin. Med., 13: 19. https://doi. org/10.1186/s13020-018-0178-9.

Zuo, T.T., Li, Y.L., Wang, Y., Guo, Y.S., Shen, M.R. et al. (2023). Distribution, speciation, bioavailability, risk assessment, and limit standards of heavy metals in Chinese herbal medicines. Pharmacol. Res. Mod. Chin. Med., 6: 100218. https://doi.org/10.1016/j. prmcm.2023.100218.

14

Applications, Safety and Cordyceps Products

Lek Teng Lim

1. Introduction

The entomophagous fungus *Cordyceps*, also called caterpillar fungus, is predominantly found in the Himalayan belt of the Indian subcontinent in Asia, encompassing regions such as Bhutan, Chin, Nepal, North-east India, and Tibet (Chen et al., 2000; Arif and Kumar, 2003). It is known to parasitize and mummify Lepidopteran larvae. As an exotic medicinal mushroom, the dead larva along with the fruit body of *Cordyceps* have a long history of use in traditional Chinese medicine (TCM). They are used to treat a variety of ailments such as severe acute respiratory syndrome, ageing, cancer, cardiovascular diseases, diabetes, hyperlipidemia, liver problems, nervous breakdown, and renal disorders (Wang and Shiao, 2000; Chen et al., 2006, 2013; Kuo et al., 2006).

Wild *Cordyceps* are used in various food products such as porridge and soup (Jiang, 1994). The required dosage depends on the type of ailment. It should be safely administered with s precautions. Several adverse effects have been reported from the consumption of raw or processed *Cordyceps*. Wild or cultivated *Cordyceps* are used in several solid and liquid food products for humans, fish, and poultry. Many cosmeceutical products, especially skin and hair care products, consist of *Cordyceps*. Under the food and drug administration (FDA), *Cordyceps* has been marketed as one of the dietary supplements (Dong and Yao, 2007). Owing to the threat of deforestation,

BioFact Life Sdn Bhd, Parit Jawa, Muar, Johor, Malaysia.
Email: ltlim@biofactlife.com

forest fire, and overharvest of *Cordyceps*, it has been designated as an endangered fungus (Winkler, 1998; 2000; 2003). To alleviate the pressure on wild *Cordyceps* natural resources, there has been a notable focus on artificial cultivation and fermentation for commercial purposes, encompassing both fruit bodies and mycelial products (Holliday et al., 2004; Phull et al., 2022). This chapter offers insights into the therapeutic potential, dosage, mode of administration, precautions, safety, adverse effects, and marketed products.

2. Therapeutic Potential

2.1 *Anti-Arrhythmic Agents*

Anti-arrhythmic agents are used to suppress abnormal rhythms in the heart. *Cordyceps* are effective in reducing the heart rate. Individuals with heart disease and those using anti-arrhythmic medications should avoid simultaneous intake of *Cordyceps*, as both may exhibit similar anti-arrhythmic effects.

2.2 *Anti-Coagulant Medications*

Anti-coagulants prevent coagulation by inhibiting the action of clotting factors or platelets. thereby preventing the formation of blood clots. It has been reported that *Cordyceps* inhibit the aggregation of human blood platelets. The effects of anti-coagulant medications may be altered, when taken with Cordyceps, potentially increasing the risk of bleeding. Anti-coagulant medications include anti-thrombin III, argatroban, enoxaparin, bivalirudin, dalteparin, danaparoid sodium, heparin, lipirudin, tinzaparin, and warfarin.

2.3 *Anti-Diabetic Medications*

Anti-diabetic medications are utilized to treat diabetic mellitus by lowering blood glucose levels. It has been reported that patients consuming *Cordyceps* for the alteration of blood glucose metabolism might need to reduce the dosages of oral or injected anti-diabetic medications (Kiho et al., 1996; Zhang et al., 2006; Dong et al., 2014). Extra caution should be taken by insulin-dependent diabetics, as hypoglycaemia can occur in patients taking insulin or other oral anti-diabetic drugs as *Cordyceps* increase insulin receptivity in cells (Kiho et al., 1993).

2.4 *Anti-HIV drugs*

A novel compound has emerged that shows high potential in the treatment of HIV infections and likely in the management of advanced cases of AIDS.

The Immune-Assist 247 combines varied species of medicinal mushrooms and an infrequent compound extracted from green tea. The original Immune-Assist formula consists of a cocktail of 1,6-β-glucans, polyphenols derived from green tea leaves, and some direct-acting antiretroviral components from *Cordyceps sinensis*. A direct immune stimulation of β-glucans, the potent antiretroviral effect of transformed nucleosides from *Cordyceps sinensis*, and the inhibition of fusion effects of polyphenols from green tea act synergistically to contribute a major role in controlling HIV.

2.5 *Anti-Platelet Medications*

An anti-platelet drug or anti-aggregator is used to decrease platelet aggregation and inhibit thrombus formation. *Cordyceps* may interact with anti-platelet drugs and affect blood clotting ability. *Cordyceps* are reported to decrease platelet aggregation. This may increase the risk of bleeding and alter the medication effects; hence, dosage adjustments may be needed. Anti-platelet drugs include abciximab, anagrelide, aspirin, cilostazol, clopidogrel, dipyridamole, eptifibatide, ticlodipine, and tinofiban.

2.6 *Anti-Retroviral Drugs*

Anti-retroviral drugs are a group of medications used to treat infections caused by retroviruses, such as HIV. Naturally occurring anti-retroviral compounds are found in *Cordyceps*. 2,3-deoxyadenosine could result in increased effectiveness of drugs, thereby requiring dosage adjustment for patients undertaking concurrent therapy with other antiretroviral drugs.

2.7 *Hormone Replacement Therapy*

Patients undergoing hormone replacement therapy or taking birth control pills should be cautious when taking *Cordyceps* as they may induce steroid-like effects. For example, there are two categories of hormone remedies for breast cancer. Medications that inhibit estrogen and progesterone,crucial in breast cancer treatment, and drugs that suppress hormone production from the ovaries might interact with the hormonal functions of *Cordyceps* (Chang et al., 2008). Special advice from the physician is required when breast cancer patients decide to take *Cordyceps*.

Cordyceps might induce interactions with other herbs and dietary supplements that possess similar or adverse functions. In short, other herbs and dietary supplements that act on blood clotting, blood sugar level, immune system, blood pressure, cholesterol, heart rate, and hormone level might reduce or increase the effectiveness of *Cordyceps*. Consult with a qualified healthcare professional before combining therapies to prevent serious illnesses from happening. To minimize the likelihood of medications affecting the potency of *Cordyceps*, take the herb approximately two hours before or after the medications.

2.8 Hypertensive Medications

Consumers should be cautious when taking medications for lowering blood pressure because *Cordyceps* may act to lower blood pressure too. Drugs used for blood pressure are angiotensin-converting enzyme inhibitors (ACE inhibitors).

2.9 Immunosuppressants

Immunosuppressive drugs inhibit and prevent activities of the immune system. *Cordyceps* possess an immunosuppressant effect. The inhibitory effects of *Cordyceps* on the immune response responsible for organ transplant rejection have been studied. *Cordyceps* may stimulate the immune system and may reduce the efficacy of immunosuppressants such as prednisolone or cyclophosphamide. Other medications that suppress the immune system include azathioprine (Imuran), basiliximab (Simulect), corticosteroids (glucocorticoids), cyclosporine (Neoral and Sandimmune), daclizumab (Zenapax), muromonab-CD3 (OKT3), mycophenolate (CellCept), orthoclone (OKT3), prednisone (Deltasone and Orasone), sirolimus (Rapamune), tacrolimus (FK506 and Prograf), and others.

2.10 Medications to Reduce Cholesterol

Several studies have confirmed that *Cordyceps* help to lower the total blood cholesterol level. Therefore, *Cordyceps* might interact with medications and further enhance the cholesterol lowering effect.

2.11 Monoamine-Oxidase Inhibitors

Monoamine oxidase inhibitors (MAOIs) are used to treat depression. MAOIs have been reserved as a last-line treatment that is used when other classes of anti-depressant drugs have failed. Studies show that *Cordyceps* may have an anti-depressant activity that will alter the effects of MAOI medications. It may inhibit monoamine oxidase type B. Some of the *Cordyceps* constituents might act as adrenoceptors and dopamine reuptake inhibitors. MAOI medications include phenelzine, tranylcypromine, and isocarboxazid.

2.12 *Prescription Drugs Containing Caffeine*

Caffeine is commonly used in many prescription and over-the-counter headache medications to increase alertness, decrease fatigue, and improve muscle coordination (Li and Li, 2009). Adenosine contained in *Cordyceps* has a direct relaxing effect on smooth vascular muscles by decreasing the excitability of nerves. Caffeine blocks the adenosine effects and interferes with the calming action of the brain. Therefore, consumers are advised not to take coffee, tea, or any food containing caffeine within 2 hours of intake.

3. Methods of Administration

3.1 *Wild Cordyceps*

There are various ways to administer wild *Cordyceps*, each with its own advantages and disadvantages. There is no definitive guideline on the best way to take *Cordyceps*; it ultimately depends on individual consumer preferences (Table 1).

Before delving into the methods of consuming wild *Cordyceps*, it is crucial to understand how to clean them. Wild *Cordyceps* are sometimes adulterated with lead wire, and precautions must be taken to mitigate the potential health impacts of lead (Chen et al., 2013).

Wild *Cordyceps* should be marinated with rice wine for two hours, followed by surface brushing and rinsing with water. The technique of fermentation to prepare fruit wine using *C. sinensis* has been explained by Wu et al. (2008). Wild *Cordyceps* can be taken with vitamin C to aid the digestion and absorption of the medicinal components of the mushrooms. The following methods of wild *Cordyceps* consumption can also be applied to cultivated *Cordyceps*.

Table 1. Traditional dietary intake of wild *Cordyceps* for different ailments (adapted from Jiang, 1994).

Types of meat used to cook	Medicinal use
Cooked with duck	For patients with cancer, asthenia, and after severe illness
Cooked with chicken	For hyposexuality (especially emission)
Cooked with black-boned chicken	For asthenia (especially Qi-Yin asthenia)
Cooked with lean pork	For fatigue, male impotence, and kidney asthenia
Cooked with sparrow	For anti-ageing or senescence
Cooked with quail	Fatigue, poor appetite, kidney asthenia, and tuberculosis
Cooked with steamed turtle	For male and female hyposexuality
Cooked with baked abalone	For chronic bronchitis, COPD, tuberculosis, arteriosclerosis, and cataracts

3.2 *Wild Cordyceps Cooked in Porridge*

The easiest way to prepare this is to add the wild *Cordyceps* together with rice and then cook them into porridge. To enhance its medicinal properties, other herbs such as *Radix astragalus*, yam (*Dioscorea batatas*), Chinese wolfberry, and Chinese red date can also be added. Different types of herbs might be beneficial for different types of illnesses. Hence, depending on the sickness you wish to cure, cook the wild *Cordyceps* porridge by selecting the most suitable herbs to mix with. Other than herbs, if you want the monotonous porridge to be tastier, meat, fish, chicken, pork, and mutton can also be added. The fat from the meat is believed to enhance the extraction of the fat-soluble active components from the herbs.

3.3 *Wild Cordyceps Soaked in Liquor*

Wild *Cordyceps* can be prepared by soaking them in food spirit. This is effective in extracting bioactive compounds from the wild *Cordyceps* into the liquor providing consumers with the benefits (Chen et al., 2013). The method is as simple as crushing the wild *Cordyceps* and mixing it with rice wine. Shaking the mixture may enhance the extraction of the bioactive ingredients. The wild *Cordyceps* is then left to soak in the mixture for about seven days, after which the liquor is filtered to remove the residue. Yang et al. (2015) explained how to prepare *Cordyceps* rice wine as a novel brewing process. Further, Lao et al. (2019) discuss the significance of enzymatic hydrolysis and fermentation for the flavor and nutritional quality of *C. militaris* beverages. Drinking this self-made wild *Cordyceps* liquor is beneficial for curing fatigue, excessive sweating, lack of appetite, insomnia, and phlegm. Furthermore, the soaked wild *Cordyceps* can also be eaten.

3.4 *Wild Cordyceps Simmered in Soup*

Wild *Cordyceps* can be consumed with a variety of meats in the form of medicinal soup. Simmering wild *Cordyceps* in the soup can extract the active components from it. The type of meat used depends on the targeted medical condition.

3.5 *Medications and Injections*

The discovery and development of suitable new medications to address emerging diseases have already consumed so much time and effort. The best way to remedy this situation is by isolating useful chemicals from medicinal plants, which can then be used as drugs or starting materials for the synthesis of many vital drugs. The isolated chemicals are useful in studies concerning the mechanisms involved in the treatment of diseases, which later are hoped to lead to their cure.

Cordycepin in cultivated *Cordyceps* possesses diverse biological properties, especially anti-cancer activity (Xu et al., 1992; Wu et al., 2007; Phull et al., 2022). Since wild *Cordyceps* are very scarce in nature, cordycepin can be extracted from the cultivation of medicinal fungi or prepared synthetically. Cordycepin from cultivated *Cordyceps* is expected to play evolutionary roles in pharmacognosy, leading to the creation of a feasible base for the pharmaceutical industries as some modern diseases such as cancer, SARS, and swine flu are yet to have proper remedies.

4. Dosage of Consumption

Cordyceps possess numerous medicinal properties. Nevertheless, if consumed in an inappropriate manner only a small fraction of its medicinal values will be obtained. Different people have different reactions to *Cordyceps*, whereas different types of sicknesses require different product dosages. Dosage is very critical in the herbal and drug industries. Wrong dosages of medications may actually be fatal. The same thing applies to *Cordyceps*. The required dose for *Cordyceps* may vary due to the source as well as the diseases (Phull et al., 2022). Clinical trials were performed using 3–5g of wild *Cordyceps* per day. For patients with certain diseases, the recommended dosage would be different. In severe liver disease conditions, the dosage will be higher, about 6–9 g per day. More details about the doses of *Cordyceps* for specific diseases are available in the review by Phull et al. (2022). Approximate wild Cordyceps dosages recommended for patients with specific diseases are given in Table 2.

Table 2. Dosage of consumption of *Cordyceps*.

Ailment	Quantity	Dose	Remarks/Reference
Chronic bronchitis and bronchial asthma	4–5 g per day	2 doses	Up to 2 months (Cai et al., 2004)
Chronic hepatitis and liver cirrhosis	2–5 g per day	3 times	Up to 3 months
Hyperlipidaemia	3 g per day	3 times	-
Sexual dysfunction	3–6 g per day	3 times	Up to 40 days
Arrhythmia	1–2 g per day	3 times	Up to 2 weeks
Lethargy and a weak immune system	0.5 g each time	2 servings per day	Up to 2 weeks
Allergic rhinitis	3–6 g each time	3 servings per day	-
Chronic nephritis	4–6 g each	2 times per day	-

Some consumers might think that the larger the amount consumed, the better the medicinal effects would be. However, it is not true. Consumers are advised not to take more than 50 g of wild *Cordyceps* per day. Consuming excessive amounts of *Cordyceps* is likely to be harmful to the body. This is because the human body can only absorb a limited amount of the bioactive ingredients present in the *Cordyceps*. Taking an overdose of *Cordyceps* does not provide extra benefits to the body. It is an uneconomical way of taking expensive herbs. Other than that, their efficacy shows only after gradual accumulation in the body and is definitely not visible after a sudden, large consumption. Hence, *Cordyceps* normally must be taken continuously for one to two months in order to gain potent medical benefits.

Cordyceps dosages are normally measured using the Chinese Traditional Measurement Units *tael* and *mace*, or simply measured in grams. Nowadays, with the advancement of technology, the dosage can be determined not only by physical weight but also by the quantity of quality markers or active components present in it. If the amount of the quality markers in *Cordyceps* is low, a higher dosage may be required in order to confer medical effects. However, most people do not practice this, as special equipment is required to determine the number of quality markers. Other than that, the dosage of cultivated *Cordyceps* is easier to formulate due to the consistency of quality as compared to wild *Cordyceps* (Phull et al., 2022). For example, the bioactive ingredient of wild *Cordyceps* (Qing Hai) bought from a local market was tested using HPLC. Adenosine content was found to be 843 µg/g, while cordycepin was not detected. Wild Tibetian *Cordyceps* bought from a local market were found to have 673 µg/g of adenosine, while cordycepin was not detected in it. If calculated, the total adenosine content in 5 g of wild *Cordyceps* will be 843 µg/g × 5g = 4215 µg. At present, there is no reliable standard by which one can compare the quality of the different types of wild *Cordyceps*. More clinical trials should be conducted so that a clearer picture of the required dosages for a specific condition can be standardized.

5. Safety and Precautions

Although *Cordyceps* are rich in bioactive ingredients, they might not be suitable universally. Healthy individuals may wish to take *Cordyceps* for health maintenance, but some may choose not to due to its incredibly high price. It is very important for us to know our body constitution and whether our body possesses tolerance for *Cordyceps* or not. Generally, below is a list of people who should exercise caution when taking *Cordyceps*:

5.1 Babies and Infants

It is not encouraged to administer *Cordyceps* to babies on a regular basis unless they have certain illnesses that can be cured by Cordyceps dosages or if they are advised to do so by their physician. *Cordyceps* contain testosterone-like hormones, which may trigger early maturity with prolonged usage. Furthermore, the internal body structures of babies and infants are not mature, making it difficult for them to digest such complex food. Thus, Cordyceps intake is not recommended for babies and infants..

5.2 *Breastfeeding and Pregnant Women*

Different people have different perspectives on the intake of *Cordyceps* by pregnant women. Traditional Chinese medicine specialists recommend that pregnant women take *Cordyceps* for general health as well as for the benefit of their offspring. Babies born to mothers who practice this are believed to be more intelligent. On the contrary, some people believe that pregnant women should be more cautious about their food and supplement intake.

Historically, most Chinese have adhered to the practice of taking traditional herbs such as ginseng and ginger during pregnancy, breastfeeding, confinement, and the postpartum period. In fact, almost all traditional herbs do not have any strong scientific evidence to prove that they are safe for consumption during these periods. There is no truth or falsehood to such a practice, but it remains a part of Chinese culture as traditional herbs normally have positive effects on the body, mostly by restoring health and alleviating sickness.

Since there is limited research on the side effects of *Cordyceps* on pregnancy, pregnant women are not encouraged to take *Cordyceps* during the first three months of pregnancy to prevent any unknown effects on the fetus. After three months, when the pregnancy is viable, pregnant women can start taking *Cordyceps* only based on the recommendation of their physician. This is due to the fact that hormone levels during pregnancy are very sensitive and easily affected by any herb or medicine. Hence, if they still want to take *Cordyceps*, pregnant women can do so at their own risk..

5.4 *Individuals Affected by Common Sickness*

Individuals who are prone to certain common illnesses or suffer from any form of disorder or disease that involves sensitivity to hormones are not encouraged to take *Cordyceps*. The common sicknesses referred to are influenza or flu, fever, and syndrome of the excess type known as "shi". The causes of influenza are pathogen infections and poor immunity. Even though *Cordyceps* can improve the immune system and are equipped with anti-virus activity, they are capable of regulating internal inflammation. Hence, it is advisable to take *Cordyceps* only after one fully recovers from influenza to prevent the sickness from getting worse.

Syndrome of the excess ('shi') type refers to the body condition caused by an overabundance of excessive exogenous pathogens when the body's resistance has not yet weakened. The symptoms of excess syndrome include dry mouth, agitation, a sonorous voice, coarse breathing, constipation, and urinary difficulty. Individuals who suffer from the syndrome of excess type would not benefit from taking *Cordyceps*, but they might experience mild side effects such as stomach upset, dry mouth, and nausea. Thus, it is advisable to avoid consuming *Cordyceps* or reduce the dosage as advised by the physician.

6. Adverse Effects

6.1 Allergic to Cordyceps

There is only a very small group of people who are hypersensitive to mushrooms. Those who are allergic to mushrooms might also have an allergic reaction to *Cordyceps* since this herb is a kind of mushroom. Although there are no major side-effects of *Cordyceps* reported in medical journals, individuals who are allergic to mushrooms should be cautious when taking *Cordyceps* (Chen et al., 2013). Before taking this herb, an individual should seek advice from a physician to ensure that *Cordyceps* will not cause any serious adverse effects due to an allergic reaction.

6.2 Drug Interactions

There is no guarantee that taking natural herbs and foods is absolutely safe. Herbs and food taken together may interact with medications and, if taken frequently, will cause serious adverse reactions. *Cordyceps* discussed in this book are no exception. It is highly advisable to consult a pharmacist, physician, or other health practitioner about the possible interactions before taking *Cordyceps*. Unusual symptoms may foretell the occurrence of a drug interaction. Caution should be exercised about possible drug interactions since *Cordyceps* can lessen or increase the impact of medications.

6.3 Individuals on Medications

Individuals who are currently on prescription medicines or have any medical condition should be aware of the possibility of drug interactions between *Cordyceps* and medicines. It is highly recommended not to take *Cordyceps* with other medicines. Consultation with a physician is necessary for a prescription prior to administering *Cordyceps*.

6.4 Low Blood Glucose Level

Individuals with low blood glucose levels should take extra caution before taking *Cordyceps*. The polysaccharide derivative of *Cordyceps* is able to lower plasma blood glucose. Individuals with low blood glucose will get dizzy due to an extremely low blood glucose level in the body.

6.5 *Other Adverse Effects*

Adverse effects might be observed after taking *Cordyceps*, depending on an individual's body constitution. Different types of body constitutions might respond differently to the *Cordyceps*. As mentioned in the previous section, *Cordyceps* might not be suitable for all kinds of people. Some of the possible side effects that might be experienced are dry mouth, dizziness, nausea, stomach discomfort, and systemic allergic reactions. These side effects reflect that the body is adapting to the *Cordyceps* that have been consumed. Consumers are advised to consume small doses of *Cordyceps* initially. Throughout time, the body can harmonize itself to the functions and reactions of *Cordyceps*, thus reducing the occurrence of adverse effects. Nevertheless, the consumption of *Cordyceps* will not cause any serious illnesses unless the product is adulterated, the method of taking *Cordyceps* is erroneous, and the body cannot adapt itself to the *Cordyceps*. If the symptoms continue or other serious symptoms occur, consuming the product should be stopped immediately and a doctor's advice be sought.

Cordyceps can be taken safely, provided consumers follow the instructions and abide by the recommended dosages. Only a small group of people may experience dry mouth, nausea, and diarrhoea. There are reports from clinical studies of slight upper gastrointestinal discomforts, like nausea, dry mouth, and discomfort of the stomach, in some patients' on consuming cultivated strains of *Cordyceps sinensis* called Cs-4 (Xu, 1992). In another study, one patient experienced a systemic allergic reaction by consuming Cs-4. But, this type of systemic allergic reaction is uncommon. All investigators opined that wild *Cordyceps* and its mycelia fermentation products are safe for clinical use (Shashidhar et al., 2013). The side effects of taking wild *Cordyceps* are very limited, especially if there is no drug interaction or underlying lead content in the wild *Cordyceps* (Wu et al., 1996).

7. Food Products

7.1 *Cordyceps as Powder*

Some traditional medicine shops will grind the wild *Cordyceps* into powder form before selling them to consumers. Some manufacturers may even encapsulate the wild *Cordyceps* powder for easier consumption. Theoretically, the powder form of wild *Cordyceps* may enhance the digestion and absorption of its medicinal components into the body. However, experts stated that taking wild *Cordyceps* directly is unhygienic because the harvested wild *Cordyceps* contain a lot of microorganisms, fungal spores, and parasite eggs on their outer layer. Nevertheless, in certain circumstances, there is no choice but to take the wild *Cordyceps* in powder form. The wild *Cordyceps* should be cleansed properly before they are ground into powder form. Some babies with breathing problems, advised to take wild *Cordyceps* to improve their condition may not know how to take it through soup, porridge, or even liquor pathways. In this case, mothers can mix the wild *Cordyceps* powder into the babies' cereal or milk to facilitate their consumption.

7.2 *Cultivated Cordyceps*

Various types of cultivated (or fermented) *Cordyceps* products are available on the market (Shashidhar et al., 2013). Other than direct consumption as a supplement, the nutritional properties of cultivated *Cordyceps* can be utilized in other applications (Chen et al., 2021). It would be helpful, especially for those who cannot afford the expensive wild *Cordyceps*. Either wild or cultivated *Cordyceps* can be used to manufacture products in the form of powder, extract, essence, or liquid, as long as their bioactive ingredients and nutrients are present (Phull et al., 2022). However, most *Cordyceps* products are made of cultivated mycelia or stroma, as the cost is lower than that of wild *Cordyceps*.

7.3 *Dietary Supplements*

Over-the-counter medicine (OTC) refers to any medication that can be bought from any health care centre without a doctor's prescription. Supplements, or OTC medicines, exist in the form of liquids, extracts, capsules, and tablets. Consumers can take the medicines by referring to the instructions stated on product labels for the purpose of curing diseases and improving body health.

Supplements formulated with cultivated *Cordyceps* and tonic herbs are now available and selling widely in the market (Shreshta et al., 2012). Cultivated *Cordyceps* may work synergistically with other tonic herbs such as *Panax ginseng, Radix astragalus, Ganoderma lucidum, Radix glycyrrhiza glabra*, and *Bulbus fritillariae cirrhosae* to offer enhanced medicinal effects. For example, health supplements combining cultivated *Cordyceps* with *Bulbus fritillariae cirrhosae* in capsule and tablet forms are beneficial to moisten the lungs, resolve phlegm, and relieve cough and asthma (Wei et al., 2002). The effectiveness might not be as prominent if you are only taking cultivated *Cordyceps* or *Bulbus fritillariae cirrhosae*. Furthermore, by taking this type of herbal combination, it is not necessary anymore to purchase both wild *Cordyceps* and *Bulbus fritillariae cirrhosae* individually from herbal shops, which may cost more than the combination supplements. For liquid forms of supplements, ingredients such as bird's nest, collagen, and cough syrup can also be added with cultivated *Cordyceps* extract so that consumers will benefit from the major and minor ingredients present in cultivated *Cordyceps*.

7.4 Enzyme and Fermented Liquids

Cordyceps products are also formulated in liquid form. Liquid-fermented herbal enzymes from complex vegetables and fruits are now available on the market. These are made with many types of precious ingredients, added with microorganisms, and fermented for a few months. This nutritional liquid normally consists of the *Cordyceps sinensis* extract hundreds of natural vegetables, fruits, alpine plants, seaweeds, and other Chinese herbs (Koh et al., 2003). Each drop of the fermented liquid is packed with natural vitamins, potent antioxidants such as the superoxide dismutase enzyme, minerals, and amino acids, which can assist in maintaining health, stamina, beauty, and nutritional balance. This would definitely be the best liquid vitamin and nutrition supplement for a healthier lifestyle.

7.5 Fruit Juices

Fruit juice, made from all kinds of natural fruits, is loaded with nutrients for health without increasing body weight. According to a finding by the researchers at the Louisiana State University Agricultural Centre and Baylor College of Medicine, 2–5-year-old kids who regularly consume fruit juice have a significantly increased intake of vitamin C, potassium, and magnesium, with a lower intake of added sugars as compared to non-fruit juice consumers.

As fruit juice is accepted globally, manufacturers have developed a new method to further enhance its nutritional value, namely, by adding the liquid extract of cultivated *Cordyceps* (Guan et al., 2010). When consumers drink *Cordyceps* fruit juice, they do not only take in Vitamin C. Additionally, active ingredients other than Vitamin C present in cultivated *Cordyceps* are also consumed. These compounds play significant roles in strengthening the immune system as well as providing high levels of antioxidant and anti-inflammatory properties to the body (Jia and Lau, 1997). Mixing the liquid extract of cultivated *Cordyceps* with the fruit juice is a practical way to provide nutrients to children since *Cordyceps* herbal soup, which tastes bitter, may be unappealing to them. The sweet flavor of the fruit can mask the unpleasant smell of the *Cordyceps* liquid extract, thereby enhancing compliance.

7.6 Pastilles and Candies

The majority of candies on the market nowadays are laden with calories and void of any nutritional value. The addition of some essential nutrients will not only add value to the candy, it will also be beneficial to health and stamina. As a result, nobody will say that eating candy is bad for our health. Since cultivated *Cordyceps* is packed with benefits, its liquid extract can be used to produce sweets, candies, and pastilles. Furthermore, with the addition of other nutrients such as minerals and vitamins, consumers can opt to buy these nutritional candies rather than non-beneficial candies, which consist of only sugar and fat. It will also be good for children who always like to eat candy.

7.7 Wine

Cordyceps wine is very popular throughout the world, especially in China. The most basic *Cordyceps* wine can be produced by just adding crushed cultivated *Cordyceps* to a certain amount of rice wine, then covering the container and leaving it to ferment for a number of days. This type of *Cordyceps* wine enables the extraction of beneficial ingredients from cultivated *Cordyceps*, which are then easily absorbed by the body. Its medicinal properties include alleviating health problems such as fatigue, excessive sweating, phlegm, and coughing. With cultivated *Cordyceps* as the major active ingredient, ginseng, astragalus, wolfberry, and other valuable traditional compounds can also be added to the *Cordyceps* wine to confer extra benefits to the body. The resulting wine is much sought-after for healing purposes as well as to provide health care.

7.8 Other Food Products

Cultivated *Cordyceps* can be applied in the food industry to produce many fortified food products like biscuits, bread, cereal-based products, cookies, and mooncake (Chen et al., 2021). Manufacturers need to consider whether or not the process of biofunctional food production will affect the bioactive ingredients in cultivated *Cordyceps* (Chen et al., 2021; Phull et al., 2022). Some of these ingredients may be degraded by high temperatures, acidity, moisture, depending on the individual characteristics of each bioactive ingredient. For instance, temperature may be one of the factors that will reduce the amount of cordycepin present in biscuits after their production. Hence, food products can be sent for analysis to identify the presence of bioactive ingredients in order to ensure the high quality and value of cultivated *Cordyceps* products (Chen et al., 2021; Phull et al., 2022).

8. Veterinary Foods

8.1 Aquaculture Feed

Cultivated *Cordyceps* can be used as supplementary feed for aquatic animals, such as fish. For this purpose, biotechnology has enabled the mycelia of cultivated *Cordyceps* to be successfully cultured using rice as the solid substrate. The natural antibiotic potential of cultivated *Cordyceps* has raised great interest as an alternative to replace the use of synthetic antibiotics in the food chain. Extensive research on beef is continuously taking place, which has enabled dairy farms to successfully utilize cultivated *Cordyceps* as an antibiotic replacement. Trials have also been executed in many countries for its use as an antibiotic replacement for fish aquaculture. It is revolutionizing the meat production industry in an organically suitable manner, enhancing animal health and the quality of the final products.

Cordycepin (3′-deoxadenosine) is one of the main components in *Cordyceps* with antifungal and antiviral activities. As it also enhances immune function, the consumption of *Cordyceps* will surely strengthen the immunity of fish and improve the quality of fish meat. The mortality of fish during the spawning period will thus be reduced. The mycelia of cultivated *Cordyceps* can also be fermented on food sources such as industrially manufactured pellet feeds, farm-made aqua feed, organic or inorganic fertilizers and other existing supplementary feed.

8.2 Pet Foods

Cultivated *Cordyceps* can be fed to pets, such as cats and dogs, for the purpose of improving immunity and stamina. Any physiological changes that may take place when cultivated *Cordyceps* are fed to other animals are still being researched. Nevertheless, the practice of feeding animals with cultivated *Cordyceps* is not widespread due to its relatively high cost. To address this issue, feed manufacturers can try to formulate special types of feeds for animals and pets by introducing certain amounts of cultivated *Cordyceps* into them to increase nutrient content while ensuring affordability for both pet owners and farmers.

8.3 Poultry Feed

We often use chicken meat and eggs in many of our daily dishes. Chicken supplies protein, iron, and zinc to our bodies. Eggs also contain large amounts of minerals (calcium, iron, phosphorus, and potassium), protein, and vitamins (choline, folic acid, riboflavin, vitamin A, vitamin B_6, and vitamin B_{12}). Although egg yolk contains a large quantity of cholesterol, which may be responsible for a higher risk of cardiovascular diseases it is desirable to reduce the cholesterol content and increase the nutritional properties of the eggs produced simultaneously.

Cultivated *Cordyceps* may be blended with poultry feed or drinking liquid from chickens or ducks as powder or paste to improve their health and general well-being. According to studies, chickens fed with cultivated *Cordyceps* will produce eggs with a lower cholesterol content as compared with ordinary eggs. Moreover, presence of cordycepin is identified in the eggs produced to the extent of approximately 0.006–0.08 mg of cordycepin per 100g of egg matter, depending on the amount of the *Cordyceps* feed. The egg yolks of these eggs contain a high amount of cordycepin, while the egg whites do not contain any cordycepin traces. In addition, chickens fed with cultivated *Cordyceps* will gain more body mass. The protein and amino acids present in the meat are increased, but at the same time, the fat content is doubled. It has been suggested that the increased fat content may contribute to the improved flavor and texture of the meat.

9. Cosmeceuticals

9.1 Cosmeceutical Products

Many product developers utilize the various functions of *Cordyceps* as key selling points in cosmetic series. The essence of cultivated *Cordyceps* is extracted and added to a wide range of cosmetic products, such as moisturizers, creams, cleansers, facial masks, and serums, to increase their effectiveness and market value. Cultivated *Cordyceps* contain substances that are useful in restoring and nourishing damaged skin, such as cordycepin, cyclopeptides, fatty acids, amino acids, polysaccharides, cordycepic acid, and various forms of sterols, minerals, and trace elements. Each of them plays different roles in skin nourishment. Cordycepin has been scientifically proven to protect skin from ultraviolet radiation, which may lead to photo ageing (Lin et al., 2007). Cultivated *Cordyceps* also function as powerful antioxidants to protect the skin from oxidative damage (Li et al., 2001). With the combination of other ingredients in these cosmetics, their targeted effect on the skin will be fully achieved.

9.2 Skin and Hair Care Products

Most shampoos and hair care products contain synthetic ingredients such as ginseng, garlic, jojoba, olive oil, and so on. These synthetic ingredients are mainly used to increase the market value of products. Some manufacturers produce *Cordyceps* shampoo and hair care products. One of the hair care producers stated on its official website that *Cordyceps* improve cellular energy, regulate endocrine function, and increase blood flow to promote the growth of hair cells. However, there are no scientific studies on the benefits of *Cordyceps* for hair. The effectiveness of *Cordyceps* on hair can thus be further studied in the future.

Conclusion

The fruit bodies of *Cordyceps* serve as functional foods as well as medicines in traditional Chinese medicine. They are used to treat several human lifestyle diseases, such as cancer, cardiovascular diseases, and diabetes. Wild as well as cultivated *Cordyceps* are used in the production of various functional foods as well as medicine and dietary supplements. The quantity of functional foods and the dosage of medicine depend on the type of food and the ailment, respectively. Caution needs to be exercised to avoid adverse effects pertaining to overdose, allergic patients, and those with low blood glucose. Small doses are advisable for the optimum impact of *Cordyceps* as nutraceuticals and/or cosmeceuticals. The use of *Cordyceps* has been extended to poultry, aquaculture, and animal food for better productivity. More research and clinical trials are required to understand the specific bioactive compounds responsible for nutraceutical impact. Due to overharvesting of wild *Cordyceps* species, they have become endangered. Hence, interest has been shifted towards artificial cultivation and fermentation to harness the benefits of mycelia, fruit bodies, and fermented products. Although *Cordyceps* has been marketed as a nutraceutical supplement in solid (e.g., bakery products) and liquid (e.g., beverages) forms, more clinical trials are needed to standardize the optimum dose necessary either in food or medicine to combat human ailments.

References

Arif, M. and Kumar, N. (2003). Medicinal insects and insect—fungus relationship in high altitude areas of Kumaon Hills in Central Himalayas. J. Exp. Zoo. India, 6(1): 45–55.

Cai, H.Y., Liu, Z.C., Pang, J.W., Liu, M.S. and Wang, X.Y. (2004). Efficacy of *Cordyceps militaris* capsules in treatment of chronic bronchitis. Chin. J. New Drugs, 13: 171–174.

Chang, Y., Jeng, K.C., Huang, K.F., Lee, Y.C., Hou, C.W. et al. (2008). Effect of *Cordyceps militaris* supplementation on sperm production, sperm motility and hormones in Sprague-Dawley rats. Am. J. Chin. Med., 36(5): 849–859.

Chen, S.J., Yin, D.H., Li, L., Zha, X., Shuen J.H. and Zhama, C. (2000). Resources and distribution of *Cordyceps sinensis* in Naqu Tibet. Zhong Yao Cai, 11: 673–675.

Chen, C., Han, Y., Li, S., Wang, R. and Tao, C. (2021). Nutritional, antioxidant, and quality characteristics of novel cookies enriched with mushroom (*Cordyceps militaris*) flour. CYTA-Journal of Food, 19(1): 137–145.

Chen, J., Zhang, W., Lu, T., Li, J., Zheng, Y. and Kong, L. (2006). Morphological and genetic characterization of a cultivated *Cordyceps sinensis* fungus and its polysaccharide component possessing antioxidant property in H22 tumor-bearing mice. Life Science, 78: 2742–2748.

Chen, P.X., Wang, S., Nie, S. and Marcone, M. (2013). Properties of *Cordyceps sinensis*: A review. J. Func. Foods, 5: 550–569. 10.1016/j.jff.2013.01.034.

Dong, C.-H. and Yao, Y.-J. (2007). In vitro evaluation of antioxidant activities of aqueous extracts from natural and cultured mycelia of *Cordyceps* sinensis. LWT—Food Sci. Technol., 41: 669–677.

Dong, Y., Jing, T., Meng, Q., Liu, C., Hu, S. et al. (2014). Studies on the antidiabetic activities of *Cordyceps militaris* extract in diet-streptozotocin-induced diabetic Sprague-Dawley rats. Biomed. Res. Int., 2014: 160980. 10.1155/2014/160980.

Holliday, J., Cleaver, P., Megan, L.P. and Patel, D. (2004). Analysis of quality and techniques for hybridization of medicinal fungus *Cordyceps sinensis* (Berk.) Sacc. (Ascomycetes). Int. J. Med. Mushrooms, 6: 151–164.

Guan, J., Yand, F.-Q. and Li, S.-P. (2010). Evaluation of carbohydrates in natural and cultured *Cordyceps* by pressurized liquid extraction and gas chromatography coupled with mass spectrometry. Molecules, 15: 4227–4241. 10.3390/molecules15064227.

Jia, T.Y. and Lau, B.H.S. (1997). The immune-enhancing effect of Chinese herbal medicine *Cordyceps sinensis* on macrophage J774. J. Chin. Pharmacol., 32(3):142–144.

Jiang S.J. (1994). A tonic application of *Cordyceps sinensis* in medicinal diets. J. Chin. Med. Mat., 17(1): 47–48.

Kiho, T., Hui, J., Yamane, A. and Ukai, S. (1993). Polysaccharide in fungi. XXXII. Hypoglycemic activity and chemical properties of a polysaccharide from the cultural mycelium of *Cordyceps sinensis*. Biol. Pharm. Bull., 16(12): 1291–1293.

Kiho, T., Yamane, A., Hui, J., Usui, S. and Ukai, S. (1996). Polysaccharide in fungi. XXXII. Hypoglycemic activity of a polysaccharide (CS-F30) from the cultural mycelium of *Cordyceps sinensis* and its effect on glucose metabolism in mouse liver. Biol. Pharm. Bull., 19(2): 294–296.

Koh, J.H., Kim, J.M., Chang, U.J. and Suh, H.J. (2003). Hypocholesterolemic effect of hot-water extract from mycelia of *Cordyceps sinensis*. Biol. Pharm. Bull., 26(1): 84–87.

Kuo, H.C., Sua, Y.L., Yang, H.L., Huang, I.C. and Chen, T.Y. (2006). Differentiation of *Cordyceps sinensis* by a PCR-single-stranded conformation polymorphism-based method and characterization of the fermented products in Taiwan. Food Biotechnol., 20: 161–170.

Lao, Y., Zhang, M., Li, and Bhandari, B. (2019). A novel combination of enzymatic hydrolysis and fermentation: Effects on the flavor and nutritional quality of fermented *Cordyceps militaris* beverage. LWT Food Sci. Tachnol., 120 (1): 108934. 10.1016/j.lwt.2019.108934.

Li, S.P., Li, P., Dong, T.T.X. and Tsim K.W.K. (2001). Anti-oxidation activity of different types of natural *Cordyceps sinensis* and cultured *Cordyceps* mycelia. Phytomedicine, 8(3): 207–212.

Li, T. and Li, W. (2009). Impact of polysaccharides from *Cordyceps* on anti-fatigue in mice. Sci. Res. Essay, 4(7): 705–709.

Lin, C.C., Pumsanguan, W., Koo, M.M.K., Huang, H.B. and Lee, M.S. (2007). Radiation protective effects of *Cordyceps sinensis* in blood cells. Tzu. Chi. Med. J., 19(4): 226–232.

Phull, A.-R., Ahmed, M. and Park, H.-J. (2022). *Cordyceps militaris* as a bio functional food source: pharmacological potential, anti-inflammatory actions and related molecular mechanisms. Microorganisms, 10: 405. 10.3390/microorganisms10020405.

Shashidhar, M.G., Giridhar, P., Shankar, K.U. and Manohar, B. (2013). Bioactive principles from *Cordyceps sinensis*: A potent food supplement—A review. J. Func. Foods, 5: 1013–1030.

Shrestha, B., Zhang, W., Zhang, Y. and Liu, X. (2012). The medicinal fungus *Cordyceps militaris*: Research and development. Mycol. Prog., 11(3): 599–614. https://doi.org/10.1007/s11557-012-0825-y.

Wang, S.Y. and Shiao, M.S. (2000). Pharmacological functions of Chinese medicinal fungus *Cordyceps sinensis* and related species. J. Food Drug Anal., 8: 248–257.

Wei, T., Jin, Z.L., Kung, X.J. and Wei, W.L. (2002). Study on functions of *Cordyceps sinensis* mycelium about cough-relief, expectorant effect, antisepsis and anti-inflammatory effect. Food Sci., 23: 126–130.

Winkler, D. (1998). Deforestation in eastern Tibet human impact - past and present. Proceedings of the 7th Seminar for Tibetan Studies (IATS), pp. 79–96.

Winkler, D. (2000). Patterns of forest distribution and the impact of fire and pastoralism in the forest region of the Tibetan Plateau. Environmental Change in High Asia, Marburger Geographische Schriften, 135: 201–227.

Winkler, D. (2003). Forest use and implications of the 1998 logging ban in the Tibetan Prefectures of Sichuan: Case Study on forestry, reforestation and NTFP in Litang county, Ganzi TAP, China. The Ecological Basis and Sustainable Management of Forest Resources, 1: 116–125.

Wu, W.C., Hsiao, J.R. and Lian, Y.Y. (2007). The apoptotic effect of cordycepin on human OEC-M1 oral cancer cell line. Canc. Chemo. Pharmacol., 60: 103–111.

Wu, T.-N., Yang, K.-Y., Wang, C.-M., Lai, J.-S., Ko, K.-N. et al. (1996). Lead poisoning caused by contaminated Cordyceps, a Chinese herbal medicine: two case reports. Sci. Tot. Environ., 182(1–2): 193–195. 10.1016/0048-9697(96)05054-1.

Wu, Z.-W., Dou, Y.-P., Zhao, X.-F. and Wang, Z.-H. (2008). Study of fermentation technique of *Cordyceps sinensis-Lycium barbarum* wine. Food Sci. Proc. Tech., 29(1): 146–149.

Xu, F. (1992). Pharmaceutical studies of submerged culture of *Cordyceps* mycelia in China. Chin. Pharma. J., 27(4): 195–197.

Xu, R.H., Peng, X.E., Chen, G.Z. and Chen, G.L. (1992). Effects of *Cordyceps sinensis* on natural killer activity and colony formation of B16 melanoma. Chin. Med J., 105(2): 97–101.

Yang, R., Gu, D. and Gu, Z. (2015). *Cordyceps* rice wine: A novel brewing process. J. Food Proc. Eng., 39(6): 581–590. 10.1111/jfpe.12251.

Zhang, G.Q., Huang, Y.D., Bian, Y., Wong, J.H., Ng, T.B. and Wang, H.X. (2006). Hypoglycemic activity of the fungi *Cordyceps militaris, Cordyceps sinensis, Tricholoma mongolicum,* and *Omphalia lapidescens* in streptozotocin-induced diabetic rats. Appl. Microbiol. Biotechnol., 72: 1152–1156.

Index